ELECTROSPINNING for ADVANCED ENERGY and ENVIRONMENTAL APPLICATIONS

ELECTROSPINNING for ADVANCED ENERGY and ENVIRONMENTAL APPLICATIONS

edited by Sara Cavaliere

CRC Press
Taylor & Francis Group
Boca Raton London New York

CRC Press is an imprint of the
Taylor & Francis Group, an **informa** business

CRC Press
Taylor & Francis Group
6000 Broken Sound Parkway NW, Suite 300
Boca Raton, FL 33487-2742

First issued paperback 2017

© 2016 by Taylor & Francis Group, LLC
CRC Press is an imprint of Taylor & Francis Group, an Informa business

No claim to original U.S. Government works

ISBN-13: 978-1-4822-1767-4 (hbk)
ISBN-13: 978-1-138-74908-5 (pbk)

Visit the Taylor & Francis Web site at
http://www.taylorandfrancis.com

and the CRC Press Web site at
http://www.crcpress.com

Contents

Preface

Energy production and environment remediation are the major societal issues of this century, in which we are facing a dramatic population growth and the associated ecological issues. Indeed, the increase in the global demand for food, energy, and chemicals is obviously interlocked with problems of depletion of natural resources and pollution. In fact, if on the one hand the intense exploitation of the fossil fuels modifies the environment and deteriorates it with problems of pollution of soil, water, and air, on the other hand their limited supplies impose the development of novel means of energy production with a sensibly lower impact on the environment. We are currently living a global environmental crisis, and we are already realizing its effects all over the world (e.g., global warming). As the Greek etymology of the word (κρισις, krisis) states, crisis means choice, which means that in facing these problems, a choice must be made and a viable strategy must be devised to tackle them. A great challenge for research scientists is thus the development of novel and sustainable methods and materials for energy conversion and storage as well as environmental monitoring and repair.

Nanomaterials and nanotechnology can be one of the keys to address these issues. In particular, electrospinning has emerged in the last years as a promising technique to obtain nanoscaled fibrous materials that can be applied in energy and environmental fields. Indeed, electrospinning is a versatile technique allowing the production of nanofibers of diverse compositions with controlled structures and multiscale and multifunctional assembly possibilities. It allows one to prepare a large variety of materials with improved and tunable properties for a wide range of applications. Furthermore, the relatively easy scaling-up of this process may allow mass production of nanofibers and the introduction of their related fabrics and devices into the market and everyday life.

For these reasons, in the last decade, numerous articles and patents have been published on applications based on electrospun fibers used in energy conversion and storage and in environmental sensing and remediation. The aim of this book is thus to take stock of these publications and give an overview of the recent advances in the application of electrospun materials in energy- and environmentally related devices. Experts of electrospinning and their specific applications have come together to contribute to this manuscript. This will include not only the monitoring of the present results but also the identification of the future challenges to be taken up to give concrete answers to the environmental issues.

After an introduction of the electrospinning technique and its origins, with an outline of the achievable 1D nanoscaled materials and their varied applications, the book is divided into two main parts.

The first one concerns the application of the electrospun materials in energy devices. In particular, means of energy conversion and storage alternative to the fossil fuel exploitation, including low- and high-temperature fuel cells, hydrogen storage, dye-sensitized solar cells, Li-ion batteries, and supercapacitors, are addressed.

The second part deals with environmentally related applications of electrospun fibers. The use of electrospinning-issued materials in membranes for water and air purification as well as in sensors and biosensors for pollution control is developed.

This overview on the application of electrospinning in energy- and environmentally related devices is addressed to a broad audience, academic as well as industrial; beginners will benefit from the basic approaches to the electrospinning technique and their advantages in the dedicated applications, but also experts in the field may appreciate a complete review of the last advances and technical challenges. Our hope is that this book will not only serve as a reference manual for newcomers and experts in the field but that it may also inspire new ideas and developments for future energy and environmental applications.

Sara Cavaliere
Montpellier, France

Editor

Dr. Sara Cavaliere is assistant professor at the University of Montpellier in the Laboratory for Aggregates, Interfaces and Materials for Energy–Charles Gerhardt Institute for Molecular Chemistry and Materials, Montpellier, France. since 2009. She received her PhD in chemistry and materials science in 2006 in Versailles, France, after graduating in chemistry from the University of Milan, Italy. She carried out postdoctoral research work at the University of Freiburg, Germany, and the University of Lyon, France. In 2012, she was awarded a European Research Council starting grant (project SPINAM) on the use of electrospinning for the elaboration of integrated membrane-electrode assemblies for energy devices. Her research interests focus on the design and preparation of nanostructured and nanofibrous materials for energy conversion and storage applications.

Contributors

Dominique C. Adolphe
Laboratory of Textile Physics and
Mechanics
National School of Engineers Sud
Alsace
University of Haute-Alsace
Mulhouse, France

Annalisa Aluigi
Institute of Organic Synthesis and
Photoreactivity
National Research Council
Bologna, Italy

Marco Ballestri
Institute of Organic Synthesis and
Photoreactivity
National Research Council
Bologna, Italy

Andrea Bearzotti
Institute of Atmospheric Pollution
Research
National Research Council
Rome, Italy

Sara Cavaliere
Institute Charles Gerhardt of
Montpellier
Aggregates, Interfaces and Materials
for Energy
National Center for Scientific Research
University of Montpellier
Montpellier, France

David Cornu
National Graduate School of Chemistry
of Montpellier
European Membrane Institute of
Montpellier
Montpellier, France

Laura Coustan
Institute Charles Gerhardt of
Montpellier
Aggregates, Interfaces and Materials
for Energy
National Center for Scientific Research
University of Montpellier
Montpellier, France

Paolo Dambruoso
Institute of Organic Synthesis and
Photoreactivity
National Research Council
Bologna, Italy

Fabrizio De Cesare
Institute of Atmospheric Pollution
Research
National Research Council
Rome, Italy
and
Department for Innovation in
Biological, Agro-Food and Forest
Systems
University of Tuscia
Viterbo, Italy

Renaud Demadrille
Institute for Nanoscience and
    Cryogenics
Structure and Properties of Molecular
    Architectures
Alternative Energies and Atomic
    Energy Commission
French National Centre for Scientific
    Research
University Grenoble Alpes
Grenoble, France

Giorgio Ercolano
Institute Charles Gerhardt of
    Montpellier
Aggregates, Interfaces and Materials
    for Energy
National Center for Scientific Research
University of Montpellier
Montpellier, France

Frédéric Favier
Institute Charles Gerhardt of
    Montpellier
Aggregates, Interfaces and Materials
    for Energy
National Center for Scientific Research
University of Montpellier
Montpellier, France

Marcus Fehse
Department of Advanced Materials for
    Energy
IREC Catalonia Institute for Energy
    Research
Sant Adrià de Besòs, Spain

Claudia Ferroni
Institute of Organic Synthesis and
    Photoreactivity
National Research Council
Bologna, Italy

Stefano Giancola
Institute Charles Gerhardt of
    Montpellier
Aggregates, Interfaces and Materials
    for Energy
National Center for Scientific Research
University of Montpellier
Montpellier, France

Andrea Guerrini
Institute of Organic Synthesis and
    Photoreactivity
National Research Council
Bologna, Italy

Amir-Houshang Hekmati
Faculty of Engineering
Department of Textile Engineering
Islamic Azad University
Tehran, Iran

Nabyl Khenoussi
Laboratory of Textile Physics and
    Mechanics
National School of Engineers Sud
    Alsace
University of Haute-Alsace
Mulhouse, France

Damien Joly
Institute for Nanoscience and
    Cryogenics
Structure and Properties of Molecular
    Architectures
Alternative Energies and Atomic
    Energy Commission
French National Centre for Scientific
    Research
University Grenoble Alpes
Grenoble, France

Deborah Jones
Institute Charles Gerhardt of
    Montpellier
Aggregates, Interfaces and Materials
    for Energy
National Center for Scientific Research
University of Montpellier
Montpellier, France

Ji-Won Jung
Department of Materials Science and
    Engineering
Korea Advanced Institute of Science
    and Technology
Daejeon, Republic of Korea

Il-Doo Kim
Department of Materials Science and
    Engineering
Korea Advanced Institute of Science
    and Technology
Daejeon, Republic of Korea

Arthur Lovell
Cella Energy
Rutherford Appleton Laboratory
Oxfordshire, United Kingdom

Antonella Macagnano
Institute of Atmospheric Pollution
    Research
National Research Council
Rome, Italy

Yannick Nabil
Institute Charles Gerhardt of
    Montpellier
Aggregates, Interfaces and Materials
    for Energy
National Center for Scientific Research
University of Montpellier
Montpellier, France

Ahsan Nazir
Laboratory of Textile Physics and
    Mechanics
National School of Engineers Sud
    Alsace
University of Haute-Alsace
Mulhouse, France

Jacques Rozière
Institute Charles Gerhardt of
    Montpellier
Aggregates, Interfaces and Materials
    for Energy
National Center for Scientific Research
University of Montpellier
Montpellier, France

Laurence Schacher
Laboratory of Textile Physics and
    Mechanics
National School of Engineers Sud
    Alsace
University of Haute-Alsace
Mulhouse, France

Giovanna Sotgiu
Institute of Organic Synthesis and
    Photoreactivity
National Research Council
Bologna, Italy

Lorenzo Stievano
Institute Charles Gerhardt of
    Montpellier
Aggregates, Interfaces and Materials
    for Energy
National Center for Scientific Research
University of Montpellier
Montpellier, France
and
Research Network on Electrochemical
    Energy Storage
Amiens, France

Surya Subianto
Ian Wark Research Institute
University of South Australia
Boulevard, Adelaide, Australia

Greta Varchi
Institute of Organic Synthesis and
   Photoreactivity
National Research Council
Bologna, Italy

Emiliano Zampetti
Institute of Atmospheric Pollution
   Research
National Research Council
Rome, Italy

Wenjing Zhang
Department of Energy Conversion and
   Storage
Technical University of Denmark
Roskilde, Denmark

1 Fundamentals of Electrospinning

Surya Subianto, David Cornu, and Sara Cavaliere

CONTENTS

1.1 NANOFIBERS IN SCIENCE AND TECHNOLOGY

Fibers are present in nature for millennia and are common materials in our day-to-day lives. The spider web is one typical example, showing in the simplicity of its very light 1D geometry its high degree of multifunctionality. It is designed to catch insects but also possesses extraordinary flexibility and toughness allowing it to resist to wind and storms: the strength/mass ratio of such 1D materials exceeds that of steel.[1,2] In order to solve complex human and technological problems, science has often mimicked the models and the systems existing in nature. In the last years nanomaterials (i.e., structures with at least one dimension in the range of 1–100 nm) have attracted a dramatically growing scientific interest due to their unique properties compared to their bulk counterparts (superparamagnetism, quantum effect, electronic effects).[3] In particular, 1D nanomaterials including fibers, rods, tubes,

1

and ribbons are of particular interest to investigate the dependence of optical, mechanical, and electronic properties on size confinement and directionality. Due to their structure already proofed in nature and their diameter in the nanometer range, nanofibrous materials have peculiar properties, thus offering unexpected opportunities for their use in many areas. Among the key features of nanofibers and related nonwovens, we can cite the high specific surface area as well as the high porosity (up to 90%) and related lightweight, the high mechanical strength and flexibility, the 1D confinement, and the high orientation of structural elements along the fibers. The great variety of monodimensional materials that can be prepared with all the possible chemical compositions offered by the periodic table and structures ranging from solid to porous and hollow fibers leads to the possibility of tuning their properties for the targeted applications. In order to further improve them and prepare novel multifunctional nanofibers with enhanced applications to reproduce the natural fibers, another opportunity is the possibility to modify the fibers by physical and/or chemical methods during or following the production process. The two mainly used approaches of nanofiber modification are functionalization and encapsulation. The first involves depositing functional molecules or metal or semiconductor nanoparticles on the inner or outer surfaces of the 1D nanomaterials.[4] That can significantly modify the surface properties, which, for instance, can be modulated between superhydrophobic and superhydrophilic. Surface functionalization can also affect the solubility and the processability of the nanofibers as well as modify their electronic properties.

The second possible approach to engineer electric, optic, catalytic, and magnetic properties of the nanofibers is the incorporation of nanoparticles into these nanostructures.[5] On the one hand, the nanoparticles provide novel or improved performance to the 1D materials, and on the other hand the nanofibers could protect the very reactive encapsulated nanomaterials from oxidation and corrosion. In both approaches, engineering the surfaces and/or the cores of nanofibers with functional groups or nano-objects provides a straightforward route to complex hierarchical nanostructures with multiple functionalities with wide applications.

Due to the immense opportunities offered by nanofibers for creating products with new properties, much effort has been devoted in the last decades to the synthesis of such materials with different nature and compositions (polymers, semiconductors, metals, and oxides).

Several bottom-up methods have been developed, consisting in the self-assembly of the targeted material from vapor, liquid, or solid phases through nucleation and growth. For instance, solvothermal, electrochemical, and chemical synthesis using supersaturation control, vapor–liquid–solid methods (VLS) (e.g., chemical vapor deposition), and self-assembly of 0D materials, templates, or capping agents to control and direct the growth have been used to prepare nanofibers and nanotubes.[6,7] Chemical methods provide an attractive way to obtain 1D materials of different nature with potential high-volume production. However, they often require multiple steps such as the preparation/removal of templates, the elaboration of the catalyst, and the synthesis of 0D building blocks. The different process conditions in each step as well as the discontinuity of the resulting materials often limit their alignment, assembly, and processing into practical applications.

Other disadvantages include the possible damage of the obtained materials during template removal, the difficulty in the choice of the adequate capping agent and in the control of dimension and dispersity, and the contamination with metals in the case of VLS methods. The assembly of matter at the molecular level into synthetic nanofibers is thus an expensive and cumbersome process with limited control over the dimensions of the produced materials.

On the other hand, top-down methods based on size reduction such as nano-lithography and etching (electron-beam or focused-ion beam writing), mechanical stretching, and nanofluidics, have also been developed for the fabrication of nano-fibers.[8] The difficulty in achieving large amounts of materials in a large range of compositions, in a rapid way, and at low cost precludes the use of this kind of tech-niques for large-scale applications. Among the top-down engineering approaches, spinning methods use spinneret extrusions to produce continuous fibers. Materials to be extruded must be converted to a plastic or liquid state ("dope") by dissolution, heating, or pressurization. These methods can be classified according to the nature of the spinning dope, namely, melt spinning, solution spinning, and emulsion spinning (Figure 1.1).[9] Among these techniques, electrospinning is very flexible allowing the processing of solutions, gels and liquid crystals, melts, as well as emulsions.

While the other conventional extrusion techniques predominantly rely on mechanical and shear forces and geometric boundary conditions, fiber formation in electrospinning is governed by self-assembly processes induced by electric charges.[10] Indeed, its principle is the application of an electric field on a polymer droplet; when the repulsion force overcomes the surface tension, a thin polymer jet is ejected, which is elongated and accelerated by the electric field, dried during flight, and finally deposited on a substrate as a nonwoven nanofiber mat. Similar to mechanical drawing, electrospinning is a continuous process, thus allowing high-volume pro-duction, as well as the control of fiber placement and the integrated manufacturing of fibrous assemblies. In addition, different from extrusion, electrospinning is suited to produce very thin fibers with diameters down to the nanoscale[11] opening the way to novel properties and consequent applications (e.g., great improvement on the fiber mechanical properties). Another intriguing advantage of electrospun fibers over 1D nanomaterials obtained with some other methods is the possibility of fiber align-ment, which represents a challenge in electronics and composite reinforcements.[12] Electrospinning is also cost effective when compared to the other described top-down and bottom-up methods. Indeed, electrospun nanofibers are uniform and do not require expensive purification.

Electrospinning easily allows the functionalization of the fibers with molecules and/or nanoparticles. This can be achieved with one-pot routes during the electros-pinning process by using presynthesized nanoparticles[5] or preparing them in situ[13] or after fiber formation by immersing them in colloidal solutions.[14] It can be underlined that electrospinning of nanoparticles with anisotropic morphologies (nanotubes, nanorods, etc.) may provoke their alignment within the fibers in order to lower their Gibbs free energy.[15] Electrospinning shows thus its feasibility as a simple and effec-tive self-assembly method with further advantages over other self-assembly methods of obtaining free-standing mats with high flexibility and large area, which is favor-able for potential commercial applications.[16]

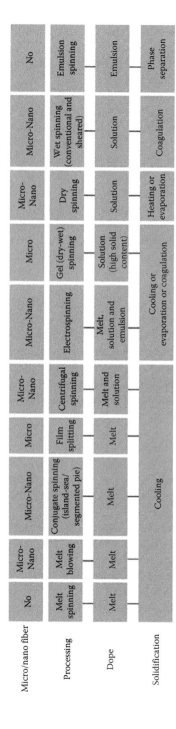

FIGURE 1.1 A concise comparison of fiber spinning processes. (Luo, C., Stoyanov, S., Stride, E., Pelan, E., and Edirisinghe, M., Electrospinning versus fibre production methods: From specifics to technological convergence, *Chem. Soc. Rev.*, 41, 4708–4735, 2012. Reproduced by permission of The Royal Society of Chemistry.)

Finally, electrospinning can be scaled up to produce nanofibers on industrial level enabling spreading of applications of the obtained fabrics.[17,18] In particular nozzle-less electrospinning[19] patented in 2005 opened the way to the high-quality and low-cost mass scale production and industrial exploitation of nanofibers generating a revolution in the field of structural materials and composites.[20] To summarize, electrospinning has the advantages of simplicity, efficiency, low cost, high yield, and high degree of reproducibility of the obtained materials as well as high versatility in the obtained compositions and architectures. The next paragraphs deal with the principle and historical development of this technique, its progress to achieve fibers with different compositions and architectures, and the wide application fields of the obtained nanomaterials.

1.2   PRINCIPLE AND THEORY OF ELECTROSPINNING

When electrostatics meets fluid dynamics, they give birth to electrospraying and electrospinning, which can be seen as their two most popular children in terms of applied technology.

The basic principle of electrospinning is simple to understand but rather complex to model. The most common electrospinning process to be considered is the needle-based technique; improved variations of this process will be depicted later. This simple system consists of a metal-tip extrusion system (a syringe coupled with a metal needle, for instance), a high-voltage supply, and a grounded collector (Figure 1.2). When a liquid is slowly pushed through a needle, a droplet of liquid is formed, exhibiting a hemispherical surface. The droplet can be then subjected to different forces like gravity. If the pressure applied on the liquid in the syringe is continuous, a liquid jet is ejected from the needle. In the case of a low-viscosity fluid, the extruded liquid is experimentally not forming a continuous and homogeneous cylindered jet due to the Rayleigh–Plateau instability, which deals with surface tensions and is related to the existence of tiny perturbations in the jet.[21] However, as it can be experimented daily with liquid honey, for instance, if the fluid is (highly) viscous, like a solution of polymer, tiny perturbations are weakened along the extruded filament. Plateau–Rayleigh instability can be therefore more or less overcome to yield a continuous jet. This is the first step to fiber formation.

If the filament formation is only governed by capillary through the needle/spinneret, even if a mechanical stretching is subsequently applied to the extruded filament, naturally by gravity or with the assistance of a mechanical setup, the resulting fiber diameter is usually in the range of mm or μm, not below. The breakthrough, which enables to come down to the nanoscale, is the contribution of electrostatics. For that purpose, high voltage (several tens of kV) is applied between the needle/spinneret and the (grounded) collector. Due to the effect of the high electric field and surface tension, the hemispherical surface of the extruded droplet is extended to create a conical shape, known as the Taylor cone in reference to the pioneer works of Sir G. Taylor on the disintegration of water drops in an electric field.[22] This cone exhibits a spherical tip and its shape tends to an oblique circular cone with an angle of 98.6°. When this value is reached, the rounded tip inverts and emits a liquid jet. This fluid

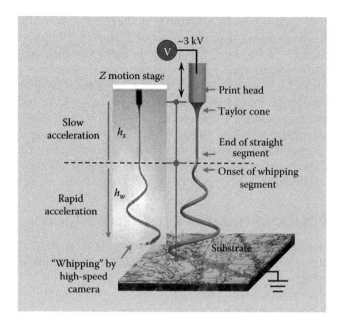

FIGURE 1.2   Image and scheme illustrating the conical path of flight of an electrospun jet. (Huang, Y., Bu, N., Duan, Y., Pan, Y., Liu, H., Yin, Z., and Xiong, Y., Electrohydrodynamic direct-writing, *Nanoscale*, 5, 12007–12017, 2013. Reproduced by permission of The Royal Society of Chemistry.)

electrodynamics phenomenon is the first step of the electrospinning and electro-spraying processes. Thus Plateau–Rayleigh instability is again to be considered. In some experimental conditions, perturbations are high enough to form droplets of liquids from the jet, which will yield in fine powders or coatings on the collector, depending on the solvent evaporation capacity. In other experimental conditions, Plateau–Rayleigh instability does not win the match, and a continuous electrospun filament is collected on the grounded collector. During the flight from the needle to the collector, solvent evaporation occurs, which permits to collect solid fibers. This phenomenon is accompanied by a specific flight path, known to result from bending instabilities.[23] In few words, the longitudinal forces induced by the external electric field are responsible for a straight path during the first part of the flight. Then, when the solvent evaporates, the fiber diameter is slowly reduced thus generating lateral repulsive forces. The motion of the jet is affected by these new contributions, which yield a flight path included in a conical volume, as illustrated by Figure 1.2.[24]

1.3   ORIGIN OF ELECTROSPINNING

The history of electrospinning is highly documented.[25–27] Up to our knowledge, the history of electrospinning really starts in 1900 with the patent deposited by J.F. Cooley[28] and entitled *Improved methods of and apparatus for electrically separating the relatively volatile liquid component from the component of relatively fixed*

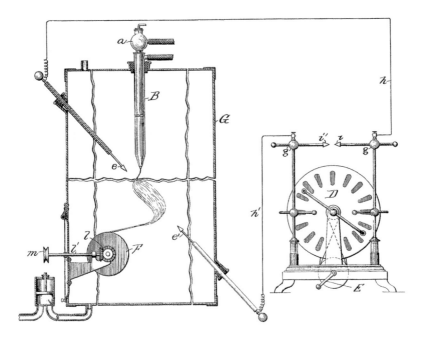

FIGURE 1.3 Cooley's apparatus: ancestor of the electrospinning/electrospraying setup. a, tube bulb; B, enclosing tube; D, electric current generator; E, crank; e and e', electrodes; F, reel; G, set-up case; g, and g', positive and negative poles; h and h', wires; i and i', supplementary electrodes; l, bevel gearing; l', counter shaft; m, driving pulley. (From Cooley, J., Apparatus for electrically dispersing fluids, U.S. Patent 692,631, 1902.)

substances of composite fluids. Pioneer and visionary, using the up-to-date high-voltage apparatus, J.F. Cooley reports the principle and system to generate powders or fibers. The influence of the viscosity on the latter was even discussed on the basis of his experiments. Figure 1.3 presents the schematic principle of Cooley's apparatus.

From this pioneer's works, there were continuous advances in electrospinning process and fundamental comprehension of the involved mechanisms. One can cite the patents of Anton Formhals on artificial threads made by electrostatic process in the beginning of the twentieth century[29,30] and the fundamental works of G. Taylor on electrohydrodynamics, as previously mentioned. To the best of our knowledge and according to the Web of Science, the word "electrospinning" in a scientific article, derived from "electrostatic spinning," was introduced in 1995 by R.H. Reneker and coworkers.[31,32] Taking benefits from all these insights, several companies are now commercializing electrospinning setups, and electrospun fibers are finding more and more useful applications since at least one decade in filtration and tissue engineering.

1.4 BEYOND THE ELECTROSPINNING OF POLYMERS

Initially, electrospinning has been exclusively applied on the production of nanofibers of natural and synthetic polymers and carbon, with targeted applications on filtration, textile, and biomedical systems.[26,27,33] Due to the dramatically growing development of the technique related to the rising interest on nanomaterials, in the last decade, it was extended to fiber formation based also on metal oxides, metals, and composite/hybrid systems.[25] In particular, the first fabrication of ceramic nanofibers, namely, titanium dioxide, has been reported in 2003.[34] The approach was based on a combination of electrospinning with sol–gel method, and since then it has been greatly expanded to produce ceramic nanofibers from hundreds of different materials.[35,36] The method consists of dissolving in a relatively volatile solvent (e.g., ethanol, isopropanol, water) the precursor of the targeted material (e.g., a metal salt or metal alkoxide) together with a carrier polymer ensuring the suitable rheological properties to obtain a fibrous material. Due to its high solubility in ethanol or water and its good compatibility with many metal precursors, poly(vinyl pyrrolidone) (PVP) is one of the most widely used carrier polymers. Other polymers, such as poly(vinyl acetate), poly(vinyl alcohol) (PVA), poly(methyl methacrylate), and poly(acrylic acid) (PAA), have also been employed for this purpose.[35] In order to avoid the rapid hydrolysis of metal precursors and control the electrospinning process, an additive such as a catalyst or a salt can be also introduced.

As catalysts, acetic or hydrochloric acid has been used to slow down hydrolysis and gelation rates, preventing the blockage of the spinneret and the formation of a nonelastic jet, thus ensuring a continuous and uniform spinning. As salts, sodium chloride or tetramethyl ammonium chloride has been used to increase the charge density on the liquid jet and thus prevent the formation of beads and obtain uniform thin fibers. Electrospinning of such solutions leads to composite nanofibers in which the salt is homogeneously dispersed into the polymer. These precursor/polymer fibers are then subjected to calcination at high temperature in order to remove the organic part and obtain pure inorganic crystalline nanofibers. A great variety of metal oxides has been prepared with this approach including ZnO, NiO, TiO_2, SnO_2, SiO_2, Fe_2O_3, $LiCoO_2$, and $LaMnO_3$.[35,37,38] When the thermal treatments are performed in the presence of reducing agents such as hydrogen, 1D metal structures (e.g., Cu, Pt, metal alloys) can be achieved.[39,40] The presence of other reactive gases (such as CH_4 or NH_3) during the thermal treatment or of a specific nitrogen/carbon/phosphorous donor in the electrospinning solution allows to further extend the range of chemical compositions of electrospun fiber toward metal carbides, oxonitrides, nitrides, and phosphates.[41–43]

The inorganic fibers issued from blended precursor/polymer solutions after removal of the carrier polymer by the thermal treatment present rough surfaces and inherently poor mechanical properties. Electrospinning of pure inorganic precursors with the suitable viscosity has thus also been introduced, in particular to obtain ceramic fibers from alkoxides.[44,45] Very recently the influence of the sol properties on the resulting silica nanofibers and the stability of the electrospinning process have been assessed.[46] Other kinds of materials such as phosphates can be prepared without the use of carrier polymers. For instance, CsH_2PO_4 can be previously polymerized giving rise to a transparent viscous solution that can be electrospun in the absence of any carrier polymer or additive to generate pure inorganic nanofibers in a one-pot route.[47]

To obtain certain categories of materials, the precursor/polymer electrospinning method is not applicable, and novel original routes have been opened thanks to the versatility of the technique. For instance, in order to prepare zirconium phosphate (ZrP) that is unsuitable for direct electrospinning due to the formation of a gel blocking the spinneret, *reactive* electrospinning has been developed.[48] This method is based on coaxial electrospinning (see Section 1.5.2), where the sheath and core solutions contain the sources of phosphorus and zirconium, respectively, resulting in the in situ formation of ZrP during jet ejection.

Dealing with inorganic nanofibers, it is worth to cite their fragility that makes them difficult to handle avoiding their direct application, for instance, in membrane technology. This issue is still not completely solved. However, several strategies including the use of specific morphologies or structures such as nanoribbons[49] or twisted yarns[50] or the formation of composites with polymers[51] can be applied to overcome the intrinsic brittleness of such ceramic materials.

Electrospinning has been further applied to the preparation of hybrids and composite nanofibers. As already pointed out, the functionalization or loading of electrospun fibers with nanoparticles and/or functional groups is a common route. In this way, polymer or ceramic fibers loaded with chromophores, biomolecules, magnetic or catalytic nanoparticles, etc., with a widespread range of applications can be easily produced.[5,13,52–54]

These few examples mentioned earlier have shown the versatility in the compositions of the 1D materials obtained by electrospinning. In addition, this technique allows the preparation of a number of other morphologies and structures than conventional solid nanofibers, including porous, hollow, fiber-in-tube, core-sheath, and multichannel fibers, nanoribbons, and nanorods. Even the architectures of the electrospun fibers can be controlled in situ by modifying processing parameters and/or setup geometries so as to obtain nonwoven, aligned, or patterned fibers, random 3D structures, submicron spring, and convoluted fibers with controlled diameters. The next paragraph will elucidate these features with more comprehensive examples.

1.5 ARCHITECTURES OF ELECTROSPUN FIBERS

Despite the apparent simplicity of the electrospinning process, electrospun nanofibers are complex materials whose morphology and microstructure are affected by numerous parameters. The applied potential, the nature of the collector, the solution parameters, the intrinsic properties of the electrospun material, as well as the electrospinning conditions all play a part in determining their morphology and microstructure. In this regard, many studies in the literature have provided the information necessary to tune the macroscopic and microscopic features of electrospun fibers.

Electrospinning can affect the morphology in both the microscale and the macroscale, as the rapid drawing of the fiber during the process has been shown to induce molecular orientation of the polymer chains, while control of macroscopic morphological features such as fiber diameter, surface porosity, fiber alignment, and also its physical form has been shown to be possible. Indeed, these macroscopic features can be controlled through solution and electrospinning parameters, and this in turn leads to further development of our understanding of the microstructure of such materials. In particular, it is well known that orientation of a polymer's molecular structure can strongly affect its properties and thus its ultimate performance in devices.

1.5.1 MORPHOLOGY AND STRUCTURE OF ELECTROSPUN FIBERS

1.5.1.1 Molecular Alignment of Electrospun Fibers

Polymer nanofibers exhibit exceptional mechanical properties compared to their macroscopic counterparts. Indeed, when the fiber dimension is comparable to orientation-correlated regions, there is a confinement of the molecular structure, resulting in a quasi-1D system with a greater ordering than the bulk material.[55] However, earlier studies have concluded that molecular orientation in the electrospun fibers was poor and not well developed. These studies have been generally conducted on multiple fibers and thus complicated by the convolution of microscopic structure orientation and macroscopic fiber alignment. Since then, however, the development of collectors that allowed the preparation of well-aligned fiber bundles has provided greater insight into the level of molecular orientation in electrospun fibers.[56] The use of a rotating collector has been shown to induce increased orientation in nanofibers. A study by Fennessey et al.[57] showed that poly(acrylonitrile) (PAN) fibers collected at progressively higher speeds on a rotating mandrel have increased crystallinity and orientation, while those on a static collector were isotropic. This was partially attributed to residual solvent allowing the polymer chains to relax when a static collector was used, but the authors also hypothesized that the high rotation speeds induced molecular orientation of the polymer in the fibers through additional stretching. Although the orientation value was relatively low, several later studies have led to similar conclusions with regard to rotating collectors.

A work using electrospun poly(ethylene oxide) (PEO) fibers[58] on charged metallic plate collectors separated by a short gap has shown that fiber alignment and molecular orientation can be significantly improved. When the fibers collected on such charged collectors are compared to those collected on a rotating drum collector, they show much higher level of molecular orientation despite having similar levels of macroscopic fiber alignment. It was initially thought that, due to the use of water as the solvent, the high level of orientation was related to the orientation of the solvent dipole in the electric field inducing orientation on the polymer backbone. However, latter studies with other solvent systems show similar level of orientation, suggesting that the high molecular orientation is not determined by the solvent relaxation time. Furthermore, similarly high orientation was later found in PEO complexes that do not form channel structure[59,60] and in pure poly(oxomethylene) fibers.[61] All these systems possess high crystallinity and a fast crystallization kinetics, which indicates that the molecular ordering in these nanofibers is driven by the formation of oriented nuclei that hinder the relaxation of the polymer chains into an isotropic state. In fibers of fully amorphous polymers with a T_g below room temperature, it can be expected that chain mobility will allow sufficient relaxation of the polymer chain back to an isotropic state even if the microstructure was initially affected by the drawing of the fibers during electrospinning.[58]

1.5.1.2 Fiber Diameter and Homogeneity

The morphology of the electrospun fibers is strongly affected by solution and electrospinning parameters such as polymer concentration, solvent, capillary size, flow rate, working distance, and applied potential. Among these parameters, solution concentration and viscosity have been generally found to be one of the most important factors affecting fiber diameter.[62] Studies with various polymer systems have found that fiber diameter increases significantly as the solution concentration increases,[26,63] and this increase follows a power law relationship.[64]

The solution viscosity is related to the extent of polymer chain entanglement[65,66] and thus plays a critical role not only in fiber size but also in its homogeneity. Below a certain critical value, the electrospinning jet breaks up and electrospraying occurs, where beads or droplets are formed. If there is sufficient density of chain entanglement, but the extent of chain overlap is less than the critical value, an intermediate, beaded fiber morphology is obtained due to the partial dampening of the Plateau–Rayleigh instability (see Section 1.2).[67]

Such beaded fiber morphology is also related to the solvent in the solution as it is driven by surface tension. During electrospinning, the fibers undergo a solidification process determined by surface tension and a relaxation process controlled by viscoelastic properties.[68] Surface tension attempts to minimize the surface area per unit mass and thus acts to counter the electrostatic charge on the jet, whereas viscoelasticity resists rapid changes in shapes.[69] Thus, decreasing the surface tension, increasing the viscosity, or increasing the net charge density tends to stabilize the electrospinning jet, resulting in smoother fibers.

Fiber uniformity has also been shown to be affected by solution conductivity,[70,71] in that increasing the conductivity through the addition of salt may produce finer, more uniform fibers through the increase in charge density, resulting in an increased elongational force exerted on the fiber jet resulting in a greater whipping instability. However, the reverse has also been observed in certain polymers such as PAA,[72] where the interaction between the ionic groups is a more significant factor affecting fiber formation than the increased charge density in solution.

Other parameters such as the dielectric constant and boiling point of the used solvent may also play a significant role in the homogeneity of electrospun fibers; however, it is more difficult to isolate these parameters and study their effect on fiber morphology. A study with polystyrene[73] found that a solvent with higher dielectric constant enables the electrospinning of the polymer; nevertheless, the overall electrospinnability depends on the combination of the solvent's dipole moment, the conductivity of the solution, the boiling point of the solvent (which determines the speed of solvent evaporation), and also the viscosity and surface tension of the solution.

1.5.1.3 Fiber Morphology

Electrospun materials are not only solid fibers; other morphologies are also possible (Figure 1.4).[74–79] In certain cases, the electrospinning process resulted in deposition of micron-sized ribbons rather than nanofibers.[74–76] This occurs when the solvent evaporates sufficiently quickly to form a skin on the polymer jet, which later collapsed to form a ribbon. Since this occurs before the instability of the jet could further stretch the fibers to the nanometer size range, the created ribbons are generally larger in terms of size compared to electrospun fibers of the same polymer.

Other morphologies such as hollow fibers have been obtained either by coaxial electrospinning (which will be discussed in the latter section) or through conventional

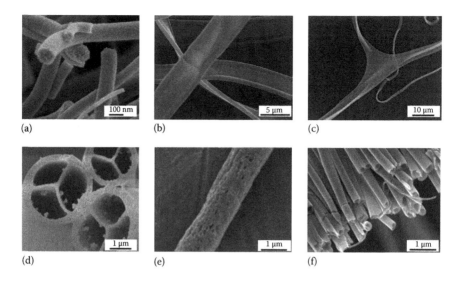

(a)   (b)   (c)

(d)   (e)   (f)

FIGURE 1.4 SEM micrographs showing some of the 1D morphologies obtainable through electrospinning (solid, ribbonlike, branched, porous, and hollow fibers). (a) Titanium dioxide solid fibers. (Reprinted from *J. Power Sources*, 257, Savych, I., Bernard d'Arbigny, J., Subianto, S., Cavaliere, S., Jones, D.J., and Rozière, J., On the effect of non-carbon nanostructured supports on the stability of Pt nanoparticles during voltage cycling: A study of TiO_2 nanofibres, 147–155, 2014, with permission from Elsevier.) (b and c) Flat and branched fibers of 10% poly(ether imide), respectively. (Koombhongse, S., Liu, W., and Reneker, D. H.: Flat polymer ribbons and other shapes by electrospinning. *J. Polym. Sci. Part B: Polym. Phys.* 2001. 39. 2598–2606. Copyright Wiley-VCH Verlag GmbH & Co. KGaA. Reproduced with permission.) (d) Titanium dioxide three-channel-tubes. (Reprinted with permission from Zhao, Y., Cao, X., and Jiang L., Bio-mimic multichannel microtubes by a facile method, *J. Am. Chem. Soc.*, 129, 764–765, 2007. Copyright 2007 American Chemical Society.) (e) Nylon-6 porous fibers. (Reprinted with permission from Gupta, A., Saquing, C.D., Afshari, M., Tonelli, A.E., Khan, S.A., and Kotek, R., Porous nylon-6 fibers via a novel salt-induced electrospinning method, *Macromolecules*, 42, 709–715, 2009. Copyright 2009 American Chemical Society.) (f) Uniaxially aligned array of anatase hollow fibers that were collected across the gap between a pair of electrodes. (Reprinted with permission from Li, D. and Xia, Y., Direct fabrication of composite and ceramic hollow nanofibers by electrospinning, *Nano Lett.*, 4, 933–938, 2004. Copyright 2004 American Chemical Society.)

electrospinning using supercritical CO_2 or through a needleless approach. It can also be obtained through electrospinning of emulsions.[80]

Under certain conditions, secondary jets may also form out of the main electrospinning jet, resulting in a "branched" fiber morphology, which has been exploited to yield morphologies resembling a net or thorns. Barakat et al.[81] have used metal salts in order to fabricate polymeric nets with many small fibers of less than 30 nm overlaying main fibers that are hundreds of nanometers in thickness. As the formation of these structures depends mostly on the ionization ability of the used salt, it was attributed to the ions creating bridges between the main fibers (which are not ionically balanced due to the salt attached to them) during electrospinning, resulting in the creation of secondary jets that form the net.

Another parameter of interest, fiber porosity has been extensively studied for applications such as filtration and tissue engineering (scaffolds) where a very high surface area may be desirable. Porosity can be increased through the addition of salts[82,83] or ice crystals[84] or by electrospinning composite fibers and removing one part of the composite. The second method relies on phase separation within the electrospun fibers that occurs with the evaporation of solvent, resulting in a binodal or spinodal structure depending on the particular system.[85] In this approach, the use of a highly swelling solvent can also result in a very high porosity after removal of the swelling agent.

1.5.2 Variations in Needle and Collector Geometry and Configuration

The earliest studies on electrospinning generally use a single needle through which the polymer solution passes through in combination with a grounded plate as the collector. Nowadays, many variations on the technique exist, in both varying the needle and collector geometries to achieve specific goals in terms of a particular fiber morphology or a larger output of fiber production, as the nature of the needle and collector significantly affects the morphology of the fibers.[86]

1.5.2.1 Coaxial Electrospinning

In addition to the conventional, single-needle electrospinning, other variations in needle geometries exist in order to take advantage of the relative simplicity of the electrospinning setup. In coaxial electrospinning, separate core and sheath solution are fed through a coaxial needle that allows for some interesting permutations to the fiber morphology[87,88] (Figure 1.5). Thus far, it has been applied for the preparation of core/sheath fibers, hollow fibers, and composite fibers and for electrospinning and immobilization of functional materials where the sheath or the core acts a carrier polymer, resulting in fibers decorated by droplets of the functional material along the axis.[89,90]

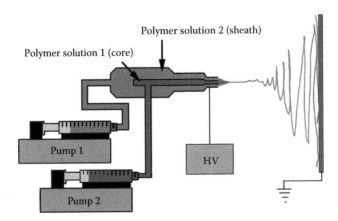

FIGURE 1.5 Schematic representation of a coaxial electrospinning setup.

Although coaxial electrospinning depends on the same parameters as conventional one, the interaction between multiple phases and their individual physical behavior adds an extra layer of complexity. The morphology and uniformity of the fiber depends on the solution properties of both the core and sheath solution, with a critical value needed on both solution concentrations and an optimized, matched flow rate needed in order to obtain uniform fibers.

Hollow fibers can be obtained if the core phase is removed after electrospinning through leaching or thermal treatment.[85] Such approach has been used in the fabrication of carbon[91,92] or inorganic hollow fibers with materials such as TiO_2[93,94] and SnO_2[95] or composite fibers,[96,97] as the core polymer can be removed during calcination, leaving an inorganic hollow fiber shell. Hollow ceramic fibers can also be obtained by using oil as the core solution[79,98,99] (Figure 1.4f), which can then be removed through washing of the fibers post-electrospinning.

1.5.2.2 Multiple Needle Electrospinning

One of the practical limitations of electrospinning is the deposition speed compared to traditional casting or extrusion methods, as in general only one jet is utilized. In order to increase throughput, some studies have investigated various approaches utilizing multiple needles in order to produce multiple fiber jets simultaneously.[18]

Studies using multiple-needle electrospinning generally found that such approach is fully feasible provided the distance between needles is sufficient so that the electrostatic fields acting on each of them are equal.[100] This is because the electrospinning jet is affected not only by the applied electric field and self-induced coulomb interactions but also by the mutual coulomb repulsion of individual charged jets.[101] As the jets are pushed away from their neighbors, closely packed arrays that will occur in the jets at the edges are bent outward, while the inner ones are squeezed along the line of the spinning nozzle.[102] Several needle geometries have been investigated, with both in-line and array-type arrangements being used to provide a greater throughput.

1.5.2.3 Needleless Electrospinning

Needleless electrospinning was another technique developed in order to achieve a greater throughput of fiber production.[103] Yarin et al.[104] utilized a two-layer system of a ferromagnetic suspension and a polymer solution. The ferromagnetic suspension is placed below the polymer solution, and a magnetic field is applied so that perturbation of the surface is created, forming spikes that are directed upward. When an electric potential is applied, these perturbations become the site of jet formation, with jets directed upward toward a collector where the nanofibers formed by these jets are deposited.

In another needleless approach developed and patented by Jirsak et al.,[103] a drum is utilized where the surface of the drum picks up the electrospinning solution as it rotates, and the potential difference is applied to the collector in order to produce the fibers (Figure 1.6). The study with poly(amic acid) found that higher concentration was required compared to conventional, single-needle electrospinning in order to produce fibers. Other subsequent studies have used polymers such as poly(vinylidene fluoride),[105] PAN[106] or PVA[107], and found that the spinneret geometry has a significant

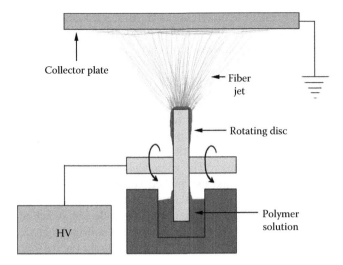

Collector plate

Fiber jet

Rotating disc

Polymer solution

HV

FIGURE 1.6 Schematics for a typical needleless electrospinning setup with a rotating disc.

influence on fiber morphology, in that disc and coil spinnerets with a narrow surface show that the electric field is more evenly distributed than cylindrical spinnerets. The coil spinneret was also shown to have a very high fiber production rate compared to other geometries due to its greater (longer) surface circumference.

1.5.2.4 Near-Field Electrospinning

Near-field electrospinning[108] is a method used to generate patterned fiber mat where a conductive tip rather than a needle is held at a close distance to the collector. At such small working distances, the electrospinning jet does not undergo a bending instability and thus are deposited in a controlled manner on a small area. When combined with a moving collector substrate, this results in the possibility of direct writing of micrometer- to millimeter-sized patterns on the substrate.[109–112] This approach has been used to fabricate patterned fibers of both organic and inorganic materials; however, the small working distances generally meant that fibers deposited in near-field electrospinning are micrometer rather than nanometer in scale.

1.5.2.5 Rotating Drum/Disc Collector

Although such simple collectors are sufficient to obtain fibers, it may not produce a homogeneous fiber mat as more fibers are deposited on the center of the target, resulting in a variation in thickness through the mat, which may also affect the fiber morphology. As such, the rotating drum collector has been developed where a cylindrical target is used with both rotational and translational motions. This allows the deposition of the fibers to occur homogeneously on the drum surface, resulting in a fiber mat with uniform thickness.

The rotation speed of the drum can also be used to control fiber alignment as at very high rotation speeds,[113,114] the fibers are aligned on the direction of rotation. This has been used in the rotating disc collector,[115,116] where a disc is used side-on to the needle as the greater circumference of the disc relative to conventional drum means that it has a greater take-up velocity of fibers, therefore achieving fiber alignment at relatively lower rpm compared to drum collectors.

1.5.2.6 Collectors for Aligned Fibers

Other configuration exists in order to obtain uniaxially aligned fibers (Figure 1.7). The parallel plate collectors consist of two strips or plates of metallic collectors, which are separated by an insulator. In this configuration the electrospinning jet is drawn in between the two plates, resulting in the deposition of uniaxially aligned fibers perpendicular to the plates.[117] The alignment of the nanofibers itself is dependent on the width of the gap and the applied voltage, as these parameters determine the direction of electrostatic forces acting on the fibers.[118] The number of fibers deposited on this type of collectors can also be controlled by switching an applied bias between the two plates.[119] Other variations of this collector include the parallel ring[120] and frame collectors[26] where conducting frames are set onto a triangular setup. By using this kind of collectors and controlling the electrospinning parameters, it is

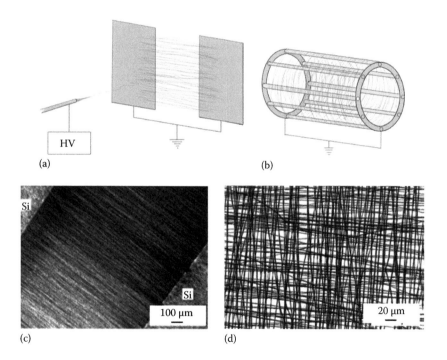

FIGURE 1.7 (a) Parallel plate and (b) rotating frame collectors used to align electrospun fibers and related examples: optical micrographs of PVP aligned fibers (c) collected on the gap formed between two silicon stripes; (d) stacked with their long axes rotated by 90°. (Reprinted with permission from Li D., Wang, Y., and Xia, Y., Electrospinning of polymeric and ceramic nanofibers as uniaxially aligned arrays, *Nano Lett.*, 3, 1167–1171, 2003. Copyright 2003 American Chemical Society.)

possible to prepare simple device structures, such as 2D arrays of crossbar junctions, 3D grid layer-by-layer structures, and individual nanofibers (Figure 1.7).[121]

1.5.2.7   Patterned Collectors

Randomly oriented fibers are generally collected on a grounded metallic plate collector; however, some studies have shown that it is possible to generate micro- and macroscale patterns of nanofibers using conductive grids or textiles.[122,123] A charged needle or knife edge can also be used to direct nanofiber deposition on a rotating drum collector.[124] A cross-grid pattern can also be obtained in a single-step process through the use of an aluminum table attached to a rotating disc collector.[115] The use of an insulating collector has also been shown to yield a bird's nest pattern,[125] where ionic liquid-doped polystyrene increases charge accumulation at the collector, only allowing a thin layer of fibers to be deposited in a particular area before depositing on other locations on the collector.

1.6   APPLICATIONS OF ELECTROSPUN MATERIALS

Until the 1990s, there was little interest in the use electrospun nanofibers, which were mainly employed in the filtration industry.[32] Since this period, the process rapidly attracted an increasing attention generated by potential applications of the obtained materials in nanotechnology. The electrospinning-related publications have nearly doubled every year, reaching about 200 articles in 2003. Since then, in the latest decade 18,600 articles dealing with electrospinning have been published with an exponential growth (2003–2013 SciFinder data: keyword electrospinning). Over a hundred synthetic and natural polymers as well as inorganic and composite materials have been electrospun into nanofibers. As demonstrated in the previous paragraphs, electrospinning is a high-yield fabrication method for nanofibers of a tremendous variety of chemical compositions and architectures.[17] This enables a huge variety of properties that can be tuned for specific purposes, such as axial strength, flexibility, porosity, and specific surface area, but also targeted catalytic, electronic, optical, and magnetic properties depending on the inherent composition and functionalization of the 1D nanomaterials. The recent years have thus witnessed the use of electrospun nanofiber mats and fabrics of several polymer, ceramic, and composite materials with different structures and architectures in disparate domains ranging from filtration,[126] protective clothing,[127] packaging,[128] self-cleaning surfaces,[129] to biomedicine (biomimetic fibers for tissue engineering and drug delivery),[130–133] (photo)catalysis,[35] electronics and photonics,[134–137] sensing,[38,138] and energy devices.[139–141] Figure 1.8 illustrates the different areas of application of the publications dealing with electrospinning with the relative percentages.

In particular, an exponential growth of the scientific research is focusing on energy and environment protection, which are the major societal issues of this century[142] (Figure 1.9). Indeed, on the one hand, the global energy demand is dramatically increasing, while on the other hand, the fossil fuel resources are limited and represent the principal contribution to the emission of greenhouse gases (Figure 1.10).[143] A great challenge in energy and environmental research is thus the development of new methods and materials for energy conversion and storage and

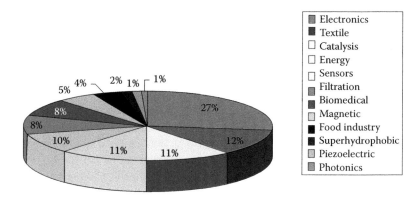

FIGURE 1.8 Relative proportion of publications in main areas of applications of electrospinning in the period 2003–2013 (SciFinder data: keywords electrospinning + type of application in the legend).

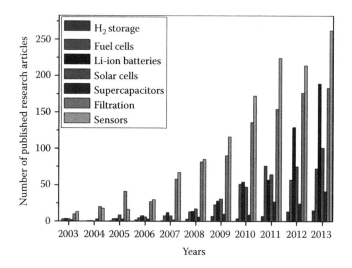

FIGURE 1.9 Number of research articles concerning electrospun nanofibers in energy and environmental applications published in the period of 2003–2013 (SciFinder data: keywords electrospinning + words in the legend).

environmental repair. Electrospinning is a promising answer to these issues as a means of fabrication and multiscale assembly of novel materials with improved and tunable properties. This explains the need for monitoring the present results and finding the future challenges within this book. Its aim is thus to present an overview of recent advances in the application of electrospun materials in energy devices including low- and high-temperature fuel cells, Li-ion batteries, solar cells, and supercapacitors as well as in environmental-related devices for water and air pollution control and biosensing.

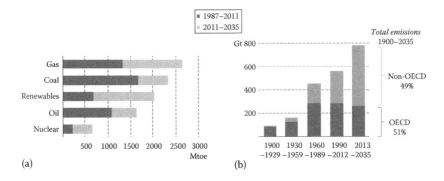

FIGURE 1.10 (a) Histogram representing the actual and foreseen growth in total primary energy demand from 1987 (toe, ton of oil equivalent); (b) actual and estimated cumulative energy-related CO_2 emissions from 1900 for OECD (Organisation for Economic Cooperation and Development) and non-OECD countries. (From OECD/IEA, World energy outlook, Presentation to Press, IEA Publishing, London, U.K., November 12, 2013, http://www.iea.org/t&c/termsandcondition.)

1.7   SUMMARY AND CHALLENGES

Electrospinning makes possible the infinite combinations of chemical compositions, offered by the periodic table, with morphologies and structures tunable with solution and process parameters. This gives access to an enormous variety of 1D nanomaterials with different targeted properties with potential application in a wide range of fields.

Nevertheless, for the full exploitation of the electrospun materials, some challenges need to be taken up. First of all, the full control on the composition and on the spatial organization of nanofibers, for example, in arrays or hierarchical architectures, is crucial to meet application in real devices. In that respect, the issues of fragility showed by ceramic fibers are still not completely solved. Specific morphologies (e.g., ribbons) or structures (e.g., yarns) or compositions (e.g., composites with polymers) can be applied with this aim.

Another essential point to be considered to further exploit the potential of nanofibers in practical applications is the improvement in the productivity of electrospinning. This implies not only the upscaling of the process that has already started, for instance, by using nozzleless spinnerets, but also the development of a technology allowing the direct integration of electrospun materials into an existing production system. The possibility of incorporating functional nanofibers in devices (for energy, electronic, textile, etc.) or directly building devices only from nanofibers is one of the major objectives to reach.

Finally, a great challenge is represented by novel developments and applications of nanofibers, which must go beyond materials chemistry and physics to reach transversal applications. This means the convergence of academia and industry in an interdisciplinary trend toward electrospun materials with technological uses.

Thanks to these developments and mass production, electrospinning may be one of the most significant technologies of this century allowing to solve global issues related to energy and environment.

REFERENCES

1. Porter, D. and Vollrath, F. Nanoscale toughness of spider silk. *Nano Today* **2**, 6 (2007).
2. Hagn, F., Eisoldt, L., Hardy, J. G., Vendrely, C., Coles, M., Scheibel, T., and Kessler, H. A conserved spider silk domain acts as a molecular switch that controls fibre assembly. *Nature* **465**, 239–242 (2010).
3. Zhou, Z.-Y., Tian, N., Li, J.-T., Broadwell, I., and Sun, S.-G. Nanomaterials of high surface energy with exceptional properties in catalysis and energy storage. *Chem. Soc. Rev.* **40**, 4167–4185 (2011).
4. Li, D., McCann, J. T., and Xia, Y. Use of electrospinning to directly fabricate hollow nanofibers with functionalized inner and outer surfaces. *Small* **1**, 83–86 (2005).
5. Cavaliere, S., Salles, V., Brioude, A., Lalatonne, Y., Motte, L., Monod, P., Cornu, D., and Miele, P. Elaboration and characterization of magnetic nanocomposite fibers by electrospinning. *J. Nanoparticle Res.* **12**, 2735–2740 (2010).
6. Xiong, Y., Mayers, B. T., and Xia, Y. Some recent developments in the chemical synthesis of inorganic nanotubes. *Chem. Commun. (Camb.)* **40**, 5013–5022 (2005).
7. Martin, C. and Van Dyke, L. Template synthesis of organic microtubules. *J. Am. Chem. Soc.* **112**(24), 8976–8977 (1990).
8. Xia, Y., Yang, P., Sun, Y., Wu, Y., Mayers, B., Gates, B., Yin, Y., Kim, F., and Yan, H. One-dimensional nanostructures: Synthesis, characterization, and applications. *Adv. Mater.* **15**, 353–389 (2003).
9. Luo, C., Stoyanov, S., Stride, E., Pelan, E., and Edirisinghe, M. Electrospinning versus fibre production methods: From specifics to technological convergence. *Chem. Soc. Rev.* **41**, 4708–4735 (2012).
10. Wendorff, J. H., Agarwal, S., and Greiner, A. *Electrospinning: Materials, Processing and Applications* (Wiley VCH, Weinheim, Germany, 2012).
11. Li, D. and Xia, Y. Electrospinning of nanofibers: Reinventing the wheel? *Adv. Mater.* **16**, 1151–1170 (2004).
12. Wang, M. C. P. and Gates, B. D. Directed assembly of nanowires. *Mater. Today* **12**, 34–43 (2009).
13. Cavaliere, S., Subianto, S., Chevallier, L., Jones, D. J., and Rozière, J. Single step elaboration of size-tuned Pt loaded titania nanofibres. *Chem. Commun. (Camb.)* **47**, 6834–6836 (2011).
14. Formo, E., Yavuz, M. S., Lee, E. P., Lane, L., and Xia, Y. Functionalization of electrospun ceramic nanofibre membranes with noble-metal nanostructures for catalytic applications. *J. Mater. Chem.* **19**, 3878–3882 (2009).
15. Ko, F., Gogotsi, Y., Ali, A., Naguib, N., Ye, H., Yang, G. L., Li, C., and Willis, P. Electrospinning of continuous carbon nanotube-filled nanofiber yarns. *Adv. Mater.* **15**, 1161–1165 (2003).
16. Zhang, C.-L. and Yu, S.-H. Nanoparticles meet electrospinning: Recent advances and future prospects. *Chem. Soc. Rev.* **43**, 4423–4448 (2014).
17. Persano, L., Camposeo, A., Tekmen, C., and Pisignano, D. Industrial upscaling of electrospinning and applications of polymer nanofibers: A review. *Macromol. Mater. Eng.* **298**, 504–520 (2013).
18. Zhou, F.-L., Gong, R.-H., and Porat, I. Mass production of nanofibre assemblies by electrostatic spinning. *Polym. Int.* **58**, 331–342 (2009).

19. Jirsak, O., Sanetrnik, F., Lukas, D., Kotek, V., Martinova, L., and Chaloupek, J. A method of nanofibres production from a polymer solution using electrostatic spinning and a device for carrying out the method. WO 2005024101 (2005).

20. Petrik, S. and Maly, M. Production nozzle-less electrospinning nanofiber technology. *MRS Proc.* **1240**, WW03–WW07 (2009).

21. Strutt, J. and Rayleigh, L. On the instability of jets. *Proc. Lond. Math. Soc.* **10**, 4–13 (1878).

22. Taylor, G. Disintegration of water drops in an electric field. *Proc. R. Soc. Lond. Ser. A: Math. Phys. Sci.* **280**, 383–397 (1964).

23. Reneker, D. H., Yarin, A. L., Fong, H., and Koombhongse, S. Bending instability of electrically charged liquid jets of polymer solutions in electrospinning. *J. Appl. Phys.* **87**, 4531–4547 (2000).

24. Huang, Y., Bu, N., Duan, Y., Pan, Y., Liu, H., Yin, Z., and Xiong, Y. Electrohydrodynamic direct-writing. *Nanoscale* **5**, 12007–12017 (2013).

25. Greiner, A. and Wendorff, J. H. Electrospinning: A fascinating method for the preparation of ultrathin fibers. *Angew. Chem. Int. Ed. Engl.* **46**, 5670–5703 (2007).

26. Huang, Z.-M., Zhang, Y.-Z., Kotaki, M., and Ramakrishna, S. A review on polymer nanofibers by electrospinning and their applications in nanocomposites. *Compos. Sci. Technol.* **63**, 2223–2253 (2003).

27. Frenot, A. and Chronakis, I. Polymer nanofibers assembled by electrospinning. *Curr. Opin. Colloid Interface Sci.* **8**, 64–75 (2003).

28. Cooley, J. Apparatus for electrically dispersing fluids. U.S. Patent 692,631 (1902).

29. Formhals, A. U.S. Patent 1,975,504 (1934).

30. Formhals, A. U.S. Patent 2,187,306 (1940).

31. Srinivasan, G. and Reneker, D. Structure and morphology of small diameter electrospun aramid fibers. *Polym. Int.* **36**, 195–201 (1995).

32. Doshi, J. and Reneker, D. Electrospinning process and applications of electrospun fibers. *J. Electrostat.* **35**, 151–160 (1995).

33. Agarwal, S., Greiner, A., and Wendorff, J. H. Functional materials by electrospinning of polymers. *Prog. Polym. Sci.* **38**, 963–991 (2013).

34. Li, D. and Xia, Y. Fabrication of titania nanofibers by electrospinning. *Nano Lett.* **3**, 555–560 (2003).

35. Dai, Y., Liu, W., Formo, E., Sun, Y., and Xia, Y. Ceramic nanofibers fabricated by electrospinning and their applications in catalysis, environmental science, and energy technology. *Polym. Adv. Technol.* **22**, 326–338 (2011).

36. Ramaseshan, R., Sundarrajan, S., Jose, R., and Ramakrishna, S. Nanostructured ceramics by electrospinning. *J. Appl. Phys.* **102**, 111101 (2007).

37. Jiang, J., Li, Y., Liu, J., Huang, X., Yuan, C., and Lou, X. W. Recent advances in metal oxide-based electrode architecture design for electrochemical energy storage. *Adv. Mater.* **24**, 5166–5180 (2012).

38. Wu, H., Pan, W., Lin, D., and Li, H. Electrospinning of ceramic nanofibers: Fabrication, assembly and applications. *J. Adv. Ceram.* **1**, 2–23 (2012).

39. Shui, J. and Li, J. C. M. Platinum nanowires produced by electrospinning. *Nano Lett.* **9**, 1307–1314 (2009).

40. Khalil, A., Hashaikeh, R., and Jouiad, M. Synthesis and morphology analysis of electrospun copper nanowires. *J. Mater. Sci.* **49**, 3052–3065 (2014).

41. García-Márquez, A., Portehault, D., and Giordano, C. Chromium nitride and carbide nanofibers: From composites to mesostructures. *J. Mater. Chem.* **21**, 2136–2143 (2011).

42. Zukalova, M., Prochazka, J., Bastl, Z., Duchoslav, J., Rubacek, L., Havlicek, D., and Kavan, L. Facile conversion of electrospun TiO_2 into titanium nitride/oxynitride fibers. *Chem. Mater.* **22**, 4045–4055 (2010).

43. Lee, J.-H. and Kim, Y.-J. Hydroxyapatite nanofibers fabricated through electrospinning and sol–gel process. *Ceram. Int.* **40**, 3361–3369 (2014).
44. Iimura, K., Oi, T., Suzuki, M., and Hirota, M. Preparation of silica fibers and non-woven cloth by electrospinning. *Adv. Powder Technol.* **21**, 64–68 (2010).
45. Lee, S. W., Kim, Y. U., Choi, S.-S., Park, T. W., Joo, Y. L., and Lee, S. G. Preparation of SiO_2/TiO_2 composite fibers by sol–gel reaction and electrospinning. *Mater. Lett.* **61**, 889–893 (2007).
46. Geltmeyer, J., Van der Schueren, L., Goethals, F., De Buysser, K., and De Clerck, K. Optimum sol viscosity for stable electrospinning of silica nanofibres. *J. Sol–Gel Sci. Technol.* **67**, 188–195 (2013).
47. Goñi-Urtiaga, A., Scott, K., Cavaliere, S., Jones, D. J., and Roziére, J. A new fabrication method of an intermediate temperature proton exchange membrane by the electrospinning of CsH_2PO_4. *J. Mater. Chem. A* **1**, 10875–10880 (2013).
48. Subianto, S., Donnadio, A., Cavaliere, S., Pica, M., Casciola, M., Jones, D. J., and Rozière, J. Reactive coaxial electrospinning of ZrP/ZrO_2 nanofibres. *J. Mater. Chem. A* **2**, 13359–13365 (2014).
49. Huang, S., Wu, H., Zhou, M., Zhao, C., Yu, Z., Ruan, Z., and Pan, W. A flexible and transparent ceramic nanobelt network for soft electronics. *NPG Asia Mater.* **6**, e86 (2014).
50. Lotus, A. F., Bender, E. T., Evans, E. A., Ramsier, R. D., Reneker, D. H., and Chase, G. G. Electrical, structural, and chemical properties of semiconducting metal oxide nanofiber yarns. *J. Appl. Phys.* **103**, 024910 (2008).
51. Seol, Y.-J., Kim, K.-H., Kim, I. A., and Rhee, S.-H. Osteoconductive and degradable electrospun nonwoven poly(epsilon-caprolactone)/CaO-SiO_2 gel composite fabric. *J. Biomed. Mater. Res. A* **94**, 649–659 (2010).
52. Ner, Y., Grote, J. G., Stuart, J. A., and Sotzing, G. A. White luminescence from multiple-dye-doped electrospun DNA nanofibers by fluorescence resonance energy transfer. *Angew. Chem. Int. Ed. Engl.* **48**, 5134–5138 (2009).
53. Sahay, R., Kumar, P. S., Sridhar, R., Sundaramurthy, J., Venugopal, J., Mhaisalkar, S. G., and Ramakrishna, S. Electrospun composite nanofibers and their multifaceted applications. *J. Mater. Chem.* **22**, 12953–12971 (2012).
54. Chronakis, I. S. Novel nanocomposites and nanoceramics based on polymer nanofibers using electrospinning process—A review. *J. Mater. Process. Technol.* **167**, 283–293 (2005).
55. Arinstein, A., Burman, M., Gendelman, O., and Zussman, E. Effect of supramolecular structure on polymer nanofibre elasticity. *Nat. Nanotechnol.* **2**, 59–62 (2007).
56. Richard-Lacroix, M. and Pellerin, C. Molecular orientation in electrospun fibers: From mats to single fibers. *Macromolecules* **46**, 9473–9493 (2013).
57. Fennessey, S. F. and Farris, R. J. Fabrication of aligned and molecularly oriented electrospun polyacrylonitrile nanofibers and the mechanical behavior of their twisted yarns. *Polymer (Guildf.)* **45**, 4217–4225 (2004).
58. Kakade, M. V., Givens, S., Gardner, K., Lee, K. H., Chase, D. B., and Rabolt, J. F. Electric field induced orientation of polymer chains in macroscopically aligned electrospun polymer nanofibers. *J. Am. Chem. Soc.* **129**, 2777–2782 (2007).
59. Liu, Y., Antaya, H., and Pellerin, C. Characterization of the stable and metastable poly(ethylene oxide)–urea complexes in electrospun fibers. *J. Polym. Sci. Part B: Polym. Phys.* **46**, 1903–1913 (2008).
60. Liu, Y., Antaya, H., and Pellerin, C. Structure and phase behavior of the poly(ethylene oxide)-thiourea complex prepared by electrospinning. *J. Phys. Chem. B* **114**, 2373–2378 (2010).
61. Kongkhlang, T., Tashiro, K., Kotaki, M., and Chirachanchai, S. Electrospinning as a new technique to control the crystal morphology and molecular orientation of polyoxymethylene nanofibers. *J. Am. Chem. Soc.* **130**, 15460–15466 (2008).

62. Zander, N. Hierarchically structured electrospun fibers. *Polymers (Basel)* **5**, 19–44 (2013).
63. Demir, M. M., Yilgor, I., Yilgor, E., and Erman, B. Electrospinning of polyurethane fibers. *Polymer* **43**, 3303–3309 (2002).
64. Deitzel, J., Kleinmeyer, J., Harris, D., and Beck Tan, N. The effect of processing variables on the morphology of electrospun nanofibers and textiles. *Polymer* **42**, 261–272 (2001).
65. McKee, M. G., Wilkes, G. L., Colby, R. H., and Long, T. E. Correlations of solution rheology with electrospun fiber formation of linear and branched polyesters. *Macromolecules* **37**, 1760–1767 (2004).
66. Shenoy, S., Bates, W., Frisch, H., and Wnek, G. Role of chain entanglements on fiber formation during electrospinning of polymer solutions: Good solvent, non-specific polymer–polymer interaction limit. *Polymer* **46**, 3372–3384 (2005).
67. Pham, Q. P., Sharma, U., and Mikos, A. G. Electrospinning of polymeric nanofibers for tissue engineering applications: A review. *Tissue Eng.* **12**, 1197–1211 (2006).
68. Reneker, D. H. and Yarin, A. L. Electrospinning jets and polymer nanofibers. *Polymer (Guildf.)* **49**, 2387–2425 (2008).
69. Fong, H., Chun, I., and Reneker, D. Beaded nanofibers formed during electrospinning. *Polymer (Guildf.)* **40**, 4585–4592 (1999).
70. Zuo, W., Zhu, M., Yang, W., Yu, H., Chen, Y., and Zhang, Y. Experimental study on relationship between jet instability and formation of beaded fibers during electrospinning. *Polym. Eng. Sci.* **45**, 704–709 (2005).
71. Zong, X., Kim, K., Fang, D., Ran, S., Hsiao, B. S., and Chu, B. Structure and process relationship of electrospun bioabsorbable nanofiber membranes. *Polymer* **43**, 4403–4412 (2002).
72. Kim, B., Park, H., Lee, S.-H., and Sigmund, W. M. Poly(acrylic acid) nanofibers by electrospinning. *Mater. Lett.* **59**, 829–832 (2005).
73. Jarusuwannapoom, T., Hongrojjanawiwat, W., Jitjaicham, S., Wannatong, L., Nithitanakul, M., Pattamaprom, C., Koombhongse, P., Rangkupane, R., and Supaphol, P. Effect of solvents on electro-spinnability of polystyrene solutions and morphological appearance of resulting electrospun polystyrene fibers. *Eur. Polym. J.* **41**, 409–421 (2005).
74. Kang, H., Zhu, Y., Jing, Y., Yang, X., and Li, C. Fabrication and electrochemical property of Ag-doped SiO$_2$ nanostructured ribbons. *Colloids Surf. A: Physicochem. Eng. Asp.* **356**, 120–125 (2010).
75. Koombhongse, S., Liu, W., and Reneker, D. H. Flat polymer ribbons and other shapes by electrospinning. *J. Polym. Sci. Part B: Polym. Phys.* **39**, 2598–2606 (2001).
76. Ner, Y., Stuart, J. A., Whited, G., and Sotzing, G. A. Electrospinning nanoribbons of a bioengineered silk-elastin-like protein (SELP) from water. *Polymer* **50**, 5828–5836 (2009).
77. Zhao, Y., Cao, X., and Jiang, L. Bio-mimic multichannel microtubes by a facile method, *J. Am. Chem. Soc.* **129**, 764–765 (2007).
78. Savych, I., Bernard d'Arbigny, J., Subianto, S., Cavaliere, S., Jones, D. J., and Rozière, J. On the effect of non-carbon nanostructured supports on the stability of Pt nanoparticles during voltage cycling: A study of TiO$_2$ nanofibres. *J. Power Sources* **257**, 147–155 (2014).
79. Li, D. and Xia, Y. Direct fabrication of composite and ceramic hollow nanofibers by electrospinning. *Nano Lett.* **4**, 933–938 (2004).
80. Xu, X., Zhuang, X., Chen, X., Wang, X., Yang, L., and Jing, X. Preparation of core-sheath composite nanofibers by emulsion electrospinning. *Macromol. Rapid Commun.* **27**, 1637–1642 (2006).
81. Barakat, N. A. M., Kanjwal, M. A., Sheikh, F. A., and Kim, H. Y. Spider-net within the N6, PVA and PU electrospun nanofiber mats using salt addition: Novel strategy in the electrospinning process. *Polymer* **50**, 4389–4396 (2009).

82. Gupta, A., Saquing, C. D., Afshari, M., Tonelli, A. E., Khan, S. A., and Kotek, R. Porous nylon-6 fibers via a novel salt-induced electrospinning method. *Macromolecules* **42**, 709–715 (2009).

83. Zhang, Q., Li, M., Liu, J., Long, S., Yang, J., and Wang, X. Porous ultrafine fibers via a salt-induced electrospinning method. *Colloid Polym. Sci.* **290**, 793–799 (2012).

84. Simonet, M., Schneider, O. D., Neuenschwander, P., and Stark, W. J. Ultraporous 3D Polymer meshes by low-temperature electrospinning: Use of ice crystals as a removable void template. *Polym. Eng. Sci.* **47**, 2020–2026 (2007).

85. Bognitzki, M., Czado, W., Frese, T., Schaper, A., Hellwig, M., Steinhart, M., Greiner, A., and Wendorff, J. H. Nanostructured fibers via electrospinning. *Adv. Mater.* **13**, 70–72 (2001).

86. Baji, A., Mai, Y.-W., Wong, S.-C., Abtahi, M., and Chen, P. Electrospinning of polymer nanofibers: Effects on oriented morphology, structures and tensile properties. *Compos. Sci. Technol.* **70**, 703–718 (2010).

87. Yarin, A. L. Coaxial electrospinning and emulsion electrospinning of core–shell fibers. *Polym. Adv. Technol.* **22**, 310–317 (2011).

88. Qu, H., Wei, S., and Guo, Z. Coaxial electrospun nanostructures and their applications. *J. Mater. Chem. A* **1**, 11513–11528 (2013).

89. Greiner, A., Wendorff, J. H., Yarin, A. L., and Zussman, E. Biohybrid nanosystems with polymer nanofibers and nanotubes. *Appl. Microbiol. Biotechnol.* **71**, 387–393 (2006).

90. Yarin, A. L., Zussman, E., Wendorff, J. H., and Greiner, A. Material encapsulation and transport in core shell micro/nanofibers, polymer and carbon nanotubes and micro/nanochannels. *J. Mater. Chem.* **17**, 2585–2599 (2007).

91. Liu, B., Yu, Y., Chang, J., Yang, X., Wu, D., and Yang, X. An enhanced stable-structure core–shell coaxial carbon nanofiber web as a direct anode material for lithium-based batteries. *Electrochem. Commun.* **13**, 558–561 (2011).

92. Lee, B.-S., Son, S.-B., Park, K.-M., Yu, W.-R., Oh, K.-H., and Lee, S.-H. Anodic properties of hollow carbon nanofibers for Li-ion battery. *J. Power Sources* **199**, 53–60 (2012).

93. Han, H., Song, T., Bae, J.-Y., Nazar, L. F., Kim, H., and Paik, U. Nitridated TiO_2 hollow nanofibers as an anode material for high power lithium ion batteries. *Energy Environ. Sci.* **4**, 4532–4536 (2011).

94. Yuan, T., Zhao, B., Cai, R., Zhou, Y., and Shao, Z. Electrospinning based fabrication and performance improvement of film electrodes for lithium-ion batteries composed of TiO_2 hollow fibers. *J. Mater. Chem.* **21**, 15041–15048 (2011).

95. Cao, J., Zhang, T., Li, F., Yang, H., and Liu, S. Enhanced ethanol sensing of SnO_2 hollow micro/nanofibers fabricated by coaxial electrospinning. *New J. Chem.* **37**, 2031–2036 (2013).

96. Park, H., Song, T., Han, H., Devadoss, A., Yuh, J., Choi, C., and Paik, U. SnO_2 encapsulated TiO_2 hollow nanofibers as anode material for lithium ion batteries. *Electrochem. Commun.* **22**, 81–84 (2012).

97. Fang, D., Li, L., Xu, W., Li, G., Luo, Z., Zhou, Y., Xu, J., and Xiong, C. Hollow SnO_2-ZnO hybrid nanofibers as anode materials for lithium-ion battery. *Mater. Res. Express* **1**, 025012 (2014).

98. Sun, Z., Zussman, E., Yarin, A. L., Wendorff, J. H., and Greiner, A. Compound core–shell polymer nanofibers by co-electrospinning. *Adv. Mater.* **15**, 1929–1932 (2003).

99. Loscertales, I. G., Barrero, A., Márquez, M., Spretz, R., Velarde-Ortiz, R., and Larsen, G. Electrically forced coaxial nanojets for one-step hollow nanofiber design. *J. Am. Chem. Soc.* **126**, 5376–5377 (2004).

100. Theron, S. A., Yarin, A. L., Zussman, E., and Kroll, E. Multiple jets in electrospinning: Experiment and modeling. *Polymer* **46**, 2889–2899 (2005).

101. Varesano, A., Rombaldoni, F., Mazzuchetti, G., Tonin, C., and Comotto, R. Multi-jet nozzle electrospinning on textile substrates: Observations on process and nanofibre mat deposition. *Polym. Int.* **59**, 1606–1615 (2010).
102. Varesano, A., Carletto, R. A., and Mazzuchetti, G. Experimental investigations on the multi-jet electrospinning process. *J. Mater. Process. Technol.* **209**, 5178–5185 (2009).
103. Jirsak, O., Sysel, P., Sanetrnik, F., Hruza, J., and Chaloupek, J. Polyamic acid nanofibers produced by needleless electrospinning. *J. Nanomater.* **2010**, 1–6 (2010).
104. Yarin, A. L. and Zussman, E. Upward needleless electrospinning of multiple nanofibers. *Polymer* **45**, 2977–2980 (2004).
105. Fang, J., Niu, H., Wang, H., Wang, X., and Lin, T. Enhanced mechanical energy harvesting using needleless electrospun poly(vinylidene fluoride) nanofibre webs. *Energy Environ. Sci.* **6**, 2196–2202 (2013).
106. Niu, H., Ph, D., Wang, X., and Lin, T. Upward needleless electrospinning of nanofibers. *J. Eng. Fibers Fabr.* **7**, 17–22 (2012).
107. Kostakova, E., Meszaros, L., and Gregr, J. Composite nanofibers produced by modified needleless electrospinning. *Mater. Lett.* **63**, 2419–2422 (2009).
108. Sun, D., Chang, C., Li, S., and Lin, L. Near-field electrospinning. *Nano Lett.* **6**, 839–842 (2006).
109. Ruggieri, F., Di Camillo, D., Lozzi, L., Santucci, S., De Marcellis, A., Ferri, G., Giancaterini, L., and Cantalini, C. Preparation of nitrogen doped TiO_2 nanofibers by near field electrospinning (NFES) technique for NO_2 sensing. *Sens. Actuators B Chem.* **179**, 107–113 (2013).
110. Lameiro, R., Sencadas, V., Lanceros-Mendez, S., Correia, J. H., and Mendes, P. M. Large area microfabrication of electroactive polymeric structures based on near-field electrospinning. *Procedia Eng.* **25**, 888–891 (2011).
111. Zhou, F.-L., Hubbard, P. L., Eichhorn, S. J., and Parker, G. J. M. Jet deposition in near-field electrospinning of patterned polycaprolactone and sugar-polycaprolactone core–shell fibres. *Polymer* **52**, 3603–3610 (2011).
112. Chang, J., Liu, Y., Heo, K., Lee, B. Y., Lee, S. W., and Lin, L. Direct-write complementary graphene field effect transistors and junctions via near-field electrospinning. *Small* **10**, 1920–1925 (2014).
113. Sundaray, B., Subramanian, V., Natarajan, T. S., Xiang, R. Z., Chang, C. C., and Fann, W.-S. Electrospinning of continuous aligned polymer fibers. *Appl. Phys. Lett.* **84**, 1222–1224 (2004).
114. Pan, H., Li, L., Hu, L., and Cui, X. Continuous aligned polymer fibers produced by a modified electrospinning method. *Polymer* **47**, 4901–4904 (2006).
115. Zussman, E., Theron, A., and Yarin, A. L. Formation of nanofiber crossbars in electrospinning. *Appl. Phys. Lett.* **82**, 973–975 (2003).
116. Xu, C. Aligned biodegradable nanofibrous structure: A potential scaffold for blood vessel engineering. *Biomaterials* **25**, 877–886 (2004).
117. Karube, Y. and Kawakami, H. Fabrication of well-aligned electrospun nanofibrous membrane based on fluorinated polyimide. *Polym. Adv. Technol.* **21**, 861–866 (2010).
118. Jalili, R., Morshed, M., and Ravandi, S. A. H. Fundamental parameters affecting electrospinning of PAN nanofibers as uniaxially aligned fibers. *J. Appl. Polym. Sci.* **101**, 4350–4357 (2006).
119. Ishii, Y., Sakai, H., and Murata, H. A new electrospinning method to control the number and a diameter of uniaxially aligned polymer fibers. *Mater. Lett.* **62**, 3370–3372 (2008).
120. Dalton, P. D., Klee, D., and Möller, M. Electrospinning with dual collection rings. *Polymer (Guildf.)* **46**, 611–614 (2005).

121. Li, D., Wang, Y., and Xia, Y. Electrospinning of polymeric and ceramic nanofibers as uniaxially aligned arrays. *Nano Lett.* **3**, 1167–1171 (2003).
122. Zhang, D. and Chang, J. Electrospinning of three-dimensional nanofibrous tubes with controllable architectures. *Nano Lett.* **8**, 3283–3287 (2008).
123. Ner, Y., Asemota, C., Olson, J. R., and Sotzing, G. A. Nanofiber alignment on a flexible substrate: Hierarchical order from macro to nano. *ACS Appl. Mater. Interfaces* **1**, 2093–2097 (2009).
124. Teo, W. E., Kotaki, M., Mo, X. M., and Ramakrishna, S. Porous tubular structures with controlled fibre orientation using a modified electrospinning method. *Nanotechnology* **16**, 918–924 (2005).
125. Ye, X., Huang, X., and Xu, Z. Nanofibrous mats with bird's nest patterns by electrospinning. *Chin. J. Polym. Sci.* **30**, 130–137 (2011).
126. Tijing, L. D., Choi, J.-S., Lee, S., Kim, S.-H., and Shon, H. K. Recent progress of membrane distillation using electrospun nanofibrous membrane. *J. Membr. Sci.* **453**, 435–462 (2014).
127. Gorji, M., Jeddi, A. A. A., and Gharehaghaji, A. A. Fabrication and characterization of polyurethane electrospun nanofiber membranes for protective clothing applications. *J. Appl. Polym. Sci.* **125**, 4135–4141 (2012).
128. Perez-Masia, R., Lopez-Rubio, A., Fabra, M. J., and Lagaron, J. M. Biodegradable polyester-based heat management materials of interest in refrigeration and smart packaging coatings. *J. Appl. Polym. Sci.* **130**, 3251–3262 (2013).
129. Sas, I., Gorga, R. E., Joines, J. A., and Thoney, K. A. Literature review on superhydrophobic self-cleaning surfaces produced by electrospinning. *J. Polym. Sci. Part B: Polym. Phys.* **50**, 824–845 (2012).
130. Rogina, A. Electrospinning process: Versatile preparation method for biodegradable and natural polymers and biocomposite systems applied in tissue engineering and drug delivery. *Appl. Surf. Sci.* **296**, 221–230 (2014).
131. Wang, X., Ding, B., and Li, B. Biomimetic electrospun nanofibrous structures for tissue engineering. *Mater. Today* **16**, 229–241 (2013).
132. Jayasinghe, S. N. Cell electrospinning: A novel tool for functionalising fibres, scaffolds and membranes with living cells and other advanced materials for regenerative biology and medicine. *Analyst* **138**, 2215–2223 (2013).
133. Saito, N., Aoki, K., Usui, Y., Shimizu, M., Hara, K., Narita, N., Ogihara, N. et al. Application of carbon fibers to biomaterials: A new era of nano-level control of carbon fibers after 30-years of development. *Chem. Soc. Rev.* **40**, 3824–3834 (2011).
134. Cho, H., Min, S.-Y., and Lee, T.-W. Electrospun organic nanofiber electronics and photonics. *Macromol. Mater. Eng.* **298**, 475–486 (2013).
135. Miao, J., Miyauchi, M., Simmons, T. J., Dordick, J. S., and Linhardt, R. J. Electrospinning of nanomaterials and applications in electronic components and devices. *J. Nanosci. Nanotechnol.* **10**, 5507–5519 (2010).
136. Morello, G., Moffa, M., Girardo, S., Camposeo, A., and Pisignano, D. Optical gain in the near infrared by light-emitting electrospun fibers. *Adv. Funct. Mater.* **24**, 5225–5231 (2014).
137. Wang, P., Wang, Y., and Tong, L. Functionalized polymer nanofibers: A versatile platform for manipulating light at the nanoscale. *Light Sci. Appl.* **2**, e102 (2013).
138. Kim, I.-D. and Rothschild, A. Nanostructured metal oxide gas sensors prepared by electrospinning. *Polym. Adv. Technol.* **22**, 318–325 (2011).
139. Thavasi, V., Singh, G., and Ramakrishna, S. Electrospun nanofibers in energy and environmental applications. *Energy Environ. Sci.* **1**, 205–221 (2008).
140. Cavaliere, S., Subianto, S., Savych, I., Jones, D. J., and Rozière, J. Electrospinning: Designed architectures for energy conversion and storage devices. *Energy Environ. Sci.* **4**, 4761–4785 (2011).

141. Dong, Z., Kennedy, S. J., and Wu, Y. Electrospinning materials for energy-related applications and devices. *J. Power Sources* **196**, 4886–4904 (2011).
142. Armaroli, N. and Balzani, V. The future of energy supply: Challenges and opportunities. *Angew. Chem. Int. Ed. Engl.* **46**, 52–66 (2007).
143. OECD/IEA. World energy outlook, Presentation to Press, November 12, 2013. London, U.K.: IEA Publishing. http://www.iea.org/t&c/termsandconditions.

2 Electrospun Nanofibers for Low-Temperature Proton Exchange Membrane Fuel Cells

Surya Subianto, Stefano Giancola,
Giorgio Ercolano, Yannick Nabil, Deborah Jones,
Jacques Rozière, and Sara Cavaliere

CONTENTS

2.1   INTRODUCTION

The trio of exhaustible fossil energies, increasing carbon dioxide emissions, and the effects of global warming calls for the implementation of new technologies for energy conversion and storage. Among these, fuel cells are electrochemical devices directly converting the chemical energy of a fuel and an oxidant into electricity and heat. Their core, the so-called membrane electrode assembly (MEA), consists of two electrodes in contact with an electrolyte conducting only ionic species (Figure 2.1). At the anode, the fuel is decomposed into ions and electrons. There are several kinds of fuel cells, working in different conditions and with different fuels and, consequently, based on different materials[1] (see also Chapter 3). Depending on the type of cell,

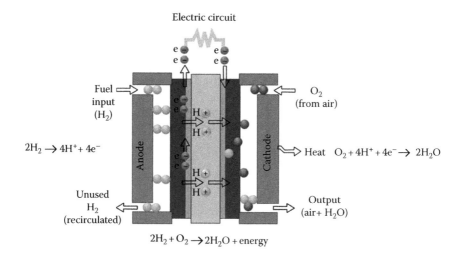

FIGURE 2.1   Schematic representation of a proton exchange membrane fuel cell.

the ionic species travel from the anode to the cathode or vice versa, while electrons travel along an external circuit to the cathode, where oxygen reduction takes place (Figure 2.1). Fuel cells are a reliable and environmentally friendly method to produce energy with higher conversion efficiencies than any other conventional thermomechanical system. Indeed, when operating with hydrogen as the fuel, they produce only water, providing a clean mechanism for energy conversion. Furthermore, hydrogen can be produced by renewable energy sources (wind, solar, biomass, etc.), positively impacting the sustainable development and energy security. Hydrocarbon fuels such as low molecular weight alcohols and bioalcohols can also be used as renewable fuel source. Despite the consequent production of carbon dioxide, these systems have a real importance for portable electronics applications. The intrinsic modularity of fuel cells in stacks allows for simple fabrication and a wide range of applications in portable, stationary, and transportation power generation.

Proton exchange membrane fuel cells (PEMFC) are one of the most promising types of fuel cells working at relatively low temperatures (80°C–120°C) and approaching commercialization. A schematic representation of this kind of system and the involved reactions is presented in Figure 2.1. Despite the significant progress in their development, further research is required to enhance their durability, reduce their costs, and further improve their performance and reliability.[2] Such research endeavors are dedicated to all core components of the cell: the electrolyte membrane,[3,4] the electrocatalyst,[5–7] and its support.[8,9] A more effective utilization of the highly priced Pt electrocatalyst, its replacement with nonnoble materials, the use of more durable materials as catalyst supports, the preparation of electrolyte membranes with elevated proton conductivity, high mechanical properties, and low gas permeability are the core challenges addressed presently by the research community for the development of novel, market-ready PEMFC. To achieve these goals, great attention is paid not only to the chemical composition of the novel materials but also to their micro- and nanostructure. The importance of the control onto the spatial organization of ionic and hydrophobic domains in the ionomer membranes[3] as well as the organized porosity within the catalyst supports and the distribution of the electrocatalyst nanoparticles onto them[5] has been clearly established. In this regard, electrospinning technique being a versatile tool in the synthesis of nanostructured materials with targeted architectures has attracted an increasing research attention for the application to PEMFC core materials in the last decade. The possibility of playing with the morphology of the electrospun fibers (solid, hollow, core–sheath), their composition (multifunctional hybrid fibers loaded with molecules or nanoparticles or chemically functionalized), and their multiscale assembly geometry (aligned or nonwoven fibers, crossbars) greatly increases the range and variety of materials that can be achieved and thus justify this intense interest. Recent advances and new directions of electrospun 1D materials in low-temperature fuel cell MEA are described in the following text.

2.2 NANOSTRUCTURED COMPOSITE IONOMER MEMBRANES

Electrospinning has been utilized for the development of nanocomposite polymer electrolyte membranes for fuel cells with the aim of modifying the morphology at both the micro- and the nanoscale so as to improve their properties.[10] Most of the studies have focused on proton exchange membrane (PEM). The latter is responsible for proton transfer from the anode to the cathode, acts as a physical barrier for reactant gases, and thus forces the electrons to pass through the external circuit, allowing the generation of electrical energy. An ideal, commercially viable PEM has to meet several requirements[11,12]: high proton conductivity, low fuel permeability, elevated electrochemical, hydrolytic, morphological and thermal stability, high dimensional stability and mechanical integrity upon hydration and dehydration, good compatibility with both electrodes, and low cost. Perfluorosulfonic acid (PFSA) membranes such as Nafion are still considered as state-of-the-art membranes for PEMFC due to their high chemical stability, high ionic conductivity and good mechanical properties in hydrated conditions, and temperatures lower than 80°C.[11] However, due to the current trend toward thin membranes (<50 μm), the issue of their durability has emerged and has to be overcome in order to facilitate a large-scale commercialization and use of PEMFC. Indeed, thin membranes are highly advantageous because of their low electrical resistance and improved water transport. Nevertheless, they suffer from poor mechanical resistance and high fuel crossover, a fundamental issue especially for direct methanol fuel cell (DMFC), where increasing the membrane selectivity to methanol (by reducing their methanol permeability) without decreasing the proton conductivity still remains a challenge.[13,14]

Among various chemical and physical[4,15,16] routes explored to increase membrane durability, embedding of ionomer into porous inert supports such as polytetrafluoroethylene (PTFE) is the most well established. The improvement in mechanical properties and dimensional stability of these membranes allows more durable and thinner (5–30 μm) membranes, which compensates for the lowered proton conductivity due to the presence of the inert support. To date, various types of porous reinforcing supports have been used like expanded PTFE sheet, micro PTFE fibrils, and polysulfone/microglass fiber fleece.[4] In this regard, macroporous, nonwoven nano-fiber mats, such as those produced by electrospinning, are advantageous because of their high volume fraction of void space resulting in a large surface area providing a large interface between the two phases, better pore interconnectivity, and a more homogeneous dispersion in the composite membrane due to their nonwoven nature. Furthermore, nanofibers have additional advantages due to their nanometric size and high directionality (see Section 1.5). In this context, electrospinning plays a key role for the realization of nanocomposite fibrous mats due to the high versatility in pre-paring 1D materials with controlled size and morphology.

In the literature, two different approaches have been pursued with electrospun mat composite membranes: one consists of embedding an ionomer into a non- or low conductive nanofiber mat (see Section 2.2.1), while the other one involves the incorporation of ion conducting nanofibers into an inert matrix (see Section 2.2.2).[10] In both cases, the properties of proton conduction and mechanical strength are dis-sociated between the electrospun mat and the matrix.

2.2.1   COMPOSITE MEMBRANES WITH THE IONOMER EMBEDDED IN A REINFORCING ELECTROSPUN MAT

In this approach, a non- or low-conducting electrospun nonwoven acts as reinforce-ment for a ionomer, which fills void spaces within the fibers and ensure the proton conduction through the membrane. Electrospun nanofibers show exceptional tensile strength and stiffness particularly because of the orientation phenomena resulting from extensional forces experienced by macromolecular chains during the process (see Section 1.5.1). Thus, they are very promising as reinforcement in nanocompos-ites. For fuel cell applications, a low-density electrospun mat is required since the volume of the pores and their interconnectivity should be maximized to ensure an efficient proton transfer. Furthermore, using this strategy a higher through plane conductivity is expected compared to blend or conventional reinforced membranes due to better interconnectivity of ionic domains along the thickness direction and a more homogenous distribution of the reinforcing fibers within the ionomer matrix.

In H_2/air PEMFC, membranes are subjected to mechanical stress due to swelling/shrinkage upon hydration/dehydration that can lead to premature failure.[4] Electrospun nanofibrous mats of robust polymers have been incorporated into the ionomer matrix in order to both increase the mechanical properties and reduce the swelling as these phenomena are strongly related.[17] For this purpose, the homogeneous fiber distribu-tion throughout the membrane achievable using electrospun mats, with a very thin layer of ionomer on both sides for electrode contact, is ideal to ensure a homogeneous swelling behavior as well as the compatibility with the ionomer-based electrodes. Nanofibers of chemically stable and mechanically strong polymers like polyvinyli-dene fluoride (PVDF),[18] polyphenylsulfone (PPSU),[19,20] polybenzimidazole (PBI),[21] and polyimide (PI)[22] have been used as reinforcing electrospun mats, although some inorganic nanofibers like ZrP/ZrO_2[23] have also been investigated. Nafion embedded with PPSU nanofibers realized by dual electrospinning (see Section 2.2.2)[19] showed a higher Young modulus and proportional limit stress than Nafion, though still main-taining good ductility. Furthermore, the mechanical properties increase with the

PPSU content, while proton conductivity decreases with decreasing Nafion volume fraction. Interestingly, this membrane shows higher in-plane proton conductivity than Nafion/PVDF solution-cast blends with the same ionomer wt% and where PVDF is a nonconducting polymer like PPSU.[19] This suggests an improved contiguity of proton pathways in nanofiber-based membranes, which facilitates proton transport. The improved interconnectivity also applies to the nonconductive phase, providing a greater improvement in mechanical properties for a given content of support material. Thus, electrospun composite membranes also show lowered mass and volumetric swelling with respect to Nafion, proving the effectiveness of this approach in improving the PEM dimensional stability. Some studies revealed that this translates to a 54% enhancement of durability for the composite membrane (30 μm) with respect to Nafion 212 (51 μm) under open-circuit voltage (OCV) humidity cycling experiment (Figure 2.2).[24] Moreover, due to its lower thickness, the nanofibrous composite membrane shows almost identical polarization curve as Nafion 212.[19]

Other attempts on reinforcing membranes have been performed using chemically cross-linked electrospun mats[21,25,26] with the aim of further improving the dimensional stability. Nafion reinforced by cross-linked and chemically functionalized (bearing sulfonic acid groups) polyvinyl alcohol (PVA) nanofibers[25] does not show a clear α-relaxation with the storage modulus, showing a slight decrease only above 150°C and then remaining constant at T higher than 200°C. On the contrary, the storage modulus of dry Nafion shows a sharp decrease at T higher than 80°C due to the α-relaxation associated with its cluster transition temperature (T_c),[27] which considering the PEMFC operating temperatures (80°C–120°C) has a very significant

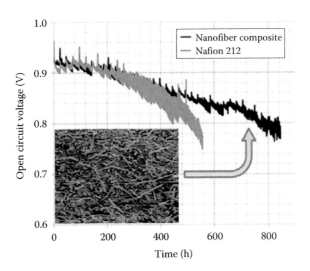

FIGURE 2.2 Results from an accelerated humidity cycling test of PEM where the OCV is measured at 80°C for MEAs with Nafion 212 or a dual-fiber electrospinning composite membrane. (Reprinted from *Curr. Opin. Chem. Eng.*, 4, Wycisk, R., Pintauro, P., and Park, J., New developments in proton conducting membranes for fuel cells, 71–78, Copyright 2014, with permission from Elsevier.)

implication on its long-term durability. This behavior for composite membranes can be ascribed to the presence of cross-linked and interconnected PVA nanofiber network ensuring good mechanical properties even at temperatures higher than the T_c of Nafion. In addition, composite membranes show not only increased stiffness and yield strength but also maintain good ductility with respect to Nafion. A cross-linkable PBI-polybenzoxazine has also been used as a PBI reinforcement for high-temperature PEMFC.[21] In this case, composite membranes show an increase of both mechanical properties and in-plane proton conductivity with respect to neat PBI, resulting in a better fuel cell performance. The increase in proton conductivity could be related to the alignment of ionic domains promoted by the interaction of PBI fibers with the PBI matrix.

Another important current challenge for PEMs is the retention of high proton conductivity under elevated temperature and low relative humidity (RH).[24] These conditions would be advantageous for various PEMFC applications, due to reduced system complexity through the elimination of water management. One promising way to improve conductivity at high temperature and low RH is the use of low equivalent weight (EW) PFSA with greater concentration of acid functionalities. In order to keep good crystallinity and mechanical properties at low EW, short-side-chain (SSC) PFSAs have been used and further reinforced by electrospun fiber mats. Recently, inorganic fibers of ZrP/ZrO_2 obtained by reactive coaxial electrospinning have been used to form nanocomposite systems with 700 EW SSC Aquivion.[23] Such composite membranes show very high in-plane proton conductivity even under dry conditions (99 mS at 110°C and 50% RH) and higher performance than the neat ionomer, suggesting an alignment of the ionic domains along the ZrP/Aquivion interface due to the hydrophilic nature of ZrP. Furthermore, the nanofibers induce an increase of membrane stiffness with respect to both cast and extruded Aquivion under humidity conditions. Another attempt to fabricate PEM with high conductivity at low RH involves the use of a glass electrolyte. The incorporation of an electrospun PI mat inside a silica network[22,28] allows thin and flexible membranes, overcoming the intrinsic fragility of glass electrolytes.

For DMFC applications, nanofibrous composite membranes have been fabricated with the main aim of achieving lower thickness and low methanol permeability. Both organic[29–35] and inorganic[36,37] nanofibers have been used as PFSA reinforcement. In particular, several studies report on membranes based on Nafion reinforced by PVA nanofibers,[32–35] due to the lower affinity of PVA for methanol and higher for water respect to Nafion.[38] In this case, a cross-linking reaction of PVA by glutaraldehyde has been performed in order to ensure fiber integrity against water and methanol exposition and to improve mechanical and thermal properties. Nanofibrous composite membranes show reduced methanol crossover and higher selectivity with respect to Nafion.[31,36,37] The membrane properties strongly depend on nanofiber loading, with a content beyond which their performance decreases due to the reduction in proton conductivity.[33,36] Furthermore, fiber distribution inside the membranes plays also a key role in blocking methanol crossover. Interestingly, Lin et al.[33] reported that such Nafion/PVA membranes show higher selectivity and thus better polarization curves with respect to cast blend membranes with same PVA content and similar thickness. This improved performance is attributed to the higher surface area and efficient distribution in the membranes of fibrous PVA. Indeed, in blended membranes, large PVA agglomerates (7–12 μm) dispersed throughout the Nafion matrix make the proton traveling pathway more tortuous than that of the interconnected micropores within the PVA fiber mat, thus resulting in a reduced conductivity. Moreover, the higher surface area of thin PVA nanofibers and their confinement within the membrane in a sandwich-like structure makes it a more effective methanol barrier, resulting in a better fuel cell performance[33] than extruded Nafion.

2.2.2 COMPOSITE MEMBRANES WITH AN INERT POLYMER EMBEDDED IN AN IONOMER ELECTROSPUN MAT

In another approach to prepare nanocomposite membranes, a highly proton-conducting polymer is electrospun into a porous fiber mat that is impregnated with a secondary polymer providing mechanical stability. In this case, the mechanical properties are improved due to the nonconducting, inert filler acting as a sheath around the ionomer nanofibers limiting their swelling. Thus, this approach requires a dense nanofiber mat that would ensure a contiguous proton-conducting path throughout the membrane.

Many studies only refer to the electrospinning of ionomers without the further impregnation step. The first part of this section will review this approach and thus the properties of such conducting nanofibers, while in the latter, the literature concerning composite membranes with double components will be discussed. Unlike conventional processing methods of ionomer membranes (solution casting and extrusion), electrospinning can introduce changes in their micro-/nanostructure. For instance, electrospun/sprayed poly(ether ether ketone) (sPEEK) membranes show a shifted and more distinct small-angle x-ray scattering ionomer peak compared to a cast membrane, indicating larger proton transport channels and better phase separation.[39] Due to the alignment of the ionic aggregates occurring during the electrospinning process, single ionomer nanofibers showed greater apparent conductivity than that of a cast film.[40–43] Indeed, it was reported that the conductivity of electrospun Nafion increases with decreasing the fiber diameter,[40] indicating that a confinement effect in thinner fibers may assist in aligning the ionic domains in the longitudinal direction. X-ray scattering under high RH confirmed such orientation in the ionic aggregates in the fibers. Similar results were reported on sulfonated polyimide (sPI) nanofibers, showing greater conductivity along the fiber axis than through the membrane[42] (Figure 2.3), and the use of single ionomer fibers in a microfuel cell showed a performance significantly higher than that of a conventional device.[41]

These results are also corroborated by specific studies on the effect of electric field–induced orientation in electrospun nanofibers. Studies on polyacrylonitrile (PAN),[44] polyethylene oxide (PEO),[45] and polyoxymethylene[46] have shown that the electric field used in electrospinning can also align the polymer chains parallel to the fiber axis. More specifically, it was found that electrospinning produced isotropic polymer fibers when the collector was grounded but contained uniaxially oriented polymer chains when negatively charged plates were used as collectors. Thus, it was postulated that the alignment of the molecular structure occurs during electrospinning, but chain relaxation promoted by residual solvent return the polymer

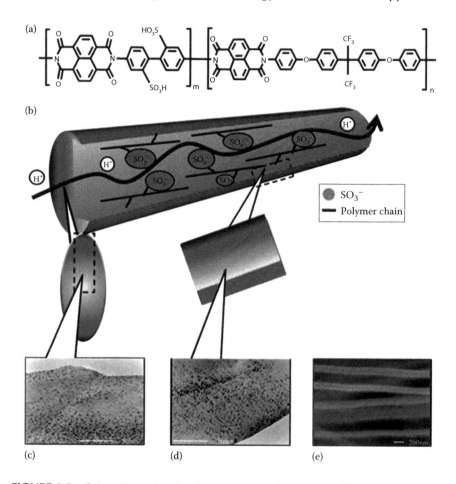

FIGURE 2.3 Orientation and molecular structure in electrospun sulfonated polyimide (a) the chemical structure of the sulfonated polyimide, (b) a schematic representation of domain orientation in the nanofibers, and TEM images of cross-sectional aligned nanofiber in (c) radial direction and (d) axial direction, and (e) SEM image of aligned nanofibers. (Reprinted with permission from Tamura, T. and Kawakami, H., *Nano Lett.*, 10, 1324–1328. Copyright 2010 American Chemical Society.)

molecular structure to the isotropic state, unless chain relaxation is hindered (e.g., through the use of charged collectors). These results suggest that the orientation of electrospun materials depends on the competing effect between the extensional forces (promoting chain orientation along the drawing direction) and orientation/relaxation (promoting a return to the isotropic state). Indeed, a study[46] shows that the electrospinning process itself induces molecular orientation parallel to the fiber axis. Both the shear force due to the flow through the needle and the columbic force during the jet formation may initiate the orientation of the polymer chain. The drawing of the solution during electrospinning alone would be sufficient to induce an orientation of the ionic domains of the ionomers along the fiber axis.[47] This may possibly be enhanced by the use of a polar solvent, as its dipole would also be preferentially aligned in the same direction. Furthermore, ionomers would be able to retain this oriented morphology through ionic bonding between charged groups locking the oriented structure and preventing relaxation.

Unlike most polymers, PFSAs such as Nafion pose a distinctive challenge when it comes to electrospinning. Ionomer solutions can be characterized into three types[48]: (1) in nonpolar solvents, they have a concentration-dependent equilibrium between intra- and interchain aggregation due to the electrostatic attraction between non-ionized ion pairs on the solvated backbone; (2) in polar solvents where the polymer backbone is soluble, they have an extended polymer chain system similar to weakly charged polyelectrolytes; and (3) in polar solvents but where the analogous neutral polymer is insoluble, they show a polymer–solvent phase separation leading to a colloidal dispersion. PFSAs fall into the third category, and as such, they lack the chain entanglement required for fiber formation, and thus the application of a potential will result in electrospraying rather than electrospinning.[49] Despite this, recent studies have electrospun Nafion[50,51] as well as other sulfonated hydrocarbon polymers[39,52] through the use of a carrier polymer such as polyacrylic acid (PAA) or PEO. In this approach, the carrier polymer imparts electrospinnability as it forms a viscous solution with sufficient chain entanglement to hold together the ionomer dispersion. The nature of the carrier polymer, then, is important in order to minimize its quantity in relation to the ionomer. Molecular weight significantly affects the threshold concentration for electrospinning, but this also depends on the nature of the solution itself as a good solvent is required.[53] The use of carrier polymers may result in reduced conductivity due to the presence of a nonconducting fraction within the fiber. Early studies used PVA, PEO,[51] or PAA[50] as Nafion carrier polymers in high amounts (10%–20% wt), dramatically lowering the proton conductivity. However, latter studies showed that using higher molecular weight carrier polymers significantly reduces the concentration needed for electrospinning (Figure 2.4), and thus Nafion nanofibers can be electrospun to a very high volume fraction (99.9%).[40] The amount of carrier polymer can be further reduced through the use of SSC, low EW PFSAs, which give greater viscosity due to the greater ionic interactions.[54] At a given PFSA concentration and carrier polymer molecular weight, there is a threshold concentration where electrospinning is possible. Below the threshold electrospraying occurs, while around this threshold, a beaded fiber morphology is obtained (Figure 2.4).

Unlike PFSAs, other ionomers such as sulfonated aromatic polymers can be electrospun without the use of a carrier. This is attributed to the fact that both the ionically functionalized and neutral segments of such polymers are soluble in polar solvents and thus do not form colloidal dispersions. Hence, their electrospinnability depends mainly on molecular weight and solution concentration. At low molecular weight, they will not form fibers due to the insufficient chain entanglement, but micellar structures rather than solvated chains, resulting in electrospraying. When the molecular weight is sufficiently high, electrospinning is possible and is dependent on ionomer concentration. Sulfonated polysulfone,[31,55] sPEEK,[39] sPI,[42] and polystyrene[52] have been electrospun into fibers without the use of carrier polymers. Fiber formation is more feasible at rather high solution concentrations.

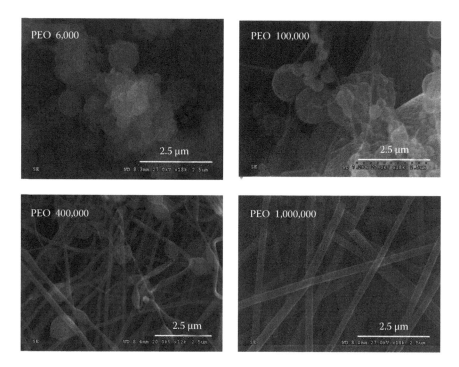

FIGURE 2.4 Effect of carrier polymer molecular weight (Poly(ethylene oxide), PEO) on the electrospinning of a fixed concentration of PEO (1% wt/v) and Aquivion™ (20% wt/v), a short-side-chain PFSA. (Subianto, S., Cavaliere, S., Jones, D.J., and Rozière, J.: Effect of side-chain length on the electrospinning of perfluorosulfonic acid ionomers. *J. Polym. Sci. Part A: Polym. Chem.* 2013. 51. 118–128. Copyright Wiley-VCH Verlag GmbH & Co. KGaA. Reproduced with permission.)

Since the pioneering work of the Pintauro group,[55] composite membranes utilizing inert polymers embedded into ionomer fibers have been fabricated by electrospinning Nafion[51] and 3M PFSA.[52] This strategy allowed lower dimensional swelling and improved mechanical properties compared to cast membranes. Electrospun composite membranes present an improvement in the organization of hydrophilic and hydrophobic domains of the ionomers compared to traditional blends. For instance, a study by Tamura et al. using the same sulfonated PI for both the matrix and the filler material showed higher conductivity and stability compared to a cast membrane,[42] due to the improved alignment of the PI microstructure within the electrospun fibers.

Generally, the fabrication of such nanofiber-based membranes comprises of the following steps: (1) electrospinning of the ionomer nanofiber mat; (2) annealing, densification, and welding of the mat; and (3) impregnation with a reinforcing filler polymer. The second step is required to increase the volume fraction of ionomer in the composite membrane to achieve a sufficient conductivity, as electrospun mats have a very high void volume. Moreover, in some cases where the ionomer mat has a low density and thus the proton conductivity is significantly reduced, other properties such as mechanical and hydrolytic stability of the membrane[26] are improved by the

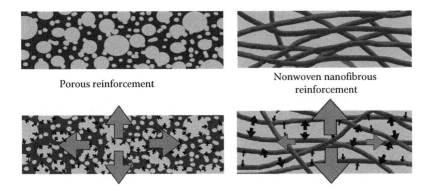

Porous reinforcement

Nonwoven nanofibrous
reinforcement

FIGURE 2.5 Schematic representation of the swelling in nonwoven-reinforced membrane.

densification procedure. For instance, in DMFC, electrospun composite membranes led to comparable performance to Nafion despite their lower proton conductivity, due to the reduced methanol crossover and lower membrane thickness.[29,30,35]

Densification and welding does, however, introduce a peculiar challenge of performing a homogeneous impregnation of the nanofiber mat. Densification would reduce pore size and may create air pockets preventing full impregnation. To overcome this issue, some studies have looked at the use of either dual electrospinning[19] or a combination of electrospinning and electrospraying.[56] The idea is that by electrospinning the ionomer and the filler polymer together, the final impregnation step will not be required as the filler polymer is already present as either entangled nanofibers or electrosprayed particles within the electrospun ionomer mat. Thus, hot pressing would be sufficient to create a dense membrane, without the need for a separate impregnation step. Composite membranes have been fabricated by dual electrospinning using Nafion[19,57] for PEMFC or functionalized polysulfone[58] for alkaline fuel cells. A comparison between composite membranes made by impregnation and dual electrospinning showed that the former have lower in-plane swelling despite having similar water uptake. This was attributed to the difference in the connectivity of the reinforcing matrix. In the impregnated composite membrane, the reinforcing matrix is introduced into an electrospun mat of Nafion as a pore-filling liquid and thus may not be fully interconnected. While in the dual electrospun membrane, the reinforcing matrix is also an electrospun mat with interlocking fibers, and thus there is a strong interconnectivity in the reinforcing matrix as well. Upon swelling, this results in a greater displacement in the thickness direction as the fibers are not welded, and thus the swelling ionomer can displace them apart in the z direction, but constraint the in-plane swelling as the reinforcing nanofibers possess strong interconnectivity in the x and y direction (Figure 2.5).

2.3 ELECTROSPUN ELECTROCATALYST SUPPORTS

A robust and efficient support material for electrocatalyst is a very fundamental prerequisite for fuel cell performance and durability. The general requirements for electrocatalyst supports are high electronic conductivity, high specific surface area, high electrochemical and chemical stability under operating conditions, and the aptitude to form a layer with the adapted porosity for an efficient gas transport. The support can affect the electrocatalytic activity and durability of the loaded catalysts by specific electronic interactions as well as by the possibility of high dispersion and a narrow size distribution of catalyst nanoparticles due to their morphology and porosity. In this respect, they can also provide a way to decrease the metal loading (as for noble metals such as Pt)[59] and thus electrode costs. Highly conductive and high surface area carbon blacks (e.g., commercial Vulcan XC-72, Shawinigan, Black Pearl 2000, Denka Black, and Ketjen Black) are currently used as conventional fuel cell catalyst supports. However, especially at high potentials they suffer from severe corrosion via the following reaction[60]:

$$C + 2H_2O = CO_2 + 4H^+ + 4e^-   (E^0 = 0.207 \text{ V vs. NHE at } 25°C)$$

High potentials (e.g., higher than 1.5 V/RHE) can arise at both anode and cathode side during fuel cell operation, for instance, due to the cell reversal phenomenon[61] or to the stopping/starting of the fuel cell stack.[62] Furthermore, high temperature and humidity[63] as well as the presence of noble metal nanoparticles[64] may accelerate the carbon oxidative degradation. The corrosion of the support leads to further issues

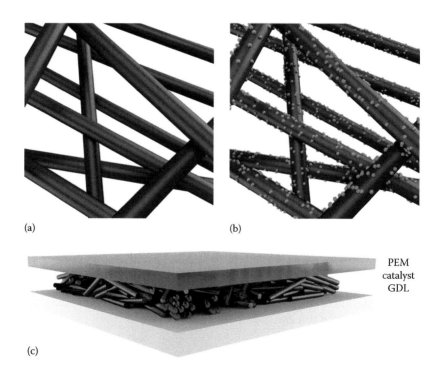

(a)   (b)

PEM
catalyst
GDL

(c)

FIGURE 2.6   Schematic representation of a MEA based on electrospun electrodes. Carbon nanofibers (a), Pt-decorated carbon nanofibers (b), and idealized MEA assembly composed of GDL—Pt-decorated segmented carbon nanofibers—PEM (c).

as the aggregation and migration of the catalyst nanoparticles and their detachment with loss of the overall performance of the fuel cell.[65,66]

One of the main objectives of the PEMFC research in the last years has thus been developing carbon-based or alternative support materials with enhanced corrosion resistance. In order to obtain faster electron transfer and high electrocatalytic activity, the interest was focused on nanostructured supports. In particular, 1D carbon materials such as nanotubes[67] and nanofibers[68] have been investigated. Indeed, fibrous materials are highly porous because of their 3D structure thus providing an efficient gas diffusion rate. Their high surface area allows an improved utilization of the electrocatalysts by offering an efficient three-phase boundary (Figure 2.6). Finally, the specific 1D morphology can improve directional properties such as the electronic conductivity by favoring the electron transfer along the fiber axe. Due to these advantages and the relatively easy and upscalable method, electrospinning has been extensively used in the last decade to prepare carbon-based and noncarbon support materials for PEMFC electrocatalysts.[10] This led to greater stability of the supports and improved dispersion of the catalysts and overall better performance of the resulting electrodes. Further improvements can be achieved by modifying the morphology and enhance the porosity and the exposed surface area, for instance, by using templating porogens or coaxial electrospinning to achieve core/shell and hollow structures.

2.3.1 CARBON-BASED SUPPORTS

Carbon nanowires were successfully deposited by electrospinning for the first time in 1999[69]; Chun et al. adapted a processing technique developed in the 1960s, where industrial carbon fibers were produced via melt spinning of PAN followed by oxidative stabilization at low temperature and carbonization. They replaced the melt spinning with electrospinning of PAN from a dimethylformamide (DMF)/PAN solution.

Alternative polymeric solutions have been since successfully electrospun and carbonized to produce both pure and composite carbon nanofibers (CNFs); the most common alternatives being PBI in dimethylacetamide (DMAc),[70] pitch in tetrahydrofuran (THF),[71,72] PVA in water.[73] Another alternative are carbonized PI electrospun fibers networks, produced by electrospinning PI in DMAc[74] followed by carbonization and by electrospinning a solution of poly(amidic acid) and THF–methanol followed by imidization first and later carbonization.[70,75,76] Despite the numerous listed alternatives, almost 75% of research papers (Web of Science v5.14 search performed using keywords Electrospinning–Electrospun–Etch CNFs) involving electrospun CNFs are based on PAN in DMF, due to its high melting point, easy stabilization and high carbon yield. Electrospun CNFs can improve over several aspects of the commercial carbon-based catalyst supports. For instance, Park et al. in 2009 focused the attention on the different superficial morphology specific to the electrospun CNFs; in particular, their PAN-based CNFs were characterized by shallower pores with larger average diameter when compared to commercial carbons.[77] The different pore morphology resulted in improved Pt utilization thanks to absence of Pt nanoparticles deposited in deep pores where it is impossible to generate the necessary triple phase and therefore use that fraction of catalyst. For the catalyst deposition on both the commercial carbon and the electrospun nanofibers, a standard polyol method was adopted, whose validity had been previously demonstrated on carbon nanotubes as well as commercial carbon.[78] In this work, the sole use of electrospun CNFs instead of commercial carbon resulted in a twofold increase of Pt usage.

The process was later improved upon by functionalizing the CNFs with aminopyrene prior to a deposition of Pt-Ru nanoparticle via the polyol method.[79] This improved catalyst/support system was characterized by smaller and better dispersed nanoparticles and showed higher active surface area and better performance toward the methanol oxidation reaction (MOR) than nonfunctionalized electrospun CNFs with Pt-Ru-based catalyst.

Another morphological difference when using electrospun carbon is the possibility to achieve continuous network of fibers as self-supporting mats. This widens the range of alternative techniques available for the Pt deposition. For example, the electrodeposition of Pt on continuous electrodes for DMFC has been investigated.[80] The electrodeposited catalyst resulted in clusters of Pt nanoparticles with diameters in the 50–200 nm range, much larger than the optimized 3–5 nm particle size usually adopted and easily achieved by the polyol method.[78] Despite the poor particle size and distribution of the electrodeposited Pt, the methanol oxidation capability of the electrospun carbon fibers resulted higher than the commercial catalyst with more than double of oxidation currents. The authors attributed this improvement in performance to the higher conductivity of the CNF mats and to a closer contact between Pt particles and carbon, which could once again increase the utilization efficiency.

In a successive work, the same research team replaced the Pt electrodeposition with a deposition based on formaldehyde vapor mediated reduction of chloroplatinic acid (H_2PtCl_6) absorbed onto the electrospun carbon mats.[81] The change in preparation technique successfully reduced the Pt nanoparticles size to an average of 5 nm and improved dispersion. Despite this dramatic change in the Pt morphology, only a minimal improvement in performance was obtained over the electrodeposited Pt; in detail, the ~420 mA mg^{-1} specific methanol oxidation currents of the electrodeposited Pt on carbon mat was increased only to ~445 mA mg^{-1}. Therefore, pivotal to the performance increase over the commercial electrocatalyst is the continuous highly interconnected nature of the self-standing electrospun CNF mats together with the absence of deep porosity. In other words, according to these works, the difference is in the carbon support more than in the Pt nanoparticles.

The versatility of the electrospinning technique allows the preparation of catalyst-loaded CNFs with an in situ method. Pt and the support can be electrospun together to obtain carbon–platinum composite nanofibers from a solution of PAN and Pt acetylacetonate.[82,83] A similar technique was applied to produce carbon–palladium composite nanofibers for alkaline DMFC thus minimizing the number of fabrication steps.[84] Despite the fact that some of the Pd particles were completely embedded in the carbon, therefore unable to participate in the reaction process, enhanced electrocatalytic performance toward methanol electrooxidation in alkaline conditions compared to the commercial Pd/C catalyst were obtained using this one-pot method.

Electrospun CNFs find application also as catalyst support in microbial fuel cells. Recently, the possibility to grow a bioelectro catalytic layer on electrospun carbon fiber derived from PAN has been demonstrated.[85,86] Excellent results have been obtained with current density values as high as 30 A m^{-2} never reported before for O_2 electrocatalytic reduction. The contribution of the CNFs results in higher performance by enabling higher loading of active enzymes and faster kinetics at the electrode surface. This concept was further improved upon by Engel et al.; they added double and multiwalled nanotubes to the surface of the electrospun nanofibers by simply dipping the carbonized mats in a water dispersion of the nanotubes.[87] They reported higher electrical conductance of the surface treated carbon mats yet similar current densities for the oxygen reduction reaction (ORR) performance.

2.3.2 NONCARBON SUPPORTS

As mentioned earlier, carbon blacks are the most commonly used PEMFC electrocatalyst supports because of their high surface area, flexible porous structure, excellent electronic conductivity, and low cost. However, they do not meet the durability requirements for PEMFC applications (e.g., automotive)[88] and need to be replaced with stable alternative materials. Noncarbon supports have been intensively investigated in the past decades as alternative supports.[89–91] The materials developed so far include metal oxides such as SiO_2,[92] WO_x,[93–95] MoO_x,[96] SnO_2,[97–100] TiO_x,[101–105] and more recently also transition metal carbides (WC, Mo_2C)[106–108] and nitrides (TiN, MoN).[109] The main advantage of oxides over carbon is their high chemical stability and corrosion resistance in PEMFC operation conditions.[90] Thermochemical calculations also shows the stability of oxides of several transition and posttransition metals (Ti, Nb, Sn, Ta, W, Sb) under PEMFC operating conditions (80°C, 1.0 V/SHE and pH = 0).[110] Furthermore, metal oxides can present promoting effects, enhancing the electrocatalytic activity by synergic electronic interactions with the metal electrocatalyst, the so-called strong metal-support interaction.[111,112] Ceramic supports have already shown promising results in terms of performance and durability with respect to conventional Pt/C electrodes.[8,89,91] For instance, a modification of the carbon support surface with titanium oxide[113] or the use of pure TiO_2 supports[102,114] improved the electrochemical stability of the electrode. Furthermore, the addition of oxides to alcohol fuel cell cathodes is known to be an efficient way to improve the tolerance of Pt to methanol for the ORR.[115] The use of oxides in direct alcohol fuel cell anodes is also growing because of the promoting effect on the alcohol electrooxidation kinetics,[116] as well as the enhanced CO tolerance.[117]

The main challenge to be addressed with metal oxides is related to their poor electronic conductivity. The strategies currently employed to enhance this property are the doping with aliovalent ions (e.g., Nb, V, Ta, Ru, Sb)[110,118,119] and the chemical reduction to nonstoichiometric compositions.[104,120] The research interest to alternative fuel cell electrocatalyst supports prepared by electrospinning is growing in the last 5 years. In particular, many studies have been reported on titanium dioxide nanofibers. Most part of them deal with Pt supports materials for the cathodic side.[121–125] Both the strategies, that is, doping with Nb and reducing to TiO_x (x < 2), have been applied to these electrospun fibers to make them conductive. For instance, Pt supported on titanium oxide fibers subjected to different treatments after electrospinning (undoped anatase, Nb doped anatase, and reduced under H_2 rutile) presented a higher stability compared to commercial Pt/C after prolonged electrochemical cycles.[121] In particular Nb doped-TiO_2 fibers exhibited the most promising results in terms of stability and ORR activity. Nevertheless, the latter was lower for the Nb doped-TiO_2 fibers compared to the carbon-based catalyst probably due to its lower electrical conductivity. Pt deposited onto Nb doped electrospun fibers in the rutile form and reduced at the surface showed also enhanced corrosion resistance and an electrocatalytic activity similar to Pt-loaded CNFs.[123] Furthermore, after potential cycling the agglomeration and growth of the catalyst particles was lower when compared to Pt/CNFs. Electrospun Ti_4O_7 was also evaluated as Pt support toward oxygen reduction and compared to the conductive oxide NbO_2.[124] Both electrocatalysts showed a low electrochemical surface area (ECSA) loss after cycling. However, their ORR mass activities dramatically dropped after the durability tests, because of the formation of electronically insulating Nb_2O_5 and TiO_2 onto the nanofiber surface, as revealed by x-ray photoelectron spectroscopy (XPS).

Titania nanofibers prepared by electrospinning found application also in direct alcohol fuel cell anodes.[126–128] Anatase nanofibers decorated with Pt nanoparticles[126] and nanowires[127] greatly promoted the MOR and the electrode durability compared to commercial Pt/C. Pd-loaded TiO_2 electrospun fibers demonstrated a higher electrocatalytic activity toward the oxidation of several alcohols such as glycerol, methanol, ethylene glycol, and 1,2-propanediol[128] in basic medium.

SnO_2 is another promising oxide that has been evaluated as a PEMFC support in 1D morphologies obtained by electrospinning. Pt/SnO_2 fibers have been used as anodes for the hydrogen oxidation reaction (HOR) with excellent performance.[129,130] However, the activity toward ORR was completely blocked. On the contrary, Nb doped tin oxide fiber in tubes (Figure 2.7) exhibited higher corrosion resistance than carbon-based electrodes and, when loaded with Pt, a comparable electroactivity to Pt/C toward oxygen reduction.[131]

WO_3 fibers are mainly investigated as PEMFC anode materials. Mesoporous electrospun WO_3 nanofibers anchored with Pt nanocubes or nanospheres were prepared

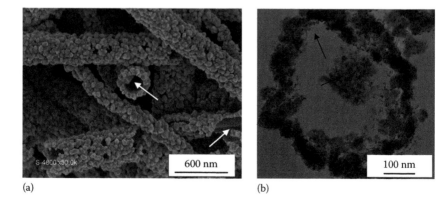

(a)   (b)

FIGURE 2.7 SEM image of Nb doped SnO_2 fibers in tubes (a) and TEM micrograph of the Pt catalyzed supports (b). (Reprinted with permission from Cavaliere, S. et al., *J. Phys. Chem. C*, 117, 18298–18307. Copyright 2013 American Chemical Society.)

by combining electrospinning and galvanic replacement. Such 1D Pt/WO_3 hetero-structures were then hybridized with carbon nanotubes and showed exceptional catalytic activity for methanol oxidation. In particular, Pt nanocubes presented higher performance than their spherical counterparts.[132]

Novel methods to improve the dispersion of Pt nanocatalysts on oxide nanofibrous materials have been studied. Approaches including in situ electrospinning-based methods[129,133] or embedding of preformed metal particles[134] or galvanic replacement[132] leading to Pt-loaded nanofibers can demonstrate their possible application in fuel cell electrodes.

A final aspect that should be underlined when dealing with metal oxide nanofibrous supports is their brittleness, avoiding their direct use in the mat form in fuel cell MEAs. Several strategies, such as pressing, combination with polymers, and preparation of twisted yarns[135] and ribbons.[136]

Transition metal carbides and nitrides are also investigated as electrocatalyst supports because of their high chemical stability and electrical conductivity.[137–139] So far, only very few papers deal with these materials prepared by electrospinning. As an example, we can cite the Pt-loaded electrospun titanium carbide nanofibers with high surface area (500 m^2 g^{-1}).[140] The obtained Pt/TiC mats presented higher ECSA and current density toward the oxygen reduction compared to Pt-loaded carbon paper. Other works on these materials concern their direct use as electrocatalysts without the need for metal nanoparticles (see Section 2.4).

2.3.3   Composite Supports

In order to overcome the low conductivity issue of noncarbon supports and the corrosion of carbon supports, various composite materials exploiting the high conductivity of carbon with the stability and promoting effect of metal oxides have been developed by electrospinning.

One approach consists in embedding oxide nanoparticles in the CNFs matrix. TiO_2 has been embedded in CNFs introducing a mixture of high-purity anatase and rutile nanoparticles in a PAN/DMF electrospinning solution.[141] Titania nanoparticles would provide an enhanced electrocatalytic activity similar to that found using pure oxide electrospun nanofibers as support for noble metals catalysts earlier discussed in this chapter (see Section 2.3.2). In this work, Pt-Ru nanoparticles have been used as electrocatalyst: this novel catalyst/support system showed the highest performance with respect to conventional Pt/C, Pt-Ru/TiO_2, and Pt-Ru/CNFs. By tuning the Ti/C ratio, four times higher mass activity than Pt-Ru on carbon was achieved. The DMFC employed this composite catalyst doubled the power output using only a fourth of the amount of Pt-Ru used in a DMFC with commercial Pt-Ru/C. Later that year, the same group demonstrated the addition of CeO_2 by electrospinning a solution of PAN/DMF and ammonium cerium nitrate $(Ce(NH_4)_2(NO_3)_6)$.[142] Ceria was chosen because Ce cations can switch between the +3 and +4 oxidation states and therefore acts as oxygen buffer; furthermore, ceria has been previously proven to enhance the catalytic performance by increasing the noble metal dispersion on the support surface and acting as a cocatalyst toward methanol electrooxidation.[143] Embedding ceria in the carbon matrix did not hinder the beneficial effects derived

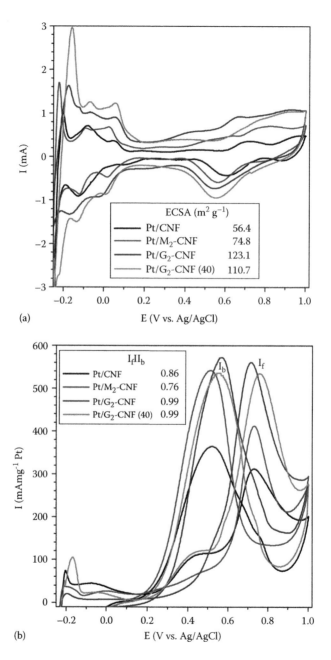

(a)

(b)

FIGURE 2.8 CV curves of Pt–CNF, Pt–G2–CNF, and Pt–M2–CNF (28 wt% Pt) and Pt–G2–CNF (40 wt% Pt) catalysts measured in 0.5 M H$_2$SO$_4$ (a) and H$_2$SO$_4$–CH$_3$OH (0.5–1 M) (b) aqueous solutions at a scan rate of 50 mV s^{-1}. (Wang, C. et al., *Nanoscale*, 6, 1377–1383, 2014. Reproduced by permission of The Royal Society of Chemistry.)

from the synergic combination of the oxide and the noble metal. On the contrary, MOR current of this novel catalyst support was three times higher than that of the commercial catalyst after prolonged reaction showing higher activity and stability. The authors were able to correlate the enhanced performance to an intimate contact between the CeO_2 nanoparticles and the carbon achieved by embedding the CeO_2 particles in the CNFs. This promoted the electron conduction between the different phases and consequently a better interaction between the methanol on the surface of the catalyst and the surrounding CeO_2.

Another approach to produce carbon–oxide composite supports is to introduce carbon in oxides nanofibers. Electrospinning a precursor solution of titanium(IV) isopropoxide, niobium ethoxide, PVP, acetic acid, and ethanol Nb doped TiO_2/PVP composite nanofibers were produced. In a subsequent thermal treatment in hydrogen atmosphere, the transformation to carbon embedded in Nb-TiO_2 was achieved.[144] Pt nanoparticles were deposited using microwave-assisted polyol deposition. They obtained ORR activity comparable with that of a commercially available carbon supported Pt catalyst and increased durability. However, the mass activity of the ceramic supported catalyst was low, most likely as a consequence of the low conductivity and low surface area of the ceramic support.

The possibility of easily combining different carbon allotropes is another advantage offered by electrospinning. In particular, graphene nanoribbons and multiwalled carbon nanotubes were successfully introduced in electrospun CNFs in 2014 by Wang et al.[145] They showed that simply electrospinning a solution of PAN/DMF and the desired carbon allotrope is possible to enhance the graphitization degree thus the conductivity without the need to increase the carbonization temperature. Therefore, it is possible to produce highly conductive CNFs without reducing the nitrogen content that would consequently result in decreased CO tolerance and durability of Pt catalysts.[146] The addition of graphene nanoribbons resulted in highly dispersed Pt nanoparticles; compared to pure CNFs and to CNFs loaded with multiwalled nanotubes, the electrodes containing graphene show much higher ECSA and better CO tolerance, indicating excellent catalytic activity and long-term stability (Figure 2.8).

2.4   SELF-SUPPORTED ELECTROCATALYSTS

In the last years, electrospinning has been used to prepare pure noble metal nanowires to be used directly as unsupported electrocatalysts. Their 1D morphology can indeed influence the performance toward the oxygen reduction by providing a facile pathway for the charge transfer and improving the accessibility to the catalyst.[147] Nevertheless, still few studies are reported on metal fibers compared to support materials, probably due to the difficulty in achieving very thin diameters by this technique. Kim et al. prepared electrospun Pt nanowires that mixed with Pt/C led to an improvement of the catalytic activity toward ORR in a fuel cell.[148] The design of bimetallic nanowires to prepare electrospun self-supported catalyst has also been described (Figure 2.9). For instance, Pt/Fe alloy nanowire displayed a high catalytic activity and durability compared to Pt/C toward oxygen reduction.[149] Concerning anodic reactions, Pt and Pt/Rh nanowires have been developed for the oxidation in cyclohexane-fueled fuel cells,[150] and Pt/Rh and Pt/Ru nanowires for methanol oxidation.[151] In all cases, bimetallic nanofibers show better performance than commercial Pt/C.

Another category of self-supported electrospun electrocatalysts is represented by metal-free fibers made of materials possessing intrinsic catalytic activity. For instance, metal carbides and in particular WC, with his Pt-like electronic structure,

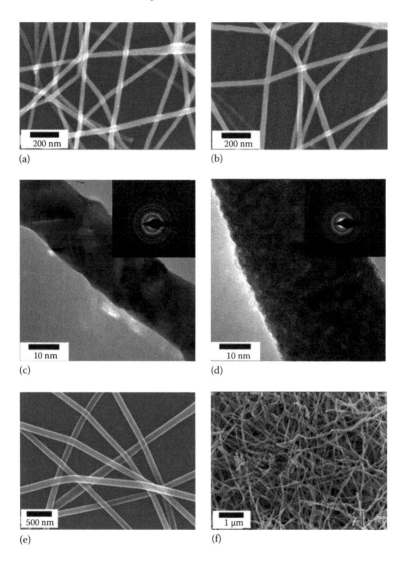

FIGURE 2.9   FESEM images of Pt (a) and Pt1Rh1 (b) nanowire. HRTEM images and corresponding SAED patterns of single nanowires of Pt (c) and Pt1Rh1 (d). FESEM image of as-spun PtRh precursor/PVP composite nanofibers (e). SEM image of the Pt1Rh1 nanowires on the carbon paper electrode (f). *(Continued)*

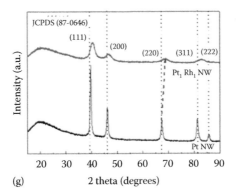

(g)

FIGURE 2.9 (Continued) XRD patterns of the electrospun Pt and Pt1Rh1 nanowires (g). (Reprinted from *Electrochem. Commun.*, 11, Kim, H., Kim, Y., Seo, M., Choi, S., and Kim, W., Pt and PtRh nanowire electrocatalysts for cyclohexane-fueled polymer electrolyte membrane fuel cell Hyung, 446–449, Copyright 2009, with permission from Elsevier.)

can catalyze the oxygen reduction.[152] Electrospun tungsten carbide nanofibers showed a catalytic activity toward ORR similar to that of the commercial 20% Pt/Vulcan XC-72R. In this case, several post treatments such as NH_3 etching were investigated to remove the thin layer of carbon formed during the synthesis so as to improve the performance.[153] Chromium carbide and nitride nanofibers have also been prepared by electrospinning.[154] The carbide is formed during a direct carburation using the carrier polymer as the carbon source. Adding a specific N donor polymer to the electrospinning solution enabled them to tune the proportion of nitride and carbide. The slow intrinsic electrocatalytic activity toward ORR is attributed to the low surface area and chromium content of the nanofibers.

Metal oxide fibers, discussed earlier as supports, have also been reported as self-supported electrocatalysts. WO_3 nanofibers have shown electrochemical activity toward the hydrogen oxidation even without metal catalyst loading.[155] Their activity can be enhanced by decreasing their diameter and by the addition of carbon black. The obtained WO_3/C hybrid materials presented HOR performance superior when compared to a commercial 20% wt Pt/C catalyst. $PdO–Co_3O_4$[156] and NiO[157] have been used as anode electrocatalysts in DMFCs in basic conditions. In particular, the latter demonstrated a positive impact of the 1D morphology on the electrocatalytic activity in comparison with NiO nanoparticles. Cobalt oxide nanoparticles incorporated into electrospun CNFs led to an increase of the electrochemical activity toward methanol oxidation of the CNFs alone.[158]

Electrospun carbon fibers doped with nitrogen and nonnoble metals such as cobalt and iron are also growing in interest as PEMFC electrocatalysts. Hollow porous electrospun N/CNFs[159] and Fe-N/CNFs[160] have shown great electroactivity toward ORR and stability in acidic medium. FeCo/CNFs obtained by electrospinning and pyrolysis presented comparable electrocatalytic activity and higher tolerance to ethanol crossover than Pt/C in alkaline fuel cells.[161]

2.5   ELECTROSPUN ELECTRODE COMPONENTS

A potential strategy to improve power density and durability of fuel cell electrodes is to use electrospinning to alter the catalyst electrode morphology. A promising technique is the electrospinning of particle/polymer (catalyst/ionomer) mixture in a nonwoven nanofiber electrode catalyst mat. It was found that replacing a conventional MEA cathode with a nanofiber mat where the fibers were composed of Pt/C catalyst with a Nafion/PAA binder performed exceptionally well in H_2/air fuel cell MEAs with a low cathode Pt loading of 0.1 mg Pt cm^{-2}. The nanofiber cathode also exhibits outstanding stability in accelerated durability tests.[162,163] Transition metal oxides and in particular cerium oxide have been conclusively shown to show peroxide degradation and radical scavenging properties that are relevant to mitigating degradation processes in PEMFC. In this context, the preparation of electrospun cerium oxide supports is of value.[164] However, the problems of handling brittle ceramic webs as described earlier with metal oxide supports impedes their direct incorporation as a self-standing mat. Carbon–cerium oxide composite electrospun mats represent a promising alternative,[142] as discussed in Section 2.3.3. A further alternative is to associate the radical scavenger with the ionomer binder in the catalyst layer. New PFSA nanofiber material enriched with cerium oxide nanoparticles has been developed as radical trap at the electrode interface with the membrane.[165] The results of OCV hold testing at low RH (50%) and high temperature (90°C) show that whereas MEAs integrating nonmodified Nafion 212 show a marked drop in OCV with time, and end of life at <200 h, an MEA comprising a radical scavenger mat at the anode gave very stable OCV and a lifetime close to 1000 h and markedly low fluoride emission rate.

2.6   CONCLUSIONS AND FUTURE CHALLENGES

The energy storage capability of hydrogen when coupled with renewable energy sources is a means of overcoming the drawbacks arising from the intermittent character of such sources with several landmark demonstration projects in operation using the hydrogen cycle. Fuel cells are essential for the conversion of hydrogen to electrical energy and heat and are part of the future technologies for energy conversion and storage, alongside electrolysers, batteries, and redox flow systems. The outlook for fuel cells for automotive applications has never been brighter, with the Mirai fuel cell vehicle under commercialization by Toyota. Challenges still remain for cost and durability nevertheless and new materials are required to meet these challenges. This chapter has shown that electrospinning is already showing great promise in the generation of materials for the PEM, catalyst support, and even the complete electrode. Despite these achievements, the full realization of the potential of the great versatility of electrospinning in fuel cell materials is probably yet to come, as they are under assessment for industrialization. Areas under active investigation include using the surface area presented by ceramic and CNFs for deposition of ultrathin (<3 nm) continuous films of platinum or platinum alloys by associating electrospinning and surface modification techniques such as atomic layer deposition to vastly increase the mass activity of the resulting electrocatalyst and bring about a step change reduction in the amount of precious metal used in current fuel cell stacks to equivalent or below that used in incumbent combustion engine vehicles. Combining electrospun heteroatom doped CNFs and nonprecious metal catalytic centers is a route to zero-platinum fuel cell catalysts. Still in the context of fuel cell electrodes, the direct integration of a catalyzed electrospun support web into a fuel cell electrode is particularly attractive but is currently feasible only for carbon, due to brittle nature of the majority of electrospun ceramic supports. This is a challenge where creativity is called for in the approaches to materials composition and stabilization approaches since the rewards are expected to be high. It is however in the fuel cell membrane component that electrospun materials have so far made greatest impact. Disassociating functions of structure and transport, controlling phase separation either by separating polymer phases by intent or associating nonmiscible polymers at designed length scales, ionically or chemically cross-linking the electrospun and matrix components all become possible and have all been demonstrated. Furthermore, interfaces can be adjusted by associating ALD, reactive plasma, electron beam, or other surface modification techniques with electrospinning. A challenge remains in the handling of ultrathin electrospun polymer nanofiber reinforcements but which can surely be overcome as the full benefits are further proven. The day is close for realization of the ambitious objective of an integrated catalyst-coated membrane process assembled by successive electrospinning of electrode and membrane components.[166]

REFERENCES

1. Stambouli, A. B. Fuel cells: The expectations for an environmental-friendly and sustainable source of energy. *Renew. Sustain. Energy Rev.* **15**, 4507–4520 (2011).
2. Borup, R. et al. Scientific aspects of polymer electrolyte fuel cell durability and degradation. *Chem. Rev.* **107**, 3904–3951 (2007).
3. Jones, D. J. and Rozière, J. *Encyclopedia of Electrochemical Power Sources*, Vol. 2 (Elsevier, Amsterdam, the Netherlands, 2009, pp. 667–679).
4. Subianto, S. et al. Physical and chemical modification routes leading to improved mechanical properties of perfluorosulfonic acid membranes for PEM fuel cells. *J. Power Sources* **233**, 216–230 (2013).
5. Rabis, A., Rodriguez, P., and Schmidt, T. Electrocatalysis for polymer electrolyte fuel cells: Recent achievements and future challenges. *ACS Catal.* **2**, 864–890 (2012).
6. Lefèvre, M., Proietti, E., Jaouen, F., and Dodelet, J. P. Iron-based catalysts with improved oxygen reduction activity in polymer electrolyte fuel cells. *Science* **324**, 71–74 (2009).
7. Thompsett, D. *Proton Exchange Membrane Fuel Cells, Materials Properties and Performance* (CRC Press, Boca Raton, FL, 2010, pp. 2–60).
8. Shao, Y., Liu, J., Wang, Y., and Lin, Y. Novel catalyst support materials for PEM fuel cells: Current status and future prospects. *J. Mater. Chem.* **19**, 46–59 (2009).
9. Balgis, R., Anilkumar, G. M., Sago, S., Ogi, T., and Okuyama, K. Nanostructured design of electrocatalyst support materials for high-performance PEM fuel cell application. *J. Power Sources* **203**, 26–33 (2012).
10. Cavaliere, S., Subianto, S., Savych, I., Jones, D. J., and Rozière, J. Electrospinning: Designed architectures for energy conversion and storage devices. *Energy Environ. Sci.* **4**, 4761–4785 (2011).

11. Li, J., Pan, M., and Tang, H. Understanding short-side-chain perfluorinated sulfonic acid and its application for high temperature polymer electrolyte membrane fuel cells. *RSC Adv.* **4**, 3944–3965 (2014).

12. Alberti, G., Narducci, R., Luisa, M., Vona, D., and Giancola, S. More on Nafion conductivity decay at temperatures higher than 80°C: Preparation and first characterization of in-plane oriented layered morphologies. *Ind. Eng. Chem. Res.* **52**, 10418–10424 (2013).

13. Neburchilov, V., Martin, J., Wang, H., and Zhang, J. A review of polymer electrolyte membranes for direct methanol fuel cells. *J. Power Sources* **169**, 221–238 (2007).

14. Deluca, N. W. and Elabd, Y. A. Polymer electrolyte membranes for the direct methanol fuel cell: A review. *J. Polym. Sci. Part B: Polym. Phys.* **44**, 2201–2225 (2006).

15. Alberti, G., Narducci, R., Di Vona, M. L., and Giancola, S. Annealing of Nafion 1100 in the presence of an annealing agent: A powerful method for increasing ionomer working temperature in PEMFCs. *Fuel Cells* **13**, 42–47 (2013).

16. Casciola, M. et al. Zirconium phosphate reinforced short side chain perflurosulfonic acid membranes for medium temperature proton exchange membrane fuel cell application. *J. Power Sources* **262**, 407–413 (2014).

17. Alberti, G., Narducci, R., and Sganappa, M. Effects of hydrothermal/thermal treatments on the water-uptake of Nafion membranes and relations with changes of conformation, counter-elastic force and tensile modulus of the matrix. *J. Power Sources* **178**, 575–583 (2008).

18. Jang, W. G., Hou, J., and Byun, H. Preparation and characterization of PVdF nanofiber ion exchange membrane for the PEMFC application. *Desalin. Water Treat.* **34**, 315–320 (2011).

19. Ballengee, J. and Pintauro, P. Composite fuel cell membranes from dual-nanofiber electrospun mats. *Macromolecules* **44**, 7307–7314 (2011).

20. Ballengee, J. B., Haugen, G. M., Hamrock, S. J., and Pintauro, P. N. Properties and fuel cell performance of a nanofiber composite membrane with 660 equivalent weight perfluorosulfonic acid. *J. Electrochem. Soc.* **160**, F429–F435 (2013).

21. Li, H. and Liu, Y. Composite membranes of polybenzimidazole and crosslinked polybenzimidazole-polybenzoxazine electrospun nanofibers for proton exchange membrane fuel cells. *J. Mater. Chem. A* **1**, 1171–1178 (2013).

22. Lim, J.-M. et al. Polyimide nonwoven fabric-reinforced, flexible phosphosilicate glass composite membranes for high-temperature/low-humidity proton exchange membrane fuel cells. *J. Mater. Chem.* **22**, 18550–188557 (2012).

23. Subianto, S. et al. Reactive coaxial electrospinning of Zrp/ZrO$_2$ nanofibres. *J. Mater. Chem. A* **2**, 13359–13365 (2014).

24. Wycisk, R., Pintauro, P., and Park, J. New developments in proton conducting membranes for fuel cells. *Curr. Opin. Chem. Eng.* **4**, 71–78 (2014).

25. Mollá, S., Compañ, V., Gimenez, E., Blazquez, A., and Urdanpilleta, I. Novel ultrathin composite membranes of Nafion/PVA for PEMFCs. *Int. J. Hydrogen Energy* **36**, 9886–9895 (2011).

26. Yun, S.-H. et al. Sulfonated poly(2,6-dimethyl-1,4-phenylene oxide) (SPPO) electrolyte membranes reinforced by electrospun nanofiber porous substrates for fuel cells. *J. Membr. Sci.* **367**, 296–305 (2011).

27. Page, K. A., Cable, K. M., and Moore, R. B. Molecular origins of the thermal transitions and dynamic mechanical relaxations in perfluorosulfonate ionomers. *Macromolecules* **38**, 6472–6484 (2005).

28. Lee, H.-J. et al. Highly flexible, proton-conductive silicate glass electrolytes for medium-temperature/low-humidity proton exchange membrane fuel cells. *ACS Appl. Mater. Interfaces* **5**, 5034–5043 (2013).

29. Choi, S. W., Fu, Y.-Z., Ahn, Y. R., Jo, S. M., and Manthiram, A. Nafion-impregnated electrospun polyvinylidene fluoride composite membranes for direct methanol fuel cells. *J. Power Sources* **180**, 167–171 (2008).

30. Hasani-Sadrabadi, M. M., Shabani, I., Soleimani, M., and Moaddel, H. Novel nanofiber-based triple-layer proton exchange membranes for fuel cell applications. *J. Power Sources* **196**, 4599–4603 (2011).

31. Shabani, I., Hasani-Sadrabadi, M. M., Haddadi-Asl, V., and Soleimani, M. Nanofiber-based polyelectrolytes as novel membranes for fuel cell applications. *J. Membr. Sci.* **368**, 233–240 (2011).

32. Lin, H.-L. et al. Preparation of Nafion/poly(vinyl alcohol) electro-spun fiber composite membranes for direct methanol fuel cells. *J. Membr. Sci.* **365**, 114–122 (2010).

33. Lin, H.-L. and Wang, S.-H. Nafion/poly(vinyl alcohol) nano-fiber composite and Nafion/poly(vinyl alcohol) blend membranes for direct methanol fuel cells. *J. Membr. Sci.* **452**, 253–262 (2014).

34. Mollá, S. and Compañ, V. Performance of composite Nafion/PVA membranes for direct methanol fuel cells. *J. Power Sources* **196**, 2699–2708 (2011).

35. Mollá, S. and Compañ, V. Polyvinyl alcohol nanofiber reinforced Nafion membranes for fuel cell applications. *J. Membr. Sci.* **372**, 191–200 (2011).

36. Thiam, H. S. et al. Nafion/Pd–SiO$_2$ nanofiber composite membranes for direct methanol fuel cell applications. *Int. J. Hydrogen Energy* **38**, 9474–9483 (2013).

37. Thiam, H. S. et al. Performance of direct methanol fuel cell with a palladium–silica nanofibre/Nafion composite membrane. *Energy Convers. Manag.* **75**, 718–726 (2013).

38. Deluca, N. and Elabd, Y. Nafion®/poly(vinyl alcohol) blends: Effect of composition and annealing temperature on transport properties. *J. Membr. Sci.* **282**, 217–224 (2006).

39. Li, X. et al. Fabrication of sulfonated poly(ether ether ketone ketone) membranes with high proton conductivity. *J. Membr. Sci.* **281**, 1–6 (2006).

40. Dong, B., Gwee, L., Salas-de la Cruz, D., Winey, K. I., and Elabd, Y. A. Super proton conductive high-purity nafion nanofibers. *Nano Lett.* **10**, 3785–3790 (2010).

41. Pan, C. et al. Nanowire-based high-performance "Micro Fuel Cells": One nanowire, one fuel cell. *Adv. Mater.* **20**, 1644–1648 (2008).

42. Tamura, T. and Kawakami, H. Aligned electrospun nanofiber composite membranes for fuel cell electrolytes. *Nano Lett.* **10**, 1324–1328 (2010).

43. Takemori, R., Ito, G., Tanaka, M., and Kawakami, H. Ultra-high proton conduction in electrospun sulfonated polyimide nanofibers. *RSC Adv.* **4**, 20005–20009 (2014).

44. Fennessey, S. F. and Farris, R. J. Fabrication of aligned and molecularly oriented electrospun polyacrylonitrile nanofibers and the mechanical behavior of their twisted yarns. *Polymer (Guildf.)* **45**, 4217–4225 (2004).

45. Kakade, M. V. et al. Electric field induced orientation of polymer chains in macroscopically aligned electrospun polymer nanofibers. *J. Am. Chem. Soc.* **129**, 2777–2782 (2007).

46. Kongkhlang, T., Tashiro, K., Kotaki, M., and Chirachanchai, S. Electrospinning as a new technique to control the crystal morphology and molecular orientation of polyoxymethylene nanofibers. *J. Am. Chem. Soc.* **130**, 15460–15466 (2008).

47. Richard-Lacroix, M. and Pellerin, C. Molecular orientation in electrospun fibers: From mats to single fibers. *Macromolecules* **46**, 9473–9493 (2013).

48. Jiang, S., Xia, K.-Q., and Xu, G. Effect of additives on self-assembling behavior of Nafion in aqueous media. *Macromolecules* **34**, 7783–7788 (2001).

49. Sanders, E. H. et al. Characterization of electrosprayed Nafion films. *J. Power Sources* **129**, 55–61 (2004).

50. Chen, H., Snyder, J. D., and Elabd, Y. A. Electrospinning and solution properties of Nafion and poly(acrylic acid). *Macromolecules* **41**, 128–135 (2008).

51. Laforgue, A., Robitaille, L., Mokrini, A., and Ajji, A. Fabrication and characterization of ionic conducting nanofibers. *Macromol. Mater. Eng.* **292**, 1229–1236 (2007).

52. Subramanian, C., Weiss, R. A., and Shaw, M. T. Electrospinning and characterization of highly sulfonated polystyrene fibers. *Polymer (Guildf.)* **51**, 1983–1989 (2010).

53. Shenoy, S., Bates, W., Frisch, H., and Wnek, G. Role of chain entanglements on fiber formation during electrospinning of polymer solutions: Good solvent, non-specific polymer–polymer interaction limit. *Polymer (Guildf.)* **46**, 3372–3384 (2005).

54. Subianto, S., Cavaliere, S., Jones, D. J., and Rozière, J. Effect of side-chain length on the electrospinning of perfluorosulfonic acid ionomers. *J. Polym. Sci. Part A: Polym. Chem.* **51**, 118–128 (2013).

55. Choi, J., Lee, K. M., Wycisk, R., Pintauro, P. N., and Mather, P. T. Nanofiber network ion-exchange membranes. *Macromolecules* **41**, 4569–4572 (2008).

56. Lavielle, N. et al. Straightforward approach for fabricating hierarchically structured composite membranes. *ACS Appl. Mater. Interfaces* **5**, 10090–10097 (2013).

57. Ballengee, J. B. and Pintauro, P. N. Preparation of nanofiber composite proton-exchange membranes from dual fiber electrospun mats. *J. Membr. Sci.* **442**, 187–195 (2013).

58. Park, A. M. and Pintauro, P. N. Electrospun composite membranes for alkaline fuel cells. *ECS Trans.* **41**, 1817–1826 (2011).

59. Chalk, S. G. and Miller, J. F. Key challenges and recent progress in batteries, fuel cells, and hydrogen storage for clean energy systems. *J. Power Sources* **159**, 73–80 (2006).

60. Eastwood, B. J., Christensen, P. A., Armstrong, R. D., and Bates, N. R. Electrochemical oxidation of a carbon black loaded polymer electrode in aqueous electrolytes. *J. Solid State Electrochem.* **3**, 179–186 (1999).

61. Taniguchi, A., Akita, T., Yasuda, K., and Miyazaki, Y. Analysis of electrocatalyst degradation in PEMFC caused by cell reversal during fuel starvation. *J. Power Sources* **130**, 42–49 (2004).

62. Reiser, C., Bregoli, L., and Patterson, T. A reverse-current decay mechanism for fuel cells. *Electrochem. Solid-State Lett.* **8**, A273 (2005).

63. Yousfi-Steiner, N., Moçotéguy, P., Candusso, D., and Hissel, D. A review on polymer electrolyte membrane fuel cell catalyst degradation and starvation issues: Causes, consequences and diagnostic for mitigation. *J. Power Sources* **194**, 130–145 (2009).

64. Passalacqua, E., Antonucci, P., and Vivaldi, M. The influence of Pt on the electro-oxidation behaviour of carbon in phosphoric acid. *Electrochim. Acta* **37**, 2725–2730 (1992).

65. Wang, F., Liu, C., Liu, C., Chao, J., and Lin, C. Effect of Pt loading order on photocatalytic activity of Pt/TiO_2 nanofiber in generation of H_2 from neat ethanol. *J. Phys. Chem. C* **113**, 13832–13840 (2009).

66. Hartl, K., Hanzlik, M., and Arenz, M. IL-TEM investigations on the degradation mechanism of Pt/C electrocatalysts with different carbon supports. *Energy Environ. Sci.* **4**, 234–238 (2011).

67. Wu, B. et al. High dispersion of platinum-ruthenium nanoparticles on the 3,4,9,10-perylene tetracarboxylic acid-functionalized carbon nanotubes for methanol electro-oxidation. *Chem. Commun. (Camb.)* **47**, 5253–5255 (2011).

68. Sebastián, D. et al. Enhanced oxygen reduction activity and durability of Pt catalysts supported on carbon nanofibers. *Appl. Catal. B: Environ.* **115–116**, 269–275 (2012).

69. Chun, I., Reneker, D., Fong, H., and Fang, X. Carbon nanofibers from polyacrylonitrile and mesophase pitch. *J. Adv. Mater.* **31**, 36–41 (1999).

70. Kim, C., Choi, Y.-O., Lee, W.-J., and Yang, K.-S. Supercapacitor performances of activated carbon fiber webs prepared by electrospinning of PMDA-ODA poly(amic acid) solutions. *Electrochim. Acta* **50**, 883–887 (2004).

71. Park, S. H., Kim, C., and Yang, K. S. Preparation of carbonized fiber web from electrospinning of isotropic pitch. *Synth. Metals* **143**, 175–179 (2004).

72. Kim, B.-H., Wazir, A. H., Yang, K.-S., Bang, Y.-H., and Kim, S.-R. Molecular structure effects of the pitches on preparation of activated carbon fibers from electrospinning. *Carbon Lett.* **12**, 70–80 (2011).

73. Zou, L. et al. A film of porous carbon nanofibers that contain Sn/SnO_x nanoparticles in the pores and its electrochemical performance as an anode material for lithium ion batteries. *Carbon N. Y.* **49**, 89–95 (2011).

74. Chung, G. S., Jo, S. M., and Kim, B. C. Properties of carbon nanofibers prepared from electrospun polyimide. *J. Appl. Polym. Sci.* **97**, 165–170 (2005).

75. Yang, K. S., Edie, D. D., Lim, D. Y., Kim, Y. M., and Choi, Y. O. Preparation of carbon fiber web from electrostatic spinning of PMDA-ODA poly(amic acid) solution. *Carbon N. Y.* **41**, 2039–2046 (2003).

76. Xuyen, N., Ra, E., and Geng, H. Enhancement of conductivity by diameter control of polyimide-based electrospun carbon nanofibers. *J. Phys. Chem. B* **111**, 11350–11353 (2007).

77. Park, J.-H. et al. Effects of electrospun polyacrylonitrile-based carbon nanofibers as catalyst support in PEMFC. *J. Appl. Electrochem.* **39**, 1229–1236 (2009).

78. Li, W. et al. Preparation and characterization of multiwalled carbon nanotube-supported platinum for cathode catalysts of direct methanol fuel cells. *J. Phys. Chem. B* **107**, 6292–6299 (2003).

79. Lin, Z., Ji, L., Krause, W. E., and Zhang, X. Synthesis and electrocatalysis of 1-aminopyrene-functionalized carbon nanofiber-supported platinum–ruthenium nanoparticles. *J. Power Sources* **195**, 5520–5526 (2010).

80. Li, M., Han, G., and Yang, B. Fabrication of the catalytic electrodes for methanol oxidation on electrospinning-derived carbon fibrous mats. *Electrochem. Commun.* **10**, 880–883 (2008).

81. Li, M., Chang, Y., Han, G., and Yang, B. Platinum nanoparticles supported on electrospinning-derived carbon fibrous mats by using formaldehyde vapor as reducer for methanol electrooxidation. *J. Power Sources* **196**, 7973–7978 (2011).

82. Lin, Z. et al. Effect of platinum salt concentration on the electrospinning of polyacrylonitrile/platinum acetylacetonate solution. *J. Appl. Polym. Sci.* **116**, 895–901 (2009).

83. Lin, Z., Ji, L., and Zhang, X. Electrocatalytic properties of Pt/carbon composite nanofibers. *Electrochim. Acta* **54**, 7042–7047 (2009).

84. Guo, Q.-H., Huang, J.-S., and You, T.-Y. Electrospun palladium nanoparticle-loaded carbon nanofiber for methanol electro-oxidation. *Chin. J. Anal. Chem.* **41**, 210–214 (2013).

85. Chen, S. et al. Electrospun and solution blown three-dimensional carbon fiber non-wovens for application as electrodes in microbial fuel cells. *Energy Environ. Sci.* **4**, 1417–1421 (2011).

86. Che, A.-F. et al. Fabrication of free-standing electrospun carbon nanofibers as efficient electrode materials for bioelectrocatalysis. *New J. Chem.* **35**, 2848–2853 (2011).

87. Engel, A. B. et al. Enhanced performance of electrospun carbon fibers modified with carbon nanotubes: Promising electrodes for enzymatic biofuel cells. *Nanotechnology* **24**, 245402 (2013).

88. Mathias, M. et al. Two fuel cell cars in every garage. *Electrochem. Soc. Interface* **14**, 24–35 (2005).

89. Antolini, E. and Gonzalez, E. R. Ceramic materials as supports for low-temperature fuel cell catalysts. *Solid State Ionics* **180**, 746–763 (2009).

90. Wang, Y.-J., Wilkinson, D. P., and Zhang, J. Noncarbon support materials for polymer electrolyte membrane fuel cell electrocatalysts. *Chem. Rev.* **111**, 7625–7651 (2011).

91. Sharma, S. and Pollet, B. G. Support materials for PEMFC and DMFC electrocatalysts—A review. *J. Power Sources* **208**, 96–119 (2012).

92. Seger, B., Kongkanand, A., Vinodgopal, K., and Kamat, P. V. Platinum dispersed on silica nanoparticle as electrocatalyst for PEM fuel cell. *J. Electroanal. Chem.* **621**, 198–204 (2008).

93. Saha, M. S. et al. Tungsten oxide nanowires grown on carbon paper as Pt electrocatalyst support for high performance proton exchange membrane fuel cells. *J. Power Sources* **192**, 330–335 (2009).

94. Kulesza, P. J. et al. Electroreduction of oxygen at tungsten oxide modified carbon-supported RuSex nanoparticles. *J. Appl. Electrochem.* **37**, 1439–1446 (2007).

95. Chhina, H., Campbell, S., and Kesler, O. Ex situ evaluation of tungsten oxide as a catalyst support for PEMFCs. *J. Electrochem. Soc.* **154**, B533 (2007).

96. Elezovic, N. R., Babic, B. M., Radmilovic, V. R., Vracar, L. M., and Krstajic, N. V. Synthesis and characterization of MoO_x-Pt/C and TiO_x-Pt/C nano-catalysts for oxygen reduction. *Electrochim. Acta* **54**, 2404–2409 (2009).

97. Kanda, K. et al. Negligible start-stop-cycle degradation in a PEFC utilizing platinum-decorated tin oxide electrocatalyst layers with carbon fiber filler. *ECS Electrochem. Lett.* **3**, F15–F18 (2014).

98. Dou, M. et al. SnO_2 nanocluster supported Pt catalyst with high stability for proton exchange membrane fuel cells. *Electrochim. Acta* **92**, 468–473 (2013).

99. Elezovic, N. R., Babic, B. M., Radmilovic, V. R., and Krstajic, N. V. Synthesis and characterization of Pt catalysts on SnO_2 based supports for oxygen reduction reaction. *J. Electrochem. Soc.* **160**, F1151–F1158 (2013).

100. Kakinuma, K. et al. Characterization of Pt catalysts on Nb-doped and Sb-doped SnO_2-δ support materials with aggregated structure by rotating disk electrode and fuel cell measurements. *Electrochim. Acta* **110**, 316–324 (2013).

101. Wang, Y.-J. et al. Synthesis of Pd and Nb–doped TiO_2 composite supports and their corresponding Pt–Pd alloy catalysts by a two-step procedure for the oxygen reduction reaction. *J. Power Sources* **221**, 232–241 (2013).

102. Huang, S.-Y., Ganesan, P., and Popov, B. N. Electrocatalytic activity and stability of titania-supported platinum–palladium electrocatalysts for polymer electrolyte membrane fuel cell. *ACS Catal.* **2**, 825–831 (2012).

103. Rajalakshmi, N., Lakshmi, N., and Dhathathreyan, K. Nano titanium oxide catalyst support for proton exchange membrane fuel cells. *Int. J. Hydrogen Energy* **33**, 7521–7526 (2008).

104. Ioroi, T., Siroma, Z., Fujiwara, N., Yamazaki, S., and Yasuda, K. Sub-stoichiometric titanium oxide-supported platinum electrocatalyst for polymer electrolyte fuel cells. *Electrochem. Commun.* **7**, 183–188 (2005).

105. Gojković, S. and Babić, B. Nb-doped TiO_2 as a support of Pt and Pt–Ru anode catalyst for PEMFCs. *J. Electroanal. Chem.* **639**, 161–166 (2010).

106. d'Arbigny, J. B., Taillades, G., Marrony, M., Jones, D. J., and Rozière, J. Hollow microspheres with a tungsten carbide kernel for PEMFC application. *Chem. Commun. (Camb.)* **47**, 7950–7952 (2011).

107. Cui, G., Shen, P. K., Meng, H., Zhao, J., and Wu, G. Tungsten carbide as supports for Pt electrocatalysts with improved CO tolerance in methanol oxidation. *J. Power Sources* **196**, 6125–6130 (2011).

108. Chen, W.-F. et al. Highly active and durable nanostructured molybdenum carbide electrocatalysts for hydrogen production. *Energy Environ. Sci.* **6**, 943–951 (2013).

109. Ham, D. J. and Lee, J. S. Transition metal carbides and nitrides as electrode materials for low temperature fuel cells. *Energies* **2**, 873–899 (2009).

110. Sasaki, K., Takasaki, F., and Noda, Z. Alternative electrocatalyst support materials for polymer electrolyte fuel cells. *ECS Trans.* **33**, 473–482 (2010).

111. Krstajic, N. V. et al. Advances in interactive supported electrocatalysts for hydrogen and oxygen electrode reactions. *Surf. Sci.* **601**, 1949–1966 (2007).

112. Tauster, S. J. Strong metal-support interactions. *Acc. Chem. Res.* **20**, 389–394 (1987).
113. Bauer, A. et al. Improved stability of mesoporous carbon fuel cell catalyst support through incorporation of TiO_2. *Electrochim. Acta* **55**, 8365–8370 (2010).
114. Chevallier, L. et al. Mesoporous nanostructured Nb-doped titanium dioxide microsphere catalyst supports for PEM fuel cell electrodes. *ACS Appl. Mater. Interfaces* **4**, 1752–1759 (2012).
115. Maheswari, S., Sridhar, P., and Pitchumani, S. Pd–TiO_2/C as a methanol tolerant catalyst for oxygen reduction reaction in alkaline medium. *Electrochem. Commun.* **26**, 97–100 (2013).
116. Zhang, H. et al. Pt support of multidimensional active sites and radial channels formed by SnO_2 flower-like crystals for methanol and ethanol oxidation. *J. Power Sources* **196**, 4499–4505 (2011).
117. Shen, Y.-L., Chen, S.-Y., Song, J.-M., and Chen, I.-G. Ultra-long Pt nanolawns supported on TiO_2-coated carbon fibers as 3D hybrid catalyst for methanol oxidation. *Nanoscale Res. Lett.* **7**, 237–241 (2012).
118. Elezović, N. and Babić, B. Synthesis, characterization and electrocatalytical behavior of Nb–TiO_2/Pt nanocatalyst for oxygen reduction reaction. *J. Power Sources* **195**, 3961–3968 (2010).
119. Chhina, H., Campbell, S., and Kesler, O. Ex situ and in situ stability of platinum supported on niobium-doped titania for PEMFCs. *J. Electrochem. Soc.* **156**, B1232 (2009).
120. Siracusano, S., Baglio, V., D'Urso, C., Antonucci, V., and Aricò, A. S. Preparation and characterization of titanium suboxides as conductive supports of IrO_2 electrocatalysts for application in SPE electrolysers. *Electrochim. Acta* **54**, 6292–6299 (2009).
121. Bauer, A. et al. Pt nanoparticles deposited on TiO_2 based nanofibers: Electrochemical stability and oxygen reduction activity. *J. Power Sources* **195**, 3105–3110 (2010).
122. Bauer, A. et al. Synthesis and characterization of Nb-TiO_2 mesoporous microsphere and nanofiber supported Pt catalysts for high temperature PEM fuel cells. *Electrochim. Acta* **77**, 1–7 (2012).
123. Savych, I. et al. On the effect of non-carbon nanostructured supports on the stability of Pt nanoparticles during voltage cycling: A study of TiO_2 nanofibres. *J. Power Sources* **257**, 147–155 (2014).
124. Senevirathne, K., Hui, R., Campbell, S., Ye, S., and Zhang, J. Electrocatalytic activity and durability of Pt/NbO_2 and Pt/$Ti4O_7$ nanofibers for PEM fuel cell oxygen reduction reaction. *Electrochim. Acta* **59**, 538–547 (2012).
125. Cavaliere, S., Subianto, S., Chevallier, L., Jones, D. J., and Rozière, J. Single step elaboration of size-tuned Pt loaded titania nanofibres. *Chem. Commun. (Camb.)* **47**, 6834–6836 (2011).
126. Long, Q., Cai, M., Li, J., Rong, H., and Jiang, L. Improving the electrical catalytic activity of Pt/TiO_2 nanocomposites by a combination of electrospinning and microwave irradiation. *J. Nanoparticle Res.* **13**, 1655–1662 (2010).
127. Formo, E. et al. Direct oxidation of methanol on Pt nanostructures supported on electrospun nanofibers of anatase. *J. Phys. Chem. C* **112**, 9970–9975 (2008).
128. Su, L., Jia, W., Schempf, A., and Lei, Y. Palladium/titanium dioxide nanofibers for glycerol electrooxidation in alkaline medium. *Electrochem. Commun.* **11**, 2199–2202 (2009).
129. Sago, S., Suryamas, A. B., Anilkumar, G. M., Ogi, T., and Okuyama, K. In situ growth of Pt nanoparticles on electrospun SnO_2 fibers for anode electrocatalyst application. *Mater. Lett.* **105**, 202–205 (2013).
130. Suryamas, A. B., Anilkumar, G. M., Sago, S., Ogi, T., and Okuyama, K. Electrospun Pt/SnO_2 nanofibers as an excellent electrocatalysts for hydrogen oxidation reaction with ORR-blocking characteristic. *Catal. Commun.* **33**, 11–14 (2013).

131. Cavaliere, S. et al. Dopant-driven nanostructured loose-tube SnO_2 architectures: Alternative electrocatalyst supports for proton exchange membrane fuel cells. *J. Phys. Chem. C* **117**, 18298–18307 (2013).

132. Zhao, Z.-G. et al. Rational design of galvanically replaced Pt-anchored electrospun WO_3 nanofibers as efficient electrode materials for methanol oxidation. *J. Mater. Chem.* **22**, 16514–16519 (2012).

133. Lee, K., Choi, J., Wycisk, R., Pintauro, P., and Mather, P. Nafion nanofiber membranes. *ECS Trans.* **25**, 1451–1458 (2009).

134. Choi, J., Lee, K., Wycisk, R., Pintauro, P. N., and Mather, P. T. Nanofiber composite membranes with low equivalent weight perfluorosulfonic acid polymers. *J. Mater. Chem.* **20**, 6282–6290 (2010).

135. Lotus, A. F. et al. Electrical, structural, and chemical properties of semiconducting metal oxide nanofiber yarns. *J. Appl. Phys.* **103**, 024910 (2008).

136. Huang, S. et al. A flexible and transparent ceramic nanobelt network for soft electronics. *NPG Asia Mater.* **6**, e86 (2014).

137. Kakinuma, K. et al. Preparation of titanium nitride-supported platinum catalysts with well controlled morphology and their properties relevant to polymer electrolyte fuel cells. *Electrochim. Acta* **77**, 279–284 (2012).

138. Lv, H., Mu, S., Cheng, N., and Pan, M. Nano-silicon carbide supported catalysts for PEM fuel cells with high electrochemical stability and improved performance by addition of carbon. *Appl. Catal. B: Environ.* **100**, 190–196 (2010).

139. Polonský, J. et al. Tantalum carbide as a novel support material for anode electrocatalysts in polymer electrolyte membrane water electrolysers. *Int. J. Hydrogen Energy* **37**, 2173–2181 (2012).

140. Liu, H., Wang, F., Zhao, Y., and Fong, H. Mechanically resilient electrospun TiC nanofibrous mats surface-decorated with Pt nanoparticles for oxygen reduction reaction with enhanced electrocatalytic activities. *Nanoscale* **5**, 3643–3647 (2013).

141. Ito, Y., Takeuchi, T., Tsujiguchi, T., Abdelkareem, M. A., and Nakagawa, N. Ultrahigh methanol electro-oxidation activity of PtRu nanoparticles prepared on TiO_2-embedded carbon nanofiber support. *J. Power Sources* **242**, 280–288 (2013).

142. Feng, C., Takeuchi, T., Abdelkareem, M. A., Tsujiguchi, T., and Nakagawa, N. Carbon–CeO_2 composite nanofibers as a promising support for a PtRu anode catalyst in a direct methanol fuel cell. *J. Power Sources* **242**, 57–64 (2013).

143. Scibioh, M. A. et al. Pt-CeO_2/C anode catalyst for direct methanol fuel cells. *Appl. Catal. B: Environ.* **84**, 773–782 (2008).

144. Bauer, A., Hui, R., Ignaszak, A., Zhang, J., and Jones, D. J. Application of a composite structure of carbon nanoparticles and Nb–TiO_2 nanofibers as electrocatalyst support for PEM fuel cells. *J. Power Sources* **210**, 15–20 (2012).

145. Wang, C. et al. Graphene nanoribbons hybridized carbon nanofibers: Remarkably enhanced graphitization and conductivity, and excellent performance as support material for fuel cell catalysts. *Nanoscale* **6**, 1377–1383 (2014).

146. Chen, Y. et al. Enhanced stability of Pt electrocatalysts by nitrogen doping in CNTs for PEM fuel cells. *Electrochem. Commun.* **11**, 2071–2076 (2009).

147. Sung, M.-T., Chang, M.-H., and Ho, M.-H. Investigation of cathode electrocatalysts composed of electrospun Pt nanowires and Pt/C for proton exchange membrane fuel cells. *J. Power Sources* **249**, 320–326 (2014).

148. Kim, H. J. et al. Highly improved oxygen reduction performance over Pt/C-dispersed nanowire network catalysts. *Electrochem. Commun.* **12**, 32–35 (2010).

149. Shui, J., Chen, C., and Li, J. C. M. Evolution of nanoporous Pt-Fe alloy nanowires by dealloying and their catalytic property for oxygen reduction reaction. *Adv. Funct. Mater.* **21**, 3357–3362 (2011).

150. Kim, H., Kim, Y., Seo, M., Choi, S., and Kim, W. Pt and PtRh nanowire electrocatalysts for cyclohexane-fueled polymer electrolyte membrane fuel cell Hyung. *Electrochem. Commun.* **11**, 446–449 (2009).

151. Kim, Y. S. et al. Electrospun bimetallic nanowires of PtRh and PtRu with compositional variation for methanol electrooxidation. *Electrochem. Commun.* **10**, 1016–1019 (2008).

152. Kim, D., Li, O. L., Pootawang, P., and Saito, N. Solution plasma synthesis process of tungsten carbide on N-doped carbon nanocomposite with enhanced catalytic ORR activity and durability. *RSC Adv.* **4**, 16813 (2014).

153. Zhou, X. S., Qiu, Y. J., Yu, J., Yin, J., and Gao, S. Tungsten carbide nanofibers prepared by electrospinning with high electrocatalytic activity for oxygen reduction. *Int. J. Hydrogen Energy* **36**, 7398–7404 (2011).

154. Garcia-Marquez, A., Portehault, D., and Giordano, C. Chromium nitride and carbide nanofibers: From composites to mesostructures. *J. Mater. Chem.* **21**, 2136–2143 (2011).

155. Zhou, X., Qiu, Y., Yu, J., Yin, J., and Bai, X. High electrochemical activity from hybrid materials of electrospun tungsten oxide nanofibers and carbon black. *J. Mater. Sci.* **47**, 6607–6613 (2012).

156. Zhang, Y., Wang, Y., Jia, J., and Wang, J. Electro-oxidation of methanol based on electrospun PdO–Co_3O_4 nanofiber modified electrode. *Int. J. Hydrogen Energy* **37**, 17947–17953 (2012).

157. Barakat, N. A., Abdelkareem, M., El-Newehy, M., and Kim, H. Influence of the nanofibrous morphology on the catalytic activity of NiO nanostructures: An effective impact toward methanol electrooxidation. *Nanoscale Res. Lett.* **8**, 402–407 (2013).

158. Al-Enizi, A. M., Elzatahry, A. A., Soliman, A. I., and Al-Theyab, S. S. Electrospinning synthesis and electrocatalytic performance of cobalt oxide/carbon nanofibers nanocomposite based PVA for fuel cell applications. *Int. J. Electrochem. Sci.* **7**, 12646–12655 (2012).

159. Qiu, Y., Yin, J., Hou, H., Yu, J., and Zuo, X. Preparation of nitrogen-doped carbon submicrotubes by coaxial electrospinning and their electrocatalytic activity for oxygen reduction reaction in acid media. *Electrochim. Acta* **96**, 225–229 (2013).

160. Yin, J., Qiu, Y., and Yu, J. Onion-like graphitic nanoshell structured Fe–N/C nanofibers derived from electrospinning for oxygen reduction reaction in acid media. *Electrochem. Commun.* **30**, 1–4 (2013).

161. Uhm, S., Jeong, B., and Lee, J. A facile route for preparation of non-noble CNF cathode catalysts in alkaline ethanol fuel cells. *Electrochim. Acta* **56**, 9186–9190 (2011).

162. Zhang, W. and Pintauro, P. N. High-performance nanofiber fuel cell electrodes. *ChemSusChem* **4**, 1753–1757 (2011).

163. Brodt, M. et al. Fabrication, in-situ performance, and durability of nanofiber fuel cell electrodes. *J. Electrochem. Soc.* **162**, F84–F91 (2015).

164. Yang, X., Shao, C., Liu, Y., Mu, R., and Guan, H. Nanofibers of CeO_2 via an electrospinning technique. *Thin Solid Films* **478**, 228–231 (2005).

165. Zaton, M., Jones, D., and Rozière, J. Mitigation of fuel cell membrane degradation with metal oxide/Nafion nanofiber interlayers. *ECS Trans.* **61**, 15–23 (2014).

166. Cavaliere, S. SPINAM—Electrospinning: A method to elaborate membrane-electrode materials for energy devices. http://www.spinam.eu (2012). Accessed in 2015.

3 Electrospinning for Solid Oxide Fuel Cells

Wenjing Zhang

CONTENTS

3.1 SOLID OXIDE FUEL CELLS

The solid oxide fuel cells (SOFCs) have dominated competing fuel cell technologies for several decades because of their high energy conversion efficiency, extremely low pollution emission, and the ability of using currently available fossil fuels.[1,2] SOFC works on the similar principles as all other fuel cells. Figure 3.1 shows schematically how a SOFC works. The cell is composed of a solid electrolyte sandwiched in between two porous electrodes. During the operation, the oxygen/air flows into the cathode, where the oxygen molecules react with electrons and split into two oxygen ions. Those oxygen ions diffuse into the solid electrolyte and migrate from the cathode side to the anode side. On the anode side, oxygen ions then oxidize the fuel and release electrons with a higher potential to an external circuit, thus providing electrical energy and water. Compared to other fuel cell technologies, SOFCs also bear certain properties, which make them particularly interesting and also challenging. First, different from polymer electrolyte membrane fuel cells, the SOFC has a solid dense electrolyte. In this case, the gas transport through the electrolyte is not governed by the size of gas molecules and the size of the electrolyte membrane. Instead, the electrically charged oxygen ions are selectively conducted through the electrolyte via solid-state conduction,[3] thus preventing the crossover of other gaseous

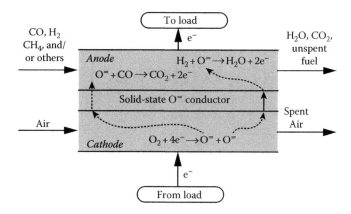

FIGURE 3.1   The principle of operation of a solid oxide fuel cell. (Reprinted with permission from Adams, T.A., Nease, J., Tucker, D., and Barton, P.I., *Ind. Eng. Chem. Res.*, 52, 3089. Copyright 2013 American Chemical Society.)

molecules such as hydrogen, carbon monoxide, and nitrogen. Therefore, SOFCs have a high flexibility to various fuels and can convert chemical energy from essentially any fuel, from hydrogen, hydrocarbon to even carbon,[3–7] to usable electrical energy. This flexibility allows SOFCs to be used with our current existing hydrocarbon fuel infrastructure.[8] Second, in contrast to other fuel cell technologies, SOFCs can be designed to work at wide temperature range (500°C–1000°C). Working at high temperature facilitates reforming and water–gas shift reaction inside the SOFCs' anode. Thus some readily available fuels (such as natural gas) can be used in SOFCs without preprocessing. The relatively high operating temperature in SOFCs also yields higher efficiency. For stand-alone applications, SOFCs can achieve chemical to electrical efficiency of 45%–65% based on the lower heating value of the fuel.[6] Moreover, high temperature exhaust gases of SOFCs bear high energy, which can be exploited in secondary power generation cycles, steam generation, or hot water production. However, high temperature operation also brings challenges to the material selection for electrodes, interconnects, and seals, which have to be able to resist high temperature (800°C–1000°C) and also have similar thermal expansion coefficient to reduce the internal stress. In addition, high operating temperature accelerates sintering and component interdiffusion.

Recently, much development has focused on the development of intermediate-temperature (IT) SOFCs, which are operated at the range of 650°C–800°C. By lowering the operating temperature, a wide range of materials can be used for interconnects and compressive nonglass/ceramic seals, thus reducing cost. Since all polarization losses increase at lower operating temperature, material selection and structure design for electrolyte and electrodes are particularly vital for IT SOFC to still achieve reasonable performance. At the cathode side, for example, oxygen reduction processes occur at the triple phase boundary (TPB), which is the interface where the electronic and ionic conducting phase coexists with oxygen. To compensate the performance loss due to the lower operating temperature, the use of mixed ionic–electronic conductor (MIEC) materials has been developed.[10] For example, lanthanum strontium cobaltite,[11,12] lanthanum strontium cobalt ferrite (LSCF),[13,14] barium strontium cobalt ferrite,[15] and strontium-doped samarium cobaltite (SSC)[16] exhibit excellent cathode performance and are widely used as cathode materials for IT SOFCs.[17] In such system, oxygen reduction takes place not only on the TPB but also on the entire cathode surface.[18] Therefore, the connectivity of cathode materials and the pore size distribution plays a crucial role in determining the performance of IT SOFCs. On the anode side, fuel oxidation occurs. The anode has to not only provide continuous electron- and ion-conducting path but also promote fuel access to the TPB and product removal. Particularly for anode-support SOFCs, a high open porosity has to be created without sacrificing too much mechanical strength. Thus, numerous efforts have been made on the design and optimization of structure and processing to achieve high performance of SOFCs.

It is well known that electrospinning is a versatile and effective method for producing nanofibers with uniform diameter, high surface area, and high porosity. Moreover, electrospinning has the possibility of generating composite networks from a rich variety of materials with the ability to control composition, morphology, and secondary structure, which allows design of optimum material characteristics for membranes. This simple and versatile method has been used to address the sizable challenges in SOFCs. This chapter is devoted to review the research work on electrospun nanofiber materials for cathode and anode in SOFCs, respectively.

3.2 ELECTROSPUN MATERIALS FOR SOFC CATHODES

As shown in Figure 3.1, an oxygen molecule is reduced on the cathode side and split into two oxygen ions, which are transported from the cathode to the electrolyte. Due to the lower operating temperature for IT SOFCs, polarization losses of the cell increase significantly due to the much slower kinetics associated with the oxygen reduction reaction and charge transport. Therefore, much attention has been drawn to develop MIEC materials and optimize their structure. For example, nanoparticles are widely used to construct an effective cathode due to its high surface-to-volume ratio and high catalytic activity. However, many challenges have emerged such as poor thermal stability and low open porosity. Incorporation of nanofibers into the cathode is an alternative approach to design a high-performance cathode for SOFCs.

3.2.1 LANTHANUM STRONTIUM COBALT FERRITE–GADOLINIA-DOPED CERIA NANOFIBER CATHODE

The $La_{1-x}Sr_xCo_{1-y}Fe_yO_{3-\delta}$ perovskite is considered as the promising cathode material because of its superior ionic–electronic conductivity and high catalytic activity for the oxygen reduction reaction compared to $La_{1-x}Sr_xMnO_3$, which is widely utilized in SOFCs operated at high temperature.[19] At lower operating temperature, the particle size and microstructure of the cathode that define pores play a vital role in LSCF cathode performance.

Zhi et al. have successfully fabricated LSCF nanofibers by sol–gel-assisted electrospinning and tested as cathode materials of an IT SOFC with yttria-stabilized

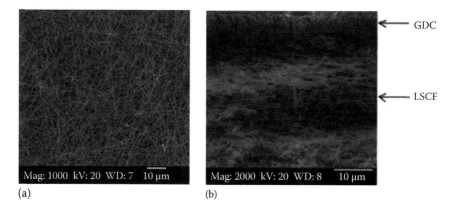

(a)   (b)

FIGURE 3.2 Scanning electron microscope (SEM) images of (a) LSCF nanofibers after calcination at 800°C for 1 h and (b) cross-section of the LSCF nanofiber cathode after GDC infiltration. (Zhi, M., Lee, S., Miller, N., Menzler, N.H., and Wu, N., *Energy Environ. Sci.*, 5, 7066, 2012. Reproduced by permission of The Royal Society of Chemistry.)

zirconia (YSZ) electrolyte.[13] Polyacrylonitrile (dissolved in *N,N*-dimethylformamide) was used as carrier polymer, which was then mixed with a solution containing La(CH$_3$COO)$_3$·1.5H$_2$O, Sr(CH$_3$COO)$_2$·0.5H$_2$O, Co(CH$_3$COO)$_2$·4H$_2$O, and Fe(CH$_3$COO)$_2$. The precursor was electrospun at an electrical field of 1.2 kV/cm with an injection rate of 0.25 mL/h. The as-spun nanofiber membrane was then sintered at 800°C. As shown in Figure 3.2a, the result is a porous LSCF perovskite nanofibers membrane with a fiber diameter of ~230 nm. Each nanofiber is composed of nanosized grains with a diameter of ~45 nm. Adhesion of LSCF nanofibers on the YSZ electrolyte is critical for cathode performance. Therefore, LSCF nanofibers were grounded with ink vehicle before they were sintered on the electrolyte. With such a cathode, the fuel cell exhibits a power density of 0.9 W/cm^2 at 1.9 A/cm^2 at 750°C. To enhance ionic conductivity, gadolinium-doped ceria (GDC) was infiltrated into LSCF nanofiber scaffold (Figure 3.2b). With 20% GDC infiltration, the performance is further improved by 19%.

Nanofiber prepared by electrospinning often has long length, more than several hundreds of micrometer. In an IT SOFC, the cathode layer is only a few micrometers thick. Therefore, electrospun LSCF nanofibers were grounded to shorter particles to shorten the continuous conducting path through cathode in IT SOFCs. Moreover, the membrane composed of shorter nanofibers has higher contact surface area with electrolyte, which facilitates adhesion between cathode and electrolyte during sintering. Zhao et al. utilized electrospinning combined with high-temperature sintering process to directly produce LSCF nanorods.[20] Similar sol–gel chemistry is used in this work with polyvinylpyrrolidone(PVP) as carrier polymer and higher electrical field (1.67 kV/cm) as compared to Zhi et al.'s work (1.2 kV/cm). More importantly, the as-spun nanofibers were calcined at 900°C for 2 h in air, resulting in a transition from long nanofibers (Figure 3.3a) to short nanorods (Figure 3.3b) with diameter in the range of 200–300 nm. After calcination at 1000°C, the nanorods are further

FIGURE 3.3 (a) SEM image of as-spun fibers; (b) SEM image of LSCF nanorod cathode calcined at 900°C for 2 h; and (c and d) SEM images of LSCF nanoparticle cathode calcined at 1000°C for 2 h. (Reprinted from *J. Power Sources*, 219, Zhao, E., Jia, Z., Zhao, L., Xiong, Y., Sun, C., and Brito, M.E., 133, Copyright 2012, with permission from Elsevier.)

reformed into nanoparticles (Figure 3.3c and d). The 1D LSCF–GDC nanocomposite cathode can be prepared by injecting nitrate aqueous solution of GDC precursor into the LSCF cathode presintered on a GDC electrolyte. Impedance analysis shows that the addition of infiltrated GDC lowers the polarization resistance of LSCF cathodes. Nanorod LSCF–GDC cathode gives five times smaller polarization resistance as compared to that of LSCF–GDC nanoparticles cathode due to larger LSCF–GDC boundaries and higher porosity.

In order to further increase TPB length and porosity, LSCF nanotubes were fabricated by sol–gel-assisted electrospinning. Zhao et al. synthesize nanofibers using nitrate aqueous solution of LSCF precursor and PVP.[21] The resulting LSCF/PVP nanofibers were then calcined at 800°C for 2 h to form LSCF nanotubes. By infiltration, GDC phase was introduced on both outer and inner surfaces of LSCF nanotubes, resulting in enlarged TPB regions and thus reducing the cathode area specific resistance.[21]

Direct electrospinning of the suspension of LSCF and GDC powders is another approach to fabricate nanofiber cathode.[22,23] In this approach, LSCF and GDC powders were first ball-milled in ethanol for 24 h, dried at 80°C for 12 h, and then mixed with PVP by vigorous stirring overnight. Compared to sol–gel-assisted electrospinning, direct electrospinning of powder suspension was carried out at relatively low electrical field (0.73 kV/cm) and fairly high feed rate (1.8 mL/h) with a much smaller needle (27 gauge). The as-spun LSCF/GDC-PVP nanofibers have smooth surface with a diameter of ~700 nm. After calcination at 800°C for 2 h, the fibrous structure is retained, and each nanofiber is roughened by LSCF and GDC nanoparticles that are around 100 nm in diameter. The pore size of LSCF–GDC network was from 1 to 2 μm. The highly increased TPB and pores in this electrospun cathode exhibits a maximum power density of 1.02 W/cm^2 at 650°C in a La$_2$Sn$_2$O$_7$–NiO–GDC anode-supported cell. It is more than double of the maximum power density of the same anode-supported cell with conventional LSCF–GDC cathode (0.49 W/cm^2). Detailed analysis of this LSCF–GDC nanofiber cathode was conducted in the work by the same group, where the electrochemical characteristics of a symmetric cell with nanofiber cathode were measured by electrochemical impedance spectroscopy (EIS).[23] The interpretation of the EIS was based on the model by Torres da Silva et al.[24] Figure 3.4 shows each polarization resistance of the symmetrical cells separated by equivalent model at different temperatures. It can be

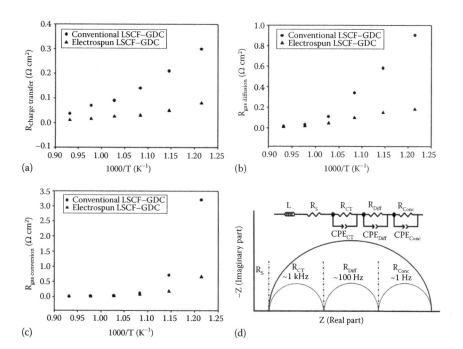

FIGURE 3.4 Plots of (a) R$_{CT}$, (b) R$_{Diff}$, and (c) R$_{Conc}$ of the symmetric cells as a function of temperature. (d) Equivalent circuit model for the interpretation of the impedance spectra obtained from the symmetrical cell tests. (Reprinted from *Ceram. Int.*, 40, Lee, J.G., Park, M.G., Park, J.H., and Shul, Y.G., 8053, Copyright 2014, with permission from Elsevier; *ECS Trans.*, 25, Torres da Silva, I.M., Nielsen, J., Hjelm, J., and Mogensen, M.B., 489, Copyright 2009, with permission from Elsevier.)

seen that R_{CT}, R_{Diff}, and R_{Conc} were reduced for the symmetric cell with LSCF–GDC nanofiber cathode. The reductions are attributed to the unique architecture of the electrospun cathode: continuous conducting path for charge transfer, larger active sites for oxygen surface exchange, and highly porous microstructure for oxygen mass transfer.

3.2.2   8 WT% Y_2O_3 STABILIZED ZrO_2–LSCF (8YSZ–LSCF) NANOFIBER CATHODE

Instead of infiltrating ionic conducting phase into LSCF nanofiber scaffold, Chou et al. designed a cathode structure where LSCF nanoparticles were mixed with 8 mol% Y_2O_3 stabilized ZrO_2 (8YSZ) nanofibers.[25] Polycrystalline 8YSZ nanofibers with a diameter of 100 nm were prepared by electrospinning a solution containing 8YSZ precursors and PVP followed by high-temperature calcination (800°C–1000°C). LSCF particle/8YSZ nanofiber cathode was fabricated by screen-printing the solution of LSCF powder, 8YSZ nanofiber, binder, and solvent followed by calcination at 1050°C. As compared to 8YSZ–LSCF powder cathode, 8YSZ–LSCF nanofiber cathode exhibits 77% higher exchange current density for O_2 reduction reaction at 800°C, due to the increased TPB region and decrease of the activation energy for oxygen ion migration.

3.2.3   STRONTIUM-DOPED SAMARIUM COBALTITE NANOFIBER CATHODE

Strontium-doped samarium cobaltite (SSC) has received much attention for its use as cathode materials for IT SOFCs due to its high activity and conductivity.[26] Choi et al. were the first group to design a $Sm_{0.2}Ce_{0.8}O_{1.9}$ (SDC)-embedded $Sm_{0.5}Sr_{0.5}CoO_{3-\delta}$ (SSC) nanofiber cathode.[27] In this work, SDC nanoparticles were mixed with SSC precursor solution ($Sm(NO_3)_3 \cdot 6H_2O$, $Sr(NO_3)$ and $Co(NO_3)_3 \cdot 6H_2O$) before electrospinning. Electrospinning of SDC nanoparticle suspension and SDC precursor gel was carried out using a water-soluble carrier polymer polyvinyl alcohol (PVA). During calcination at 800°C, three processes take place inside the nanofiber: degradation of PVA, formation of SSC phase, and sintering of SDC nanoparticles. The results are continuous fibrous nanofiber structure with a diameter of 300 ± 80 nm. Scanning transmission electron microscopy (STEM) image with the corresponding energy-dispersive spectroscopy elemental mapping confirms that the SSC and SDC nanoparticles coexist and are well mixed within each nanofiber. Figure 3.5 shows the cross-section of a single cell after the cathode was deposited and sintered on the anode-supported single cell at 1000°C for 2 h. Good adhesion between the cathode and GDC electrolyte and high porosity can be observed. Moreover, shorter nanofibers and coral-like structure can also be seen inside the cathode. The authors attribute the excellent SOFC performance to the highly porous and continuous structure that facilitates mass transport and charge-transfer reaction.

SSC nanofibers were also used by Fan et al. to construct cathode for IT SOFCs.[28] In their work, SSC nanofibers were first sintered on the GDC electrolyte at 1000°C and then mixed with GDC by multiple infiltration steps followed by calcination at

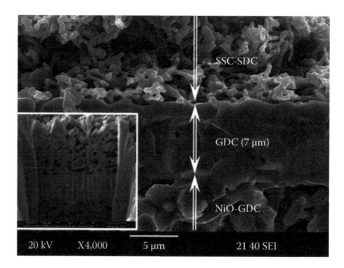

FIGURE 3.5 SEM images of the cross-section of a single cell that is composed of a strontium-doped samarium cobaltite–strontium-doped samarium cobaltite nanofiber cathode, a gadolinium-doped ceria (GDC) electrolyte, and a NiO–GDC anode. The inset is the image of front milling with focused ion beam (FIB). (Reprinted from *J. Eur. Ceram. Soc.*, 33, Choi, J., Kim, B., and Shin, D., 2269, Copyright 2013, with permission from Elsevier.)

800°C. As compared to nanoparticle-based cathode, SSC–GDC nanofiber composite cathode exhibits 2.6 times smaller polarization resistance at 650°C when SSC–GDC mass ratio of 1:0.869 for infiltration is used.

3.2.4 OTHER MATERIALS

Recently, an oxygen-deficient perovskite $Sr_xY_{1-x}CoO_{2.65-\delta}$ (SYCO) was discovered as an excellent cathode material for IT SOFCs due to its good activity for oxygen reduction reaction.[29] SYCO nanofibers were prepared by conventional sol–gel-assisted electrospinning. PVP was used as carrier polymer and the nitrate salts used as precursors. GDC was introduced into SYCO nanofiber scaffold by infiltration to further enhance ionic conductivity of the cathode. It is of particular interests that the pure SYCO perovskite can be formed at 900°C and above. It is worth noticing that SYCO prepared by the solid-state reaction requires calcination temperature above 1000°C and calcination time up to 96 h.[29–32] Authors ascribe much easier calcination for SYCO nanofibers to their large surface area and high porosity. At 1000°C, SYCO nanofibers tend to agglomerate. The SYCO–GDC nanofiber cathode offers a reduction of polarization resistance and excellent long-term stability as compared to the performance reported in the literature.[17,33]

Electrospinning was first used by Jiang et al. to produce $GdBaCo_2O_{5+\delta}$ (double-layered perovskite-type gadolinium barium cobaltate [GBCO]) nanofibers.[34] Again, sol–gel-assisted electrospinning was adopted with PVP as carrier polymer. Interestingly, a unique cathode fabrication method was proposed in this work. Figure 3.6 shows the illustration of the fabrication process of GBCO/GDC/GBCO

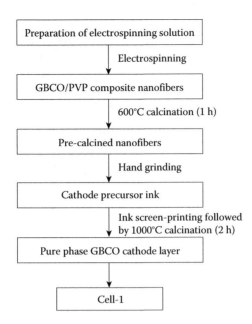

FIGURE 3.6 Illustration of fabrication process of double-layered perovskite-type gadolinium barium cobaltate (GBCO)/gadolinium-doped ceria (GDC)/GBCO symmetric cells using GBCO/PVP composite nanofibers. (Reprinted from *J. Alloy Compd.*, 557, Jiang, X., Xu, H., Wang, Q., Jiang, L., Li, X., Xu, Q., Shi, Y., and Zhang, Q., 184, Copyright 2013, with permission from Elsevier.)

symmetric cells. Unlike most of the other works, as-spun GBCO/PVP nanofibers were precalcined to remove most of the PVP polymer component and then mixed with α-terpineol and ethyl cellulose to make the precursor ink. The ink was screen-printed onto both sides of the GDC pellet followed by calcination at 1000°C for 2 h in air. Precalcination at 600°C is a method to prevent the peeling of the GBCO nanofiber cathode from GDC electrolyte due to excessive gases from PVP degradation. A coral-like structure is formed by broken and agglomerated GBCO nanorods (~30 nm in length) after calcination at 1000°C. This coral-like structure with large surface area was used in this work to enhance the electrocatalytic activity for oxygen reduction reaction and thus improve cathode performance.

La$_{1.6}$Sr$_{0.4}$NiO$_4$ (LSN) nanofibers are prepared by Li et al. using sol–gel-assisted electrospinning.[35] In this work, a rapid sintering process was discovered. After sintering of LSN nanofibers on GDC electrolyte at 900°C for 15 min, the LSN nanofibers were transformed into a hollow structure as shown in Figure 3.7, where the nanofiber is ~500 nm in diameter roughened by ~50 nm LSN nanoparticles. The enhanced electrochemical performance of LSN nanofiber cathode was ascribed to the use of rapid sintering strategy that exhibits a porous structure, relatively high surface area, and a good adhesion between the cathode and the GDC electrolyte.

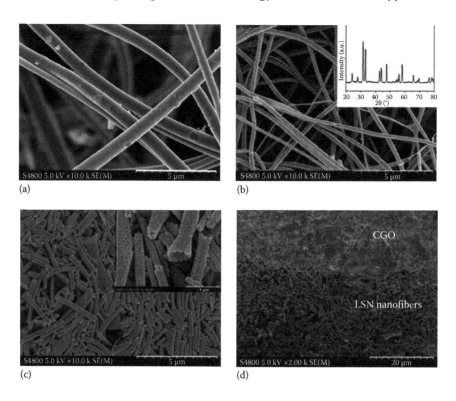

FIGURE 3.7 SEM images of PVP/$La_{1.6}Sr_{0.4}NiO_4$ (LSN) composite nanofibers (a) before and (b) after calcination at 800°C (the inset in (b) is the x-ray diffraction spectra for the nanofibers in (b)); (c) SEM image of LSN nanofiber cathode on gadolinium-doped ceria (CGO) electrolyte; and (d) SEM image of the cross-section of LSN nanofiber cathode supported on CGO electrolyte after sintering at 900°C for 15 min. (Reprinted from *J. Power Sources*, 263, Li, Q., Sun, L., Zhao, H., Wang, H., Huo, L., Rougier, A., Fourcade, S., and Grenier, J.-C., 125, Copyright 2014, with permission from Elsevier.)

3.3   ELECTROSPUN MATERIALS FOR SOFC ANODE

3.3.1   YSZ Nanofiber Anode

The anode of SOFCs (as shown in Figure 3.1) is the electrode where oxygen ions oxidize the fuel and release electrons with a higher potential to an external circuit, thus providing electrical energy. Therefore, the anode must have (a) high porosity to facilitate fuel access and product removal, (b) high electronic conductivity and catalyst activity to catalyze the reaction and efficiently transfer electrons to the external circuit, and (c) sufficient ionic conductivity to transport ions within the anode. So far, Ni–YSZ is the most used anode material for SOFCs. In such an anode, Ni acts as the catalyst to catalyze the fuel oxidation reaction and meanwhile transport electrons to the current collector. Therefore, the connectivity among Ni particles plays an important role to reduce the ohmic resistance. The YSZ not only forms a scaffold to facilitate Ni dispersion but also provides ionic conductivity. Unfortunately, this

Ni and YSZ composite formation is also facing challenges. First, the formation of contiguous conducting pathway for electrons favors a fully percolated Ni anode. However, Ni is also responsible for the carbon formation when SOFCs use hydrocarbon fuels.[36] Second, Ni has much higher TEC than that of YSZ. The mismatch TEC may result in internal stress and poor mechanical properties. Therefore, different processing techniques were designed to optimize the structure of Ni–YSZ and tailor its performance.[37–40] However, most of those processing techniques require high Ni concentration.[41]

Sol–gel-assisted electrospinning was first used by Azad et al. to produce YSZ nanofibers.[42] They used $ZrOCl_2$ and $Y(NO_3)\cdot6H_2O$ as the precursors and PVP ($M_w = 1.3$ M) as carrier polymer. The precursors and PVP were mixed with deionized water. After calcination of as-spun nanofibers in air at 1500°C for 1 h, the phase pure YSZ nanofibers were obtained. Later, Li et al. used similar chemistry to electrospin YSZ precursor-embedded PVP nanofibers.[43] However, a hollow nanofiber structure rather than conventional solid nanofiber was obtained after the calcination of as-spun nanofibers. The fibers calcined at 600°C for 12 h is 2–4 μm in diameter with the wall thickness of 300 nm. With the increase of the calcination temperature, the diameter of the hollow fibers decreases with an increase in both crystallinity and grain size. Authors claimed that the hollow structure is due to the CO and other gases released during organic binder (PVP) decomposition that induces ballooning of the fibers. It is worth mentioning that the temperature ramp rate is 4°C/min during calcination in this work, which is much higher than that used in Azad's work (0.5°C/min). The molecular weight of PVP in this work is undefined. Therefore, the influence of PVP on the formation of hollow nanofiber is unclear.

Recently, Li et al. proposed a novel route to fabricate Ni-coated YSZ nanofiber anode for SOFCs.[44] In their method, the electrospinning precursor contains the suspension of Y_2O_3 and ZrO_2 and the solution of PVP dissolved in ethanol. After electrospinning, the as-spun nanofiber mat was calcined in air at 1000°C for 3 h. Afterward, 30 wt% Ni was deposited on YSZ nanofibers by electroless plating. Figure 3.8 shows

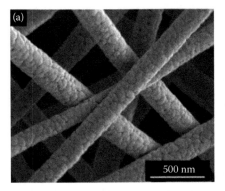
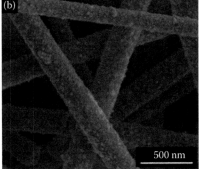

FIGURE 3.8 Field emission scanning electron microscopy (FESEM) images of the nanofibers (a) after sintering at 1000°C and (b) after Ni plating on YSZ nanofibers. (Reprinted from *J. Power Sources*, 96, Li, L., Zhang, P., Liu, R., and Guo, S.M., 1242, Copyright 2011, with permission from Elsevier.)

the SEM images of YSZ nanofibers before and after Ni plating, where an even distribution of Ni particles on YSZ nanofiber scaffold is clearly seen. This Ni–YSZ nanofiber membrane was tested as the anode in SOFCs. The power density for the SOFCs with nanofiber-derived anode is twice of that with the powder-derived anode that has also 30 wt% Ni in YSZ nanopowder matrix. Authors attribute the improvement in performance to (a) continuous ion-conducting path along the YSZ core of the fibers, (b) high electron-conducting path through adjoining and connected Ni–Ni chains on the fiber surface, and (c) larger TPB sites.

3.3.2 LA-DOPED SrTiO₃ NANOFIBER ANODE

The development of Ni-free anode materials has drawn much attention due to the disadvantages caused by Ni (carbon formation, sulfur poisoning, etc.). Recently, $La_xSr_{1-x}TiO_3$ (La-doped SrTiO₃ [LST], x = 0.1, 0.2, 0.3, 0.35, and 0.4) was discovered as a promising material for SOFC's anode due to its high electron conductivity, comparable to TEC with YSZ, and dimensional and chemical stability when subjected to redox cycling.[45,46] To compensate low catalytic activity of LST, ceria and doped ceria were often added into LST to enhance electrochemical performance.[47–49]

$Gd_{0.2}Ce_{0.8}O_{1.9}$-infiltrated $La_{0.2}Sr_{0.8}TiO_3$ nanofiber anode was prepared by sol–gel-assisted electrospinning.[46] A PVP acidic solution in DMF was mixed with $La(NO_3)\cdot6H_2O$, $Sr(NO_3)_2$, and $Ti(OC_3H_7)_4$ with molar ratios of 0.2:0.8:1. After electrospinning, the nanofibers were calcined at 900°C for 2 h and crumbed into short rodlike LST nanofibers. In Figure 3.9a, people can clearly see the uniform LST nanofibers with an average diameter of ~200 nm. The x-ray powder diffraction (XRD) confirmed a single phase with cubic perovskite structure. The calcination temperature and duration are significantly lower than those required by other processing techniques, thus lowering fabrication cost. $Gd_{0.2}Ce_{0.8}O_{1.9}$ (GDC) was deposited on

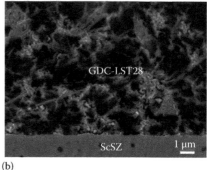

(a)   (b)

FIGURE 3.9 (a) SEM image of La-doped SrTiO₃ (LST)28 nanofibers after calcined as-spun nanofibers at 900°C for 2 h and (b) SEM image of the cross-section of gadolinium-doped ceria-LST28 nanofiber composite anode on one side of 1 mol% CeO₂–10 mol% Sc₂O₃–89 mol% ZrO₂ (ScSZ) electrolyte disk. (Reprinted from *J. Power Sources*, 265, Fan, L., Xiong, Y., Liu, L., Wang, Y., Kishimoto, H., Yamaji, K., and Horita, T., 125, Copyright 2014, with permission from Elsevier.)

LST nanofibers by infiltration. SEM images of a cross-section of GDC infiltrated LST nanofiber membrane as shown in Figure 3.9b, where LST nanofibers are randomly oriented with many junction points among the fibers. The elemental analysis on the cross-section confirms uniform distribution of GDC on all LST nanofibers. Moreover, the porous structure of LST scaffold provides access to fuels and exits to products. All those structure-related properties can enlarge TPB inside the anode to improve electrochemical performance. An optimized mass ratio of GDC to LST was found in this work. The impedance spectra show that the GDC–LST (0.92:1) nanofiber anode gives the R_p value of 0.95, 0.63, 0.38, and 0.27 Ω cm^2 at 800°C, 850°C, 900°C, and 950°C, respectively. This anode also showed no degradation in performance after 3120 min 26 cycles, indicating a good thermal cycling stability.

3.3.3 Nanofiber Pore Former for Anode

For anode-support SOFCs, anode is not only the place where electrochemical oxidation occurs but also the thickest component to maintain mechanical robustness of the cell. On one hand, sufficient open porosity is required to generate continuous channels to transport fuels in and remove the products and other gases out. On the other hand, too high porosity also induces the decrease of the mechanical strength. Therefore the design of an anode with appropriate pore structure is of great importance in determining SOFC's performance.

Pore formers such as flour, rice starch, and graphite are widely used to generate porous structure inside the anode.[50–53] These particles are mixed with anode materials and other binders. By high-temperature treatment, the particles are burnt away and leave pores inside of the anode. Pan et al. proposed to use polymer nanofibers as the pore former for the anode of SOFCs.[54] In their design, a thin PVA nanofiber membrane was electrospun on a rotating drum collector. Then the anode powder (Ni/YSZ) was sprinkled evenly on the PVA nanofiber membrane. This PVA electrospinning and powder sprinkling process was repeated for several times to obtain the desired thickness. Afterward, the mixture was compressed to form a pellet and presintered at 1000°C for 2 h. Figure 3.10 shows the SEM

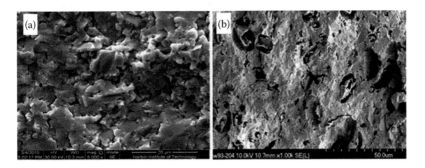

FIGURE 3.10 SEM images of anode using (a) polyvinyl alcohol nanofiber as pore former and (b) using wheat flour as pore former. (Reprinted from *Electrochim. Acta*, 55, Pan, W., Lue, Z., Chen, K., Huang, X., Wei, B., Li, W., Wang, Z., and Su, W., 5538, Copyright 2010, with permission from Elsevier.)

images of presintered anode with PVA nanofiber as the pore former and wheat flour as the pore former. The author claimed that PVA nanofiber with the diameter of 0.05–2 μm generated large amounts of wirelike pores uniformly distributed in the anode matrix, which produces 17% open porosity. This is much higher than that for the anode with wheat flour as the pore former (10%). The consequences of using PVA nanofiber as pore former are the improved gas transport inside the anode and higher electrocatalytic activity, thus increasing cell performance significantly.

3.4 CONCLUSION

Electrospun electrodes carry unique properties that are especially vital for high-performance IT SOFCs: (1) continuous electron- and/or ion-conducting path, (2) enlarged TPB regions when electrospun electrodes are covered with second materials (e.g., electrospun LSCF-GDC nanofiber/nanotube cathodes,[13,20,21] Ni–YSZ nanofiber anode[44]), (3) high open porosity that promotes mass transports, and (4) large surface area that can lower calcination temperature of the electrodes. Therefore significant performance improvements were achieved by using electrospun cathode/anode in IT SOFCs.

To the best of my knowledge, there are only few publications that focus on electrospun materials for SOFCs. More research work is required to explore the potential of electrospinning in SOFC technology. First, a better understanding is needed on the structural evolution of as-spun nanofibers during calcination/sintering. Based on previous work, we can see both temperature and ramping rate play important role in triggering the transition from nanofibers to nanorods, nanotubes, or coral-like structure. Second, most as-spun nanofiber membranes (containing precursors/ nanoparticles and carrier polymers) were calcinated into ceramic nanofibers, grounded into powders and then sintered on the electrolyte at a high temperature. Those treatments inevitably alter the original three-dimensional structures of nanofiber membranes after electrospinning and calcination. Moreover, compared to the current cell production process (e.g., tape casting, cofiring), the grinding and sintering methods introduce extra steps in the cell fabrication. Therefore, more efforts are required on the optimization of the electrospinning and sintering process that can be integrated into current SOFC's production technology.

REFERENCES

1. Visco, S. J.; Wang, L. S.; Souza, S.; Dejonghe, L. C. *Mater Res Soc Symp Proc* 1995, *369*, 683.
2. Zhu, Q. S.; Fan, B. *Solid State Ionics* 2005, *176*, 889.
3. Williams, M. C.; Strakey, J. P.; Surdoval, W. A. *Int J Appl Ceram Technol* 2005, *2*, 295.
4. Minh, N. Q. *Solid Oxide Fuel Cells 10 (SOFC-X), Pts 1 and 2* 2007, *7*, 45.
5. Burke, A. A.; Carreiro, L. G. *ECS Trans* 2011, *35*, 2815.
6. Wachsman, E. D.; Lee, K. T. *Science* 2011, *334*, 935.
7. Lo Faro, M.; Stassi, A.; Antonucci, V.; Modafferi, V.; Frontera, P.; Antonucci, P.; Arico, A. S. *Int J Hydrogen Energy* 2011, *36*, 9977.
8. Steele, B. C. H.; Heinzel, A. *Nature* 2001, *414*, 345.

9. Adams, T. A.; Nease, J.; Tucker, D.; Barton, P. I. *Ind Eng Chem Res* 2013, *52*, 3089.

10. Mai, A.; Haanappel, V. A. C.; Uhlenbruck, S.; Tietz, F.; Stöver, D. *Solid State Ionics* 2005, *176*, 1341.

11. Hayd, J.; Dieterle, L.; Guntow, U.; Gerthsen, D.; Ivers-Tiffee, E. *J Power Sources* 2011, *196*, 7263.

12. Hayd, J.; Guntow, U.; Ivers-Tiffee, E. *Ionic Mixed Conduct Ceram 7* 2010, *28*, 3.

13. Zhi, M.; Lee, S.; Miller, N.; Menzler, N. H.; Wu, N. *Energy Environ Sci* 2012, *5*, 7066.

14. Sumi, H.; Yamaguchi, T.; Hamamoto, K.; Suzuki, T.; Fujishiro, Y. *J Power Sources* 2013, *226*, 354.

15. Subramania, A.; Saradha, T.; Muzhumathi, S. *J Power Sources* 2007, *165*, 728.

16. Wu, T.; Zhao, Y.; Peng, R.; Xia, C. *Electrochim Acta* 2009, *54*, 4888.

17. Fan, L.; Liu, L.; Wang, Y.; Huo, H.; Xiong, Y. *Int J Hydrogen Energy* 2014, *39*, 14428.

18. Fleig, J.; Maier, J. *J Eur Ceram Soc* 2004, *24*, 1343.

19. Qiu, L.; Ichikawa, T.; Hirano, A.; Imanishi, N.; Takeda, Y. *Solid State Ionics* 2003, *158*, 55.

20. Zhao, E.; Jia, Z.; Zhao, L.; Xiong, Y.; Sun, C.; Brito, M. E. *J Power Sources* 2012, *219*, 133.

21. Zhao, E.; Ma, C.; Yang, W.; Xiong, Y.; Li, J.; Sun, C. *Int J Hydrogen Energy* 2013, *38*, 6821.

22. Lee, J. G.; Lee, C. M.; Park, M.; Shul, Y. G. *RSC Adv* 2013, *3*, 11816.

23. Lee, J. G.; Park, M. G.; Park, J. H.; Shul, Y. G. *Ceram Int* 2014, *40*, 8053.

24. Torres da Silva, I. M.; Nielsen, J.; Hjelm, J.; Mogensen, M. B. *ECS Trans* 2009, *25*, 489.

25. Chou, C.-C.; Huang, C.-F.; Yeh, T.-H. *Ceram Int* 2013, *39*, S549–S553.

26. Ishihara, T.; Honda, M.; Shibayama, T.; Minami, H.; Nishiguchi, H.; Takita, Y. *J Electrochem Soc* 1998, *145*, 3177.

27. Choi, J.; Kim, B.; Shin, D. *J Eur Ceram Soc* 2013, *33*, 2269.

28. Fan, L.; Xiong, Y.; Liu, L.; Wang, Y.; Brito, M. E. *Int J Electrochem Sci* 2013, *8*, 8603.

29. Rupasov, D. P.; Berenov, A. V.; Kilner, J. A.; Istomin, S. Y.; Antipov, E. V. *Solid State Ionics* 2011, *197*, 18.

30. Kobayashi, W.; Yoshida, S.; Terasaki, I. *Prog Solid State Chem* 2007, *35*, 355.

31. Fukushima, S.; Sato, T.; Akahoshi, D.; Kuwahara, H. *J Appl Phys* 2008, *103*, 07F705.

32. Son, J. Y.; Shin, Y. H.; Park, S. B.; Park, C. S.; Kim, H.; Cho, J. H.; Ali, A. I. *J Cryst Growth* 2008, *310*, 3649.

33. Li, Y.; Kim, Y. N.; Cheng, J.; Manthiram, A.; Goodenough, J. B.; Alonso, J. A.; Hu, Z. et al. *Chem Mater* 2011, *23*, 5037.

34. Jiang, X.; Xu, H.; Wang, Q.; Jiang, L.; Li, X.; Xu, Q.; Shi, Y.; Zhang, Q. *J Alloy Compd* 2013, *557*, 184.

35. Li, Q.; Sun, L.; Zhao, H.; Wang, H.; Huo, L.; Rougier, A.; Fourcade, S.; Grenier, J.-C. *J Power Sources* 2014, *263*, 125.

36. Takeguchi, T.; Kani, Y.; Yano, T.; Kikuchi, R.; Eguchi, K.; Tsujimoto, K.; Uchida, Y.; Ueno, A.; Omoshiki, K.; Aizawa, M. *J Power Sources* 2002, *112*, 588.

37. Fukui, T.; Murata, K.; Ohara, S.; Abe, H.; Naito, M.; Nogi, K. *J Power Sources* 2004, *125*, 17.

38. Ringuedé, A.; Bronine, D.; Frade, J. R. *Solid State Ionics* 2002, *146*, 219.

39. Itoh, H.; Yamamoto, T.; Mori, M.; Horita, T.; Sakai, N.; Yokokawa, H.; Dokiya, M. *J Electrochem Soc* 1997, *144*, 641.

40. Mosch, S.; Trofimenko, N.; Kusnezoff, M.; Betz, T.; Kellner, M. *Solid State Ionics* 2008, *179*, 1606.

41. Pratihar, S. K.; Das Sharma, A.; Maiti, H. S. *Mater Chem Phys* 2006, *96*, 388.

42. Azad, A. M.; Matthews, T.; Swary, J. *Mater Sci Eng B—Solid State Mater Adv Technol* 2005, *123*, 252.

43. Li, J. Y.; Tan, Y.; Xu, F. M.; Sun, Y.; Cao, X. Q.; Zhang, Y. F. *Mater Lett* 2008, *62*, 2396.

44. Li, L.; Zhang, P.; Liu, R.; Guo, S. M. *J Power Sources* 2011, *196*, 1242.

45. Marina, O. A.; Canfield, N. L.; Stevenson, J. W. *Solid State Ionics* 2002, *149*, 21.

46. Fan, L.; Xiong, Y.; Liu, L.; Wang, Y.; Kishimoto, H.; Yamaji, K.; Horita, T. *J Power Sources* 2014, *265*, 125.

47. Yoo, K. B.; Choi, G. M. *Solid State Ionics* 2009, *180*, 867.

48. Yoo, K. B.; Choi, G. M. *Solid State Ionics* 2011, *192*, 515.

49. Savaniu, C. D.; Irvine, J. T. S. *Solid State Ionics* 2011, *192*, 491.

50. Hu, J.; Lue, Z.; Chen, K.; Huang, X.; Ai, N.; Du, X.; Fu, C.; Wang, J.; Su, W. *J Membr Sci* 2008, *318*, 445.

51. Haslam, J. J.; Pham, A. Q.; Chung, B. W.; DiCarlo, J. F.; Glass, R. S. *J Am Ceram Soc* 2005, *88*, 513.

52. Clemmer, R. M. C.; Corbin, S. F. *Solid State Ionics* 2004, *166*, 251.

53. Chen, K.; Lu, Z.; Ai, N.; Huang, X.; Zhang, Y.; Xin, X.; Zhu, R.; Su, W. *J Power Sources* 2006, *160*, 436.

54. Pan, W.; Lue, Z.; Chen, K.; Huang, X.; Wei, B.; Li, W.; Wang, Z.; Su, W. *Electrochim Acta* 2010, *55*, 5538.

4 Electrospun Materials for Hydrogen Storage

Arthur Lovell

CONTENTS

4.1 INTRODUCTION

4.1.1 HYDROGEN STORAGE

Of the challenges surrounding the widespread introduction of hydrogen as a renewable, carbon-free energy carrier, the problem of storage is one of the most acute (Schlapbach and Züttel 2001). In particular, the low energy density of gaseous hydrogen requires extreme compaction of the gas in order to obtain suitably low-volume storage for portable applications such as road transport (Ahluwalia et al. 2012) and consumer electronics. The new generation of fuel cell light duty vehicles (Figure 4.1) incorporates composite pressure tanks with up to 700 bar (70 MPa) of hydrogen in close proximity to the passenger compartment, raising questions of public acceptance. The weight of these tanks and the associated balance of plant detract from the

FIGURE 4.1   A hydrogen fuel cell car, the Toyota Mirai (launched 2015). Hydrogen storage is provided by high-pressure tanks. (Reproduced from Toyota. With permission.)

overall system-specific energy (often expressed as weight% of available hydrogen), while significant energy penalties are incurred in compressing the gas to that high density. These problems are equally acute with liquid hydrogen tanks, with the additional disadvantage that boil-off unavoidably reduces the available hydrogen over time (Schlapbach and Züttel 2001).

The benefits of switching to hydrogen as an energy carrier are potentially great and justify the large resources that have been expended on research and development to date: hydrogen can be burned in internal combustion engines or consumed efficiently to generate electricity in fuel cells, in the latter case generating water vapor as the sole exhaust emission. It represents a way to store surplus renewable electricity through the electrolysis of water and thus provides a way to match supply and demand across an energy economy with large penetration of wind and solar power generation (Jacobson 2009). Hydrogen storage can help by storing electricity on a timescale of hours (as with off-grid renewable generator systems), days (e.g., by inserting hydrogen as a carbon-free component into the natural gas grid [Melaina et al. 2013]), months, or years in the case of compressed gas backup power systems and in the case of solid-state hydrogen materials, effectively indefinitely.

The inherent drawbacks of high pressure or liquid hydrogen storage have spurred the development of alternative storage technology (Yang et al. 2010). Solid-state hydrogen storage materials comprise many families of materials that operate in two main ways: physical sorption and chemical storage.

Physisorptive materials typically rely on low-energy interactions between hydrogen (H_2) molecules and a substrate, typically a porous surface or particle. H_2 is not a polar molecule and typical interaction energies through quadrupolar or Van der Waals (VdW)-type forces are low, only a few kJ/mol. This interaction can be

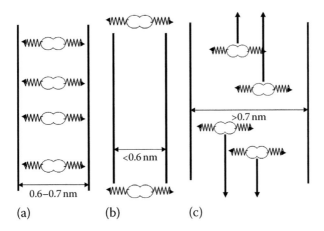

FIGURE 4.2 Diagram indicating the favorable pore width (a) from 0.6 to 0.7 nm for hydrogen. (b) The pore width is too narrow to allow H$_2$ entry; (c) The pore width is large so the hydrogen interaction with the walls is weakened. Hydrogen molecules are less confined and may escape more readily. (Reprinted from *J. Colloid Interface Sci.*, 318, Im, J.S., S.-J. Park, T.J. Kim et al., The study of controlling pore size on electrospun carbon nanofibers for hydrogen adsorption, 42–49, Copyright 2008a, with permission from Elsevier.)

enhanced through catalytic or chemical adjustment or doping (Deng et al. 2004). The quantity of hydrogen that can be stored depends strongly on temperature and generally on the specific surface area (SSA) of the substrate (Panella et al. 2005) and the internal volume accessible to hydrogen. In practice, these are porous materials, in which the pore dimensions have to be geometrically favorable for the adsorption of H$_2$. Theoretical and empirical studies have suggested that a favorable pore width for hydrogen exists in the range 0.6–0.9 nm (and more specifically around 0.7 nm). This is the scale where the hydrogen molecule, of dynamic diameter 0.4059 nm (Schlapbach and Züttel 2001), can physically insert itself, while the potential fields from more than one wall of the pore overlap, thus increasing the otherwise weak attractive force on the H$_2$ molecule (Broom 2011). The geometric reasoning is shown in Figure 4.2. However the pure physisorption of hydrogen is still limited to either cryogenic temperatures or high pressures or both. Monolayer surface coverage can generally be achieved at temperatures below 100 K or pressure in excess of 3 MPa; large excess adsorption of H$_2$ is rare at ambient temperatures and pressures. The main advantages of physisorptive systems are that they can increase the volumetric density of hydrogen storage even over pure compressed hydrogen, and that the hydrogen is readily released or refilled with high flow and very long cycle lifetime. Hybrid cryocompression systems have been designed to maximize this density by placing high-surface-area adsorbents in cold, compressed hydrogen tanks (Ahluwalia et al. 2010). The overall storage capacity generally remains low: <5 wt% H$_2$ even at 77 K. This raises questions about the ultimate suitability for pure physisorptive hydrogen storage for technological use.

Chemical hydrogen storage uses hydrides and composites where hydrogen is bonded chemically within a solid (or ionic liquid, or slurry). The strength of the hydrogen bonding, in the range of some 50–500 kJ/mol, determines the desorption of hydrogen; typically, desorption occurs at elevated temperatures and can be strongly endo- or exothermic in nature. As hydrogen is desorbed from a bulk hydride, kinetics can be slow, and in the case of exothermic rehydrogenation, huge amounts of heat can be released in the repressuring of hydrogen into the system. This requires complex engineering to implement in a vehicle (Aardahl and Rassat 2009). The types of release reaction, and the purity of the resulting hydrogen, can vary widely (Ahluwalia et al. 2012).

Hydrogen release from hydrides is governed by reaction rate, the recombination of H atoms within the material or at the surface, the kinetics of diffusion of hydrogen through the material, and the energy cost of advancing the hydride phase across the bulk. In bulk materials, these factors can often be slow or energetically expensive. Various methods to inculcate micro- and nanostructures within hydrides have been proposed and tested as a way to improve kinetics of hydrogen release and reversibility (Reardon et al. 2012). For example, nanoparticles of Mg have shown accelerated hydrogen release; however, the reversibility over many cycles is poor due to cumulative recombination or sintering of particles back toward bulk properties (Herley et al. 1985). Other structural components can be added to hold the Mg nanoparticles apart, discouraging agglomeration and improving reversibility, but at cost of reduced hydrogen weight capacity. Several methods for improving the release of hydrogen, the temperature, kinetics, and reversibility of other hydride families have been discussed, including scaffolding and templating with silica structures, metal–organic frameworks (MOFs), high-SSA carbons, and aerogels. These are often multistep processes, involving chemical etching or heat treatment and subsequent filling with hydride materials.

It should be clear that the strength of the interaction by which the hydrogen is retained within the materials is a critical parameter that governs its utility as a hydrogen store. Cyclability, the release kinetics, and thermodynamics all depend heavily on the behavior of this interaction. Consequently, recent hydrogen storage research has sought materials and methods in which this parameter might be reliably tuned and, in particular, to obtain hydrogen bonding interactions in a weak chemisorptive or strong physisorptive regime of about 20–50 kJ/mol (Bérubé et al. 2007). A large part of the literature records efforts to improve the pure physisorption interaction strength with doping or catalytic effects and, from the chemisorption side, to destabilize the strong interactions of hydrogen in chemical hydrides with catalytic, chemical, or structural alterations to make hydrogen release more favorable.

Many lightweight complex and metal hydrides have been identified in recent years, following the priorities of the U.S. Department of Energy (DoE), hydrogen and fuel cells program. This program identified the most widely used technological targets for storage systems (Satyapal et al. 2007). A rigorous down-selection process was used to reject candidate materials where the properties appeared insufficiently promising for mobile hydrogen storage (Klebanoff and Keller 2013). The more traditional families of pure metal hydrides (e.g., MgH_2; AB5-type compounds) generally have some combination of too low-hydrogen capacity, poor release kinetics, and inconvenient operating temperatures. However, more recently, families of complex hydrides have been studied, including borohydrides (e.g., $LiBH_4$)

(Nakamori et al. 2008; Rude et al. 2011), aluminum hydrides (e.g., AlH_3 and $NaAlH_4$) (Graetz et al. 2006; Ahluwalia et al. 2009), and the family of amidoboranes (Bérubé et al. 2007; Xiong et al. 2008). These eschew heavy metals in order to maximize the stored hydrogen weight as a proportion of the overall compound. These materials have quite varied characteristics, which defy collective summary. However, they have been selected and tested for their high hydrogen content and potential reversibility under moderate conditions of pressure and temperature. Amidoboranes, and in particular derivatives and blends of ammonia borane (NH_3BH_3, AB), are some of the most promising materials for on-board storage in cars. These materials release hydrogen through polymerization to form a B-N backbone with successive release of 2 H_2 per AB molecule up to 150°C (Stephens et al. 2007), giving a usable hydrogen content of 13 wt%. However, competing reactions produce unwelcome impurities, principally ammonia (NH_3) and borazine ($HNBH)_3$, and the material decomposes into a foam after the hydrogen release, with condensing of solid residues on nearby surfaces.

4.1.2 ELECTROSPINNING AND HYDROGEN STORAGE

Electrospinning is now a mature synthesis route for the continuous production of high surface-to-volume ratio micro- and nanofibers from a variety of polymeric solutions or melts (Li and Xia 2004; Greiner and Wendorff 2007). The resulting fibers and nonwoven mats can be used to make a variety of morphologies by postprocessing. The as-spun fibers can be used as scaffolds or impregnated with other materials to make composites. Active materials such as hydrides or catalytic nanoparticles can be included by suspension in the precursor solution or by cospinning a second fluid inside the first using a coaxial feed or by other means. In these cases, particles can be finely and evenly dispersed in a matrix. Spontaneous core–shell structures can be formed from a single-phase spinning process. As-spun fibers can also undergo heat treatment to be turned into ceramic or carbon fibers, with porosity or active sites included through chemical activation or etching, among other possibilities.

The inculcation of meso-, micro-, and nanostructures into specific hydrogen storage materials and composites has been shown variously to improve their production and operation (Nielsen et al. 2011; Reardon et al. 2012), and so the use of electrospinning has been studied in this field, as it can obtain homogeneous, scalable, and reproducible microstructures either in active hydrogen storage materials or their precursors, and sophisticated scaffolds and substrates for formation and processing of composite hydrogen stores. Electrospinning has not been used as widely in this field as for other energy and environmental applications (Thavasi et al. 2008; Laudenslager et al. 2010; Dong et al. 2011), but it has shown promise in a number of cases (Sundaramurthy et al. 2014), not always closely related except under the most general banner of hydrogen storage. This chapter aims to overlay a summary of the state of the art in electrospun hydrogen storage materials with an assessment of the opportunities where electrospinning is yet to make an impact, to identify the advantages electrospun materials have over other structural synthesis methods, outline areas of research (and also the gaps), and provide a short outlook for the progression of the field.

4.2 CARBON FIBERS FOR HIGH-PRESSURE HYDROGEN TANKS

Carbon fibers have been produced from electrospun precursors for a wide variety of applications in the energy area, including battery and supercapacitor electrodes and catalyst support for fuel cells. Thus, the electrospun production of high-SSA carbon-based materials for hydrogen storage is a clear extension. The high-pressure type IV composite hydrogen gas tanks produced by companies such as Quantum (Sirosh 2002) are now being rolled out in the first generation of consumer fuel cell cars. The strength in these tanks is typically provided by carbon fiber produced by traditional means from polyacrylonitrile (PAN), produced and heat-treated continuously. The carbon fiber–epoxy composite layer accounts for >70% of the total tank system cost (Hua et al. 2011).

Studies on the potential use of electrospun carbon fibers specifically to reduce the cost of production or the weight proportion of fiber required in these tanks do not seem to be available. However, the industry-leading high-strength fibers such as Toray T700S have a strength in the region of 5000 MPa. Electrospun fibers tend to have poor alignment of polymer chains and hence a relatively low strength (Yao et al. 2014). PAN-derived electrospun fibers have attained tensile strengths of 320–900 MPa (Zussman et al. 2005; Zhou et al. 2009). With further strengthening steps, such as a higher carbonization temperature, postdrawing of fibers, and stabilization and carbonization under tension, tensile strengths of 3500 MPa, approaching the high-strength commercial fibers, were reported (Arshad et al. 2011). There is at least potential for electrospinning to become a manufacturing technique for high-strength carbon nanofibers (CNFs), and an assessment of the projected costs should be made.

4.3 ACTIVATED CARBON NANOFIBERS

In the field of solid-state hydrogen storage, carbon substrates have the advantage of tunable porosity, SSA, and the ability to accept a wide variety of active and functional species in order to improve the interaction with hydrogen. A significant number of carbon materials have been produced from electrospun fibers, and these have typically been tested for their physisorptive hydrogen capacity. Catalytic materials such as vanadium have been impregnated into fibers to create hydrogen "spillover" storage. Spillover is observed in systems where a catalytic species such as Pt or Pd is available to dissociate hydrogen molecules, which then diffuse atomically onto a substrate such as a high-SSA carbon (Lueking and Yang 2002; Lachawiec et al. 2005; Yang and Wang 2009). Microporous carbons with measured SSA as high as 3400 m^2/g have been manufactured (Jiang et al. 2011) using an MOF as a template, but the hydrogen storage capacity reached only 2.77 wt% at 77 K and 1 bar: excellent for an undoped carbon but showing the difficulties of reaching the higher storage targets with this type of material.

Im et al. prepared a direct comparison of the hydrogen storage properties of chemically activated electrospun CNFs (EACNFs) with other types of carbon (Im et al. 2009c). The methodology is typical for the production of high-SSA CNFs. The fibers were electrospun from 10 wt% PAN in N,N-dimethylformamide (DMF) solution. Typically, the production of carbon fibers from the electrospun PAN was carried out in a two-stage process; stabilization in air at 523 K for 8 h followed by carbonization under nitrogen flow at 1323 K for 1 h. Sodium hydroxide (NaOH) and potassium carbonate (K_2CO_3) were used as activation agents. Activation was carried out after immersion in solution of the activation agent followed by heating at 1023 K in a N_2 atmosphere for 3 h, followed by washing and drying. Nitrogen adsorption isotherms were carried out with Brunauer–Emmett–Teller (BET) analysis (Brunauer et al. 1938) on the activated samples to determine the SSA and total pore volume. The pore size distribution was carried out using Horvath–Kawazoe method (Horvath and Kawazoe 1983) to determine the fraction of microporous volume. NaOH-activated samples showed higher hydrogen adsorption at 303 K and 5 MPa than K_2CO_3-activated samples, but the increase was less than directly proportional to the increase in SSA and overall pore volume, suggesting that micropore volume fraction has the largest positive correlation with hydrogen adsorption. Comparison of the EACNFs with hydrogen uptake in single-walled carbon nanohorn, single-walled carbon nanotube, graphitic CNF, and activated carbon (AC) (Xu et al. 2007) showed that the EACNFs have a higher uptake both as a function of SSA and micropore volume. Overall storage at room temperature was a maximum of about 1 wt% H_2 for the NaOH-activated sample with highest micropore volume (0.751 cm^3/g) and SSA (1933.2 m^2/g), although to provide a base comparison, no further modification of the EACNFs were carried out to increase the H_2 storage capacity. The advantage of the EACNFs over other forms of porous, high-SSA carbon is the large fraction of micropores with sizes in the optimum 0.6–1 nm range. Geometric and modeling (Im et al. 2007) considerations have suggested that 0.6–0.7 nm pore widths are best for trapping H_2 (Im et al. 2008a); lower widths prevent hydrogen (dynamic radius = 0.4059 nm (Schlapbach and Zuttel 2001)) from intercalating into pores except at cryogenic temperatures. Larger pore diameters reduce the overlap of adsorption potential from the opposite pore walls that an H_2 molecule interacts with and leave volumes in the center of the pore that are inaccessible to adsorbed hydrogen.

The means of generating the high-SSA fibers after spinning and stabilization varied for the production of the activated EACNFs. Other activation agents used included KOH, Na_2CO_3, H_3PO_4, $ZnCl_2$, SiO_2, or water vapors. The method for activation depended on the activation agent. KOH activation is based on carbon oxide species being burned off from the fiber surface. Nanofibers produced from activation with $ZnCl_2$ showed about 1.5 wt% hydrogen uptake at 30 bar, which was greater than KOH-activated fibers: although the KOH-activated fibers had a higher SSA, they had a lower fraction of micropores in the region of interest, of size range 0.6–0.7 nm (Im et al. 2008a).

Carbide-derived carbons with extremely high SSA were prepared from electrospun polycarbomethylsilane in tetrahydrofuran (Rose et al. 2010). First, silicon carbide was formed by pyrolysis, and then the SiC was etched through chlorination to leave carbon fibers with some residual chlorine content. BET SSAs between 1740 and 3116 m^2/g were measured, with resulting hydrogen capacity for the sample with the highest SSA found to be 3.86 wt% at 77 K and 17 bar. This value is among the highest measured at low pressures in a sorption system. The pore size distribution by low-pressure N_2 adsorption showed a large spread of small mesopores, around

2–5 nm, and a significant fraction of micropores (<1 nm). The amorphous nature of the pyrolyzed SiC may lead to the large increase in SSA compared with carbide derived from crystalline SiC.

4.4 ELECTROSPUN FIBERS AS A BASE FOR DOPED HIGH-SURFACE-AREA CARBONS

4.4.1 METAL- AND METAL OXIDE–FUNCTIONALIZED NANOFIBERS

The addition of metal or metal oxide sites to pure carbon structures has the ability to increase the hydrogen storage potential. In physisorption, the weak VdW forces of the H_2 interaction with the surface of the carbon can be enhanced by the greater interaction of hydrogen with the metal center. Much effort has gone also into combining the catalytic effect of certain species, chiefly transition metals, with the sorption potential of carbon. Hydrogen molecules are dissociated by the catalyst species and can combine with defects, dangling bonds, and other chemical interactions on the surface as atomic H and can also diffuse into the bulk medium. These metal additives are particularly amenable to an electrospinning process as they can be included either at the point of spinning or by doping or formation reaction once the fibers have been created. Im et al. have studied transition metal species as additives in a hydrogen storage system (Im et al. 2008b, 2009a) using EACNFs as a base material. 10 wt% PAN in DMF is used as the spinning solution but with the catalyst added as an additional 3.3 wt%. Fibers containing magnesium oxide (MO), copper oxide (CO), iron oxide, and iron and magnesium (Im et al. 2009a) and vanadium pentoxide (Im et al. 2008b) were spun using typical spinning parameters: solution delivery rate 1.5 mL/h; potential 15 kV; nozzle-to-collector distance 150 mm. After stabilization and carbonization steps, potassium hydroxide was used as activation agent. Samples immersed in 8 M KOH solution were activated at 1023 K in an argon atmosphere before washing and drying. SEM, energy dispersive spectroscopy (EDS), and absorption studies of the activated carbon fibers (ACFs) and their pore size distribution were carried out, followed by hydrogen uptake measurements using a volumetric method. In this, samples were held at 673 K for 1 h under hydrogen gas to obtain the reduction of the metal (oxide) and then outgassed at 473 K for 2 h. Finally, hydrogen gas was allowed onto the sample in pressures from 0 to 100 bar at 303 K using a pressure–composition–temperature (PCT) apparatus. The fiber morphology and defect structure appear to be heavily dependent on the type of catalyst present. The magnesium oxide fibers (ACF-MO) show a wrinkled outer surface and all the MO appears to be contained within the fiber. Not so the iron oxide, where clear clusters can be seen on the outer surface of fibers (ACF-IO). The copper oxide fibers (ACF-CO) have defects apparently formed by the dissociation of the CO, the melting temperature of which is below the carbonization temperature (1573 K). The activated samples contain both micro- and mesopores, with the CO fibers, the ones with the greatest microporosity. Released oxygen from the CO over melting was burned off to provide a much more microporous structure. The same effect was attributed to the formation of activated fibers with vanadium pentoxide (ACF-VO). In this study (Im et al. 2008b), the elemental content of the fibers was measured by EDS after carbonization at 1373 K, before and after KOH activation at 1073 K, for both "pure" carbon fibers and the V_2O_5-containing carbon fibers. The pure carbon fibers have >98% carbon content. The vanadium–carbon fibers have 25% V content before activation and almost 50% afterward due to removal of C during the activation. However, the oxygen content is 0.4% before and after. Thus, the V_2O_5 additive loses the majority of its oxygen in the carbonization.

The same group also studied the effect of fluorination of an ACF-VO sample by exposing the post-activated sample to fluorine gas at 303 K and 1 bar for 10 min (Im et al. 2009b) to create a metal–carbon–fluorine ACF system (ACF-VF). In this case, the fluorine and vanadium introduce a charge gap to enhance the physisorptive potential of the H_2 molecule. X-ray photoelectron spectroscopy was used to determine that the fluorine and vanadium are chemically attached to the carbon surfaces rather than each other.

The SSA and the overall pore volume for the ACF-VF was somewhat lower than for the unfluorinated ACF-VO, with less development of pores below 0.6 nm size. The micropore volume fraction remained approximately 40%. However, a hydrogen storage capacity of 3.2 wt% was measured at room temperature and 100 bar, significantly above the ACF-VO value of 2.5%.

The SSA and pore volume for these materials were determined and are shown in Table 4.1.

All the samples, but particularly the ACF and ACF-CO samples, show an enhanced pore structure in the range 0.7–0.9 nm (Figure 4.3). ACF-CO and ACF-VO have developed pores in the 0.6 nm range as well, as a function of the oxygen release from the CO and V_2O_5 in the carbonization step. The ACF-VF fibers show a further enhancement in the hydrogen uptake, owing to the charge environment between metal (δ^+) and fluorine (δ^-) sites in the carbon matrix, where the weak VdW binding

TABLE 4.1
Summary of Hydrogen Storage Properties of Metal- and Metal Oxide–Doped Activated Electrospun Carbon Fibers

Sample	SSA (m²/g)	Pore Volume (mL/g)	Fraction of Micropore Volume	Hydrogen Uptake at 303 K and 100 Bar, wt%
ACF	2880	2.46	0.4	1.78
ACF-CO	2820	2.39	0.45	2.77
ACF-MO	2130	1.78	0.45	1.69
ACF-IO	1480	1.24	0.5	1.33
ACF-M	1230	1.16	0.45	2.29
ACF-I	1040	1.04	0.45	2.05
ACF-VO	2780	2.67	—	2.41 (2.5) (Im et al. 2009b)
ACF-VF	2700	2.6	0.4	3.2 (Im et al. 2009b)

Sources: From Im, J.S. et al., *Micropor. Mesopor. Mater.*, 115, 514, 2008b; Im, J.S. et al., *Int. J. Hydrogen Energy*, 34, 3382, 2009a; Im, J.S. et al., *Int. J. Hydrogen Energy*, 34, 1423, 2009b.

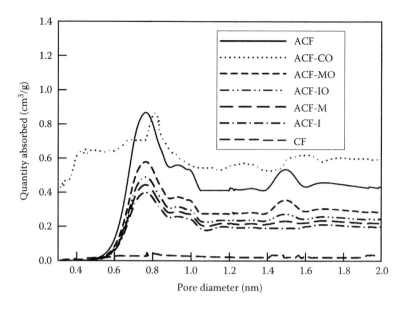

FIGURE 4.3   Pore size distribution using Horvath–Kawazoe method (Horvath and Kawazoe 1983) for metal- and metal oxide–doped carbon fibers. The unactivated carbon fibers show a low porosity, but all the activated fibers have a peak at around 0.7 nm. Only the activated carbon fiber (ACF)–copper oxide (CO) of the samples shown here has significant porosity in the <0.6 nm size range; it also has the highest surface area of the doped ACFs (2820 m²/g) in the study (Im et al. 2009a) and the highest hydrogen uptake (2.77 wt% at 100 bar and 303 K). (Reproduced from Im, J.S. et al., *Int. J. Hydrogen Energy*, 34, 3382, 2009a. With permission from International Association for Hydrogen Energy.)

of H$_2$ can be strengthened by induced quadrupolar interactions resulting from the unshielded surface charge.

Palladium-coated CNFs were developed from carbonized PAN fibers, activated with water vapor in argon at 1073 K (Kim et al. 2011). The Pd was coated using a coelectrospinning process with palladium chloride. A 1 wt% PdCl$_2$ dissolved in DMF was delivered through the outer ring of a coaxial nozzle around PAN-DMF solution. The spun fibers were stabilized and then carbonized at 1073 K, then the Pd ions were reduced using a 4% H$_2$ in argon gas mix at 523 K to form palladium nano-fibers (ACF-Pd). The resulting fibers were found to have a SSA of 815.6–1120.8 m²/g by TEM and BET analysis. The CNFs activated with water vapor showed improvement in hydrogen storage over the nonactivated fibers, and the ACF-Pd showed a greater uptake again: 2.36 wt% at 77 K and 0.82 wt% at 298 K and 1 bar. This last consisted of 0.37 wt% reversible uptake and 0.45 wt% irreversible uptake, caused by stable H bonding with carbon defects, via spillover from Pd.

Metal oxide nanowires have been derived from electrospun fibers and their hydrogen storage potential assessed (Yaakob et al. 2012). High-molecular-weight polyvinylpyrrolidone (PVP) was spun in ethanol. The zinc nitrate precursor was mixed in DMF, and separate solutions were made with 3 wt% magnesium nitrate

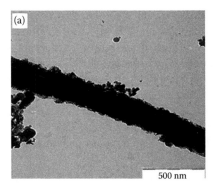
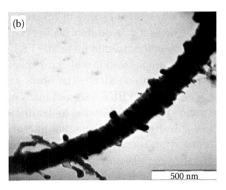

FIGURE 4.4 ZnO nanofibers from electrospun polyvinylpyrrolidone doped with (a) Al and (b) Mg nanoparticles. (From Yaakob, Z. et al., *Int. J. Hydrogen Energy*, 37, 8388, 2012. With permission from International Association for Hydrogen Energy.)

(for Mg doping) or with 3 wt% aluminum nitrate (for Al doping) under inert atmosphere. Each precursor solution was added to separate PVP/ethanol solutions dropwise and spun. As-spun nanofibers were calcined at 773 K in air, resulting in a reduction of fiber diameter to 50–150 nm in the zinc oxide fibers. The Mg-doped ZnO nanofibers show Mg particles inside and on the surface of the polymer, while the Al-doped ZnO nanofibers have Al nanoparticles chiefly on the surface (Figure 4.4). Hydrogen sorption measurements were taken at 70 bar and room temperature and compared with pure ZnO nanoparticles available commercially. The electrospun ZnO absorbed 1.4 wt% hydrogen compared with ~0.6 wt% for the nanoparticles. The Mg- and Al-doped ZnO nanofibers absorbed up to 2.29 and 2.81 wt%, respectively, a probable result of the catalytic effect of the doping and induced defects in the crystalline ZnO structure arising.

4.5 TURBOSTRATIC GRAPHITIC NANOFIBERS AND INTERCALATION COMPOUNDS WITH HYDROGEN

The manufacture of electrospun graphitic nanofibers has been reported by a number of routes, and their hydrogen capacity has been reported. Compared with the ACF, however, their SSA was low and the hydrogen uptake similarly limited. Polyvinylidene fluoride (PVDF) electrospun fibers were carbonized in a nitrogen atmosphere from 1073 to 2073 K with iron (III) acetylacetonate (IAA) to act as a graphitization promoter (Hong et al. 2007). Transition metals have been observed to increase graphitization in polyimide films (Kaburagi et al. 2001), and IAA was used to this effect with both PAN and polyimide electrospun fibers to make GNFs (Chung et al. 2005a; Park et al. 2005). Furthermore, GNFs based on PVDF without the IAA promoter had a higher nitrogen-BET SSA than PAN-based GNFs but a lower hydrogen capacity, with maximum 0.39 wt% at 100 bar and 303 K (Chung et al. 2005b), compared with ~1 wt% in the PAN-based GNFs (Kim et al. 2005). This was attributed to a lack of microporosity in the PVDF-based GNFs.

PVDF was spun at 11 wt% in 7:3 acetone/DMF containing 1,8-diazabicyclo[5.4.0] undec-7-ene as a dehydrofluorination (DHF) activator and 5.5 wt% IAA. DHF was followed by carbonization up to 2073 K. The resulting fibers showed good levels of turbostratic graphitization with the d-spacing (interlayer distance) of 0.333–0. 343 nm by x-ray diffraction (XRD). The fibers were often hollow with a high proportion of mesopores, but the BET SSA and micropore volume decreased with increasing carbonization temperature owing to densification of the sample. High levels of DHF led to low-porosity carbon fibers, but intermediate levels produced high-SSA CNFs as fluorinated gas species are given off during the carbonization process, etching pores within the fiber structure.

The hydrogen storage capacity of the PVDF-based GNFs was low, in the range 0.1–0.2 wt% at 80 bar and 303 K. CNFs carbonized at 2073 K showed higher uptake, at 0.39 wt%. As this does not correlate well with surface areas of 377–473 m²/g, it seems likely that the pore structures were also not optimized for hydrogen storage. Furthermore, the PAN-based GNFs made with IAA catalyst showed up to 1.01 wt% hydrogen storage capacity even with lower measured SSA (in the region 60–250 m²/g) (Kim et al. 2005).

4.5.1 GNF-Based Intercalation Compounds

A method to increase the adsorption capacity and molecular sieving for gas can be obtained via the intercalation of graphite with electronic acceptor or donor species. The 0.34 nm interlayer d-spacing of natural or artificial graphite can be levered open and stable graphite intercalation compounds (GICs) made with the insertion of a wide variety of single atoms, molecules, and compounds (Dresselhaus and Dresselhaus 1981; Solin and Zabel 1988). The ability of GICs to maintain very uniform sequences of filled and unfilled graphene layers over long range (called staging) has attracted researchers for hydrogen storage (Enoki et al. 1993). An intercalation process that increases the d-spacing of every layer to the 0.6–0.9 nm distance highlighted as giving the best hydrogen adsorption characteristics would provide in principle a material with the whole volume as slit pores suitable for hydrogen physisorption, compared with ACF, where a majority of pore volume is in mesopore sizes >10 nm. Moreover, the intercalant species can be chosen and tuned to provide catalytic or charge donation to the graphite suitable for enhancement of the hydrogen–GIC interaction (Patchkovskii et al. 2005). Systems tried include rubidium, cesium (Beaufils et al. 1981), and potassium compounds (Watanable et al. 1973). Divalent metal species such as calcium may increase the strength of the hydrogen interaction through greater charge difference; however, its intercalation with hydrogen into graphite has proved unstable (Srinivas et al. 2010).

The volume available to hydrogen in an intercalated layer is generally not significant owing to the presence of the intercalant species, thus the hydrogen physisorption is limited to layers that through the intercalation process are not fully intercalated. An example is the second stage potassium graphite, which has stoichiometry KC_{24} and can accept up to two H_2/K at 77 K, but every second layer is unintercalated and inaccessible to hydrogen (Lovell et al. 2008). An ideal system of this sort would remove the majority of the intercalant to provide a pillared gallery with large internal volume accessible to hydrogen (Deng et al. 2004). This possibility has also been studied in other layered structures, for example, clays, where the layers are of intrinsically greater weight and the presence of water is required to maintain the interlayer structure (Edge et al. 2014).

Electrospun graphitic nanofibers can be used for intercalation. For example, turbostratic graphitic nanofibers (TGNFs) have been produced from electrospun PAN precursor fibers (Kurban et al. 2010a) and used as the basis for potassium intercalation (Kurban 2011). The spun fibers underwent a stabilization step at 523 K in air before being carbonized and graphitized at successive temperatures from 773 to 3273 K. The resulting stabilized TGNFs had lost about 70% of the starting polymer mass. TGNFs down to 49 nm diameter were obtained from 3.8% PAN (150 kDa) spun in dimethyl sulfoxide (DMSO). TEM showed that the fibers were made of a microfibrillated structure with up to 40 turbostratic graphite layers forming interweaving ribbons within the fiber. Following potassium intercalation using a standard one-zone procedure (Dresselhaus and Dresselhaus 1981) at 573 K, hydrogen adsorption was tested at 77 K. There was no increase in the uptake of hydrogen over the bulk $KC_{24}(H_2)_2$ stoichiometry (~1.2 wt%); however, its structure was more amorphous than the bulk (turbostratic graphite) GICs and some increase in kinetics of hydrogen release was observed. The possibility of true pillared GICs engineered for hydrogen uptake remains an open question. However, the structural integration of different length scales and surface areas, together with many edge states in the electrospun-fiber GICs, shows some potential for accelerated sorption and a relatively good physisorptive prospect if a low concentration of intercalant can be achieved.

4.6   OTHER ELECTROSPUN MATERIALS FOR HYDROGEN STORAGE

4.6.1   Boron Nitride Nanofibers for Hydrogen Storage

Boron nitride nanostructures have been considered for hydrogen storage. Their properties look likely to be similar to nanocarbons. Defects and doping can increase the limited uptake of the pure material as for carbon. Titania-coated and uncoated boron nitride nanofibers (BNNFs) were examined for their hydrogen storage properties (Shahgaldi et al. 2012a). A standard solution of 10% PAN in DMF was electrospun and the resulting fibers stabilized at 250°C in air for 4 h. The titania-coated fibers were produced using a 1% solution of TiO_2 in DMF as an outer solution in a concentric nozzle before stabilization. A solution of 0.4% boron oxide (B_2O_3) was used to coat stabilized fibers. Then a treatment was carried out in O_2/NH_3 (1:9) atmosphere at 800°C and afterward at 1100°C in pure NH_3 and 1500°C in pure N_2 at 100 mL/min to obtain both pure and titania-coated BNNFs. BNNFs were also produced with variation in heating temperature of 600°C for the O_2/NH_3 mix, 900°C for NH_3, and 1300°C for pure N_2 at 100 mL/min.

Titania nanoparticles were observed to be well dispersed on the PAN nanofiber surface. After nitridation at the higher series of temperatures, rough surfaces and an average diameter of 100–150 nm were observed. The fibers produced at the lower heating temperatures had much rougher morphologies. The SSA went from 86 to 316 m^2/g for the pure BNNFs, to 416 m^2/g for the lower-temperature-synthesized titania-doped BNNFs, and to 365.5 m^2/g for the higher-temperature Ti-BNNFs. The titania-doped BNNFs synthesized at the lower temperatures had a higher hydrogen uptake than the BNNFs and the higher-temperature titania-doped BNNFs; 2.1 wt% hydrogen at 70 bar by pressure–composition isotherm versus 1.9 wt% for the higher-temperature Ti-BNNFs. This correlates with the higher SSA and the visible surface morphology by SEM.

4.6.2 POLYANILINE FIBERS FOR HYDROGEN STORAGE

Polyaniline (PANI) has become a well-studied polymer as its electrical conductivity makes it of interest for a number of nanoelectronic and bioengineering applications including biosensors and tissue engineering scaffolds (Huang et al. 2003). It is readily made into nanofibers by a number of approaches (Wang and Jing 2008). However, its property of intrinsic microporosity (IM) has also interested hydrogen storage researchers (Jurczyk et al. 2007).

Early reports of high storage capacity in HCl-treated PANI and polypyrrole (Cho et al. 2002) were not reproducible (Panella et al. 2005). Later, however, reversible hydrogen storage of 3–10 wt% was reported in PANI nanofibers (PANI-NFs) produced by standard synthesis in aqueous solution and then electrospun (Srinivasan et al. 2010). The fibers were compared with bulk PANI and found to be stable up to 150°C. PCT profiles of the PANI-NFs under hydrogen pressures of up to 8 MPa showed a reversible hydrogen uptake of approximately 3 wt% at 50°C and 6–8 wt% at 100°C–125°C. Hydrogen uptake is accompanied with significant changes in the observable microstructure (Figure 4.5); fibrillar swelling implying that the hydrogen is physically incorporated inside the polymer, with some fiber breakage and precipitated material on the fiber surface. However the mechanism for strong sorption in the PANI is not clear.

The interaction strength of the hydrogen with PANI is typically that of physisorption, and so the development of hydrogen storage systems parallels the approaches taken to improve the carbons; the surface area of the PANI has been increased; its porosity optimized; and dopant species such as ionic lithium used to increase the physisorption interaction strength. Catalytic species such as Ti, Fe, and Pd can be added to dissociate the H_2 and allow chemisorption of H atoms into the PANI matrix (Skowroński and Urbaniak 2008) as for carbons (Yildirim and Ciraci 2005). PANI nanostructures can be readily made through a number of approaches, often with the oxidative polymerization of aniline carried out in conjunction with a template. A particularly useful approach uses interfacial polymerization at the boundary of an immiscible aqueous/organic system (Huang et al. 2003), which produced uniform nanofibers in the 30–50 nm diameter range. Other researchers synthesized PANI with vanadium pentoxide (Kim et al. 2010), by analogy with the ACF-VO fibers discussed earlier (Im et al. 2009b). However, these were not electrospun.

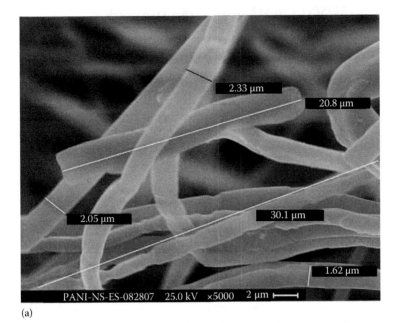

(a)

(b)

FIGURE 4.5 Polyaniline nanofibers shown (a) before and (b) after hydrogen sorption cycles. Swelling and breaking of the fiber components are a consequence of the hydrogen sorption in polyaniline. The 3–10 wt% quoted is a high figure for a purely physisorptive system. (Reproduced from Srinivasan, S.S. et al., *Int. J. Hydrogen Energy*, 35, 225, 2010. With permission from International Association for Hydrogen Energy.)

4.7 ELECTROSPUN MATERIALS FOR CHEMICAL HYDROGEN STORAGE

4.7.1 CARBON NANOFIBERS AS CATALYSTS FOR COMPLEX AND CHEMICAL HYDRIDES

The high surface area, light weight, and morphologies of CNFs can allow them to act as catalytic agents to improve the speed and temperature of dehydrogenation events in complex and chemical hydrides, with certain benefits over the more commonly studied metal catalysts such as titanium (Boganovic and Schwickardi 1997). In particular, transition and rare-earth metal catalysts are costly, add more weight and can have selectivity and durability issues. A commonly studied system is $NaAlH_4$, sodium aluminum hydride. This has limited reversibility without catalysis, and on dehydrogenation, agglomeration of the alloy takes place, reducing the speed and amount of the reversible hydrogen uptake. Using carbon as a support scaffold for nanoscale particles of $NaAlH_4$ shows improvement in this, with a number of carbon materials tried, among them CNFs (Balde et al. 2006) and nanoporous carbon composite (Stephens et al. 2009; Adelhelm et al. 2010). Scaffold supports have a large extra mass compared with catalytic carbon materials, reducing the hydrogen storage weight percentage by 30%–40%. These latter may incorporate a catalytic effect of metallic impurities such as Ni, retained as a result of the fiber growth process. Fibers produced using electrospinning may not retain similar metal particles; however, their inclusion brings significant advantages to the headline hydrogen storage properties. With graphitic nanofibers (GNFs), the lack of the metal impurities lowered the hydrogen storage performance by 10%–15%. Although a large number of test carbons have been used in this work (Hudson et al. 2012), including single and multiwalled carbon nanotubes and helical and planar GNFs, none was produced by electrospinning. The helical GNFs were found to have the best catalytic effect on the $NaAlH_4$, lowering the desorption temperature from 170°C to 143°C and increasing the rehydrogenation kinetics so that 1.75 wt% H_2 was reabsorbed in 3 h.

4.7.2 ELECTROSPINNING OF LANI₅

The AB5 series of alloy-based hydrides has long been known, and hydrogen stores based on this reversible hydriding process are commercially available. $LaNi_5$ hydrides to $LaNi_5H_6$, but traditional synthesis methods of the amorphous parent alloy are time and energy intensive, requiring heat, pulverization, and activation steps by repeated hydrogenation and dehydrogenation under vacuum. A number of methods have been tried to improve the synthesis, including reduction–diffusion processes, mechanical alloying, melting, and hydriding combustion synthesis. $LaNi_5$ samples made from self-ignition combustion synthesis combined with reduction–diffusion, using grains of calcium as heat source and reducing agent, provided 1.54 wt% hydrogen storage at 1.7 bar (Yasuda et al. 2010). This matches the commercial samples but with a more efficient process. In an attempt to find a less energy-intensive synthesis, Shahgaldi and coworkers synthesized $LaNi_5$ nanofibers by electrospinning, using PVP as the polymer and a sol–gel process (Shahgaldi et al. 2012b). Lanthanum nitrate and nickel

(II) nitrate hexahydrate precursors were mixed in 1:5 ratio and were added to a 10% PVP solution in ethanol. Spinning took place at 22 keV across a 10 cm distance. Spun PVP/LaNi$_5$ fibers were treated by heating at 450°C in air before being activated at 600°C under a hydrogen atmosphere using CaH$_2$ as reduction agent for 3 h. The sample was washed and dried. At the end of this process, x-ray peaks corresponding to the crystalline LaNi$_5$ phase were seen. The PVP was removed by temperatures of around 450°C, which was traced using FT-IR spectroscopy. Surface areas in LaNi$_5$ powders are reportedly low, up to 5.5 m^2/g. However, the surface area of the electrospun heat-treated nanofibers was checked and found to be 120 m^2/g, a large increase in materials of this type. The hydrogen storage properties of these fibers were not reported.

4.7.3 ELECTROSPINNING OF AMMONIA BORANE

The hydrogen storage properties of AB have been discussed in the introduction section. The overall hydrogen content is one of the highest known in a solid material, 19.6% by weight. The AB molecule bears a passing resemblance to ethane (C_2H_6) but because charge is offset between the hydridic N and the protic B sides of the molecule, weak dihydrogen bonds exist between neighboring molecules, and thus AB forms a waxy solid rather than a gas at ambient temperature. The low-temperature structure is shown in Figure 4.6. The dehydrogenation of AB is known in detail, with 2–2.5 of three hydrogen molecules per AB unit released at temperatures up to 150°C (Stephens et al. 2007; Wolstenholme et al. 2012), after an extended incubation time that depends on temperature. This gives a typical available hydrogen content of 13 wt%. This reaction scheme competes with other pathways, which release

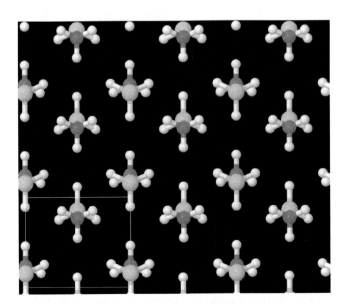

FIGURE 4.6   The structure of ammonia borane at low temperature. (Courtesy of A. Lovell.)

borazine (HNBH)$_3$, ammonia, and diborane (B$_2$H$_6$). These impurities reduce the effective hydrogen content, are toxic, and most critically degrade the fuel cells that the hydrogen stream is intended to supply. The onset of fast release of hydrogen at the AB melting temperature of 110°C also causes undesirable expansion of the solid and foaming. The combination of the dehydrogenated compounds leaves an N–B–H residue, which cannot be easily reversed back to AB. Regeneration of AB through chemical digestion and recombination routes are known, though need to be demonstrated at scale (Sutton et al. 2011).

The prospect of nano- and microstructuring of AB in order to improve the hydrogen release rate, prevent foaming, shorten or remove the incubation step, and suppress the impurities has been of great interest in developing AB into a practical hydrogen storage technology. AB combined with a nanoporous silica scaffold, SBA-15, at 1:1 weight ratio was shown to increase the rate of hydrogen release at 70°C over bulk AB and suppress the formation of borazine (Gutowska et al. 2005). The nanoconfinement and the low release temperature avoid the uncontrolled expansion of the AB-melt gas release; however, the hydrogen capacity was halved. Studies followed to show similar beneficial effects for AB supported by carbon cryogels (Feaver et al. 2007). Pore size effects were suggested to explain the lowering of hydrogen release temperature and similar results between the silica and carbon scaffolds. The AB content was only 24 wt% of the composite, so the overall hydrogen content was a quarter of the original 13 wt%. Other studies include introducing nanosized Co and Ni particles to AB (He et al. 2009); AB confined in silica hollow nanospheres (Zhang et al. 2011), in AC (Moussa et al. 2012), in MOF (Srinivas et al. 2011, 2012), and porous aromatic frameworks (Peng et al. 2012). In similar vein, nanoconfinement and scaffolding optionally combined with catalytic doping have had positive results for other hydrides including lithium borohydride (Gross et al. 2008), magnesium borohydride (Wahab et al. 2013), lithium aluminum hydride (Wahab and Beltramini 2014), and others (Nielsen et al. 2011).

Polymers present a class of low-weight and ubiquitous materials with the potential to form scaffolds or porous confinement for metal and chemical hydrides, although there are few studies published to date (Zhao et al. 2010; Jeon et al. 2011; Li et al. 2012). Electrospinning presents a way to provide a tunable single-step synthesis process for nanoscale polymer-scaffolded hydride materials. However, there are some drawbacks with the suitability of hydrides for electrospinning. They have to be combined with a compatible polymer of sufficient viscoelasticity in order to spin. They must be introduced to a solution in which they remain stable either as solute or in suspension: no easy feat for many of the more reactive hydrides, especially if spinning is to take place in air. AB composite nanofibers have been spun successfully in a number of studies. In a work that this author contributed to, a solution selection model was designed to formalize the spinning of unusual additives such as hydrogen storage materials (Kurban et al. 2010b). To the advantages of confinement of AB in a polymer fiber was added an encapsulation sheath intended to selectively emit hydrogen from the AB but delay or prevent the release of the impurities, retaining them within the fiber where possible in order to purify the gas stream and readily enable regeneration. To enable this, a coelectrospinning technique (Sun et al. 2003; Huang et al. 2006; Moghe and Gupta 2008) was used, with a nonviscoelastic AB solution in the core and a polystyrene (PS) shell polymer. PS was chosen because it is permselective to hydrogen and has a melt temperature (240°C) that would in principle keep it intact through the AB thermolysis process at 150°C. The highly polar nature of AB limited the choice of shell solutions, so to aid this, a selection model using the Hansen solubility model (Hansen 2007) was used to optimize the core–shell solution interaction. The requirement for the outer solution in core–shell electrospinning to be more viscous was met, but the AB core solution had a higher electrical conductivity than the PS shell solution even with careful selection of a PS solvent mix (Moghe and Gupta 2008). Miscible, immiscible, and semimiscible core–shell solutions were spun using a coaxial nozzle, but fully miscible core–shell solution sets did not produce fibers, instead precipitating the PS. Nor was it possible to spin PS-AB solutions through a single nozzle. The shell solution was delivered at 0.5 mL/h and the core solution flow rate varied from 0.05 to 0.5 mL/h. The immiscible solution sets used AB dissolved in water as core solution, and 20 wt% PS dissolved in a combination of toluene, dichloroethane (DCE), and pyridinium formate (PF) (solution set IS1) or DCE, nitrobenzene, and PF for the shell (IS2). The PF salt was to increase the conductivity of the solution. The semimiscible solution set sMS-1 used AB dissolved in DMSO for the core and 18–20 wt% PS dissolved in toluene/DMF for the shell. These solutions mixed but remained cloudy.

The resulting electrospun fibers, seen in Figure 4.7, showed some interesting morphological differences. Fibers from the semimiscible solution combinations (sMFs) showed high internal porosity, large (200 nm) chambers within the fiber with porous openings to the fiber surface. A hollow core was seen in the fibers with 20 wt% PS. The fibers spun from immiscible solution combinations (IFs), on the other hand, had larger hollow cores and nonporous outer walls, but had a tendency to collapse. The collapsing could be countered by adding 3 wt% of a high-molecular-weight polyethylene oxide (PEO) to the 20 wt% AB–water core solution. The location and state of the encapsulated AB was not clear from the analytical techniques used. The fibers were heated to 200°C at 1°/min to study the hydrogen release. The sMFs released hydrogen with onset at 60°C and peak at 85°C–100°C, compared with pure AB at 110°C. It was suggested that the AB was plated on the inside of interior pores in the fibers, on a length scale similar to that found in the other nanoconfinement systems. Hydrogen release from the IFs with the solid shell and little visible nanostructure was very similar to bulk AB. The mass loss was higher than expected for pure hydrogen release from both types of fiber, implying solids or impurities were also released. Although it was expected that the impurities would be reduced, it was not clear to what extent this was a function of the confinement of AB in the sMFs or the selection effect of the outer PS layer in the IFs.

The AB-PVP system has been investigated as well, using a single-phase spinning system with methanol as solvent (Tang et al. 2011). Varying concentrations of magnesium chloride ($MgCl_2$) were added to the spinning solution in order to chemically remove the resulting NH_3 on thermolysis of the AB. The same group has previously used $MgCl_2$, $ZnCl_2$, and $CaCl_2$ in AB–polyacrylamide composites (Li et al. 2012) made by mixing in aqueous solution. In the case of the electrospun AB-PVP (ES-AB-PVP), fibers of average diameter 80 nm were spun. In the case of the AB-$MgCl_2$-PVP (ES-AB-PVP+$MgCl_2$) system, the nanofibers were of the order

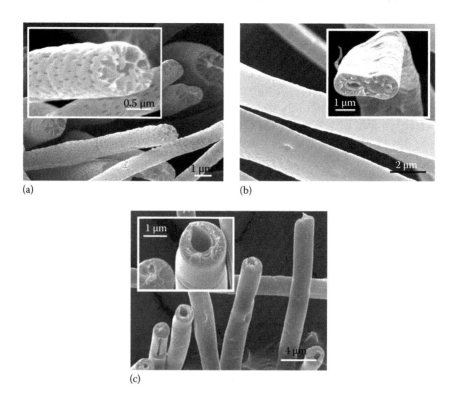

(a)

(b)

(c)

FIGURE 4.7 Electrospun polystyrene–ammonia borane (AB) core–shell fibers produced by coelectrospinning: (a) semimiscible solution fibers from solution set sMS-1 with 10 wt% AB in dimethyl sulfoxide as core solution, showing large (200 nm scale) porous structure with routes to fiber surface; (b) immiscible fibers (IFs) from solution set IS1, with 20 wt% AB in water as core solution and collapsed fibers; (c) IFs from solution set IS2, with 20% AB in water as core solution, showing noncollapsed hollow fibers with solid shells. (Reprinted with permission from Kurban, Z., Lovell, A., Bennington, S.M. et al., A solution selection model for coaxial electrospinning and its application to nanostructured hydrogen storage materials, *J. Phys. Chem. C*, 114, 21201–21213, 2010b. Copyright 2010 American Chemical Society.)

100 nm in diameter. For comparison, a ball-milled 25 wt% AB-PVP (BM-AB-PVP) sample was made. The overall AB content in the AB-PVP fibers ranged from 5 to 20 wt% but with higher AB concentrations, beads appeared in the electrospun fibers. The hydrogen and impurity release was investigated on thermolysis using mass spectrometry as shown in Figure 4.8.

The XRD data of the electrospun fibers showed an amorphous signal with no sign of the neat AB crystal phase. On thermal decomposition (Figure 4.8), the ES-AB-PVP+MgCl$_2$ showed the lowest onset of dehydrogenation, at 40°C, with peak at 92°C. The ES-AB-PVP sample also demonstrated depressed release temperatures, with the peak release at 97°C. The BM-AB-PVP sample showed some depression of the release onset but otherwise was colocated with the release of H$_2$ from neat AB, with peaks at 113°C and 140°C–150°C. Remarkably, the ES-AB-PVP+MgCl$_2$ sample

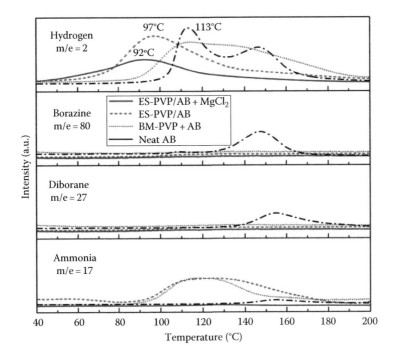

FIGURE 4.8 Mass spectrometry data of thermolysis of polyvinylpyrrolidone–ammonia borane (PVP-AB) fibers (electrospun PVP-AB; electrospun PVP-AB with added MgCl₂), ball-milled PVP-AB, and neat AB for comparison, showing release of hydrogen and three common AB impurities: borazine, diborane, and ammonia. (Tang, Z., Li, S., Yang, Z. et al., Ammonia borane nanofibers supported by poly(vinyl pyrrolidone) for dehydrogenation, *J. Mater. Chem.*, 21, 14616, 2011. Reproduced by permission of The Royal Society of Chemistry.)

showed negligible release of ammonia as well as borazine and diborane, although the ball-milled and electrospun AB-PVP samples showed a slight increase in ammonia generation over neat AB. This has been postulated as a result of the suppression of boracic by-products owing to preferential hydrogen desorption from B–H···H–B bonds leaving the NH₃ moiety to be released from the confined AB (Tang et al. 2011; Wolstenholme et al. 2012). The MgCl₂ chemically fixes the ammonia and retains it within the material. However, the MgCl₂ required for this is stoichiometric to the amount of ammonia given off, requiring a significant fraction of the AB mass and the overall low AB wt% in the fibers means that the highest system hydrogen capacity for these fibers was 2.3 wt%. Electrospun polymethyl methacrylate has also been used as support for an AB composite material, although less of a decrease in dehydrogenation temperature from that of neat AB was observed for this system (Alipour et al. 2014). Finally, high-molecular-weight PEO has been used to electrospin AB in water and acetonitrile (Nathanson 2014). Up to 75 wt% AB in AB-PEO by weight was spun, although the high conductivity and low viscosity of the solutions with most AB led to unstable spinning. Figure 4.9 shows the hydrogen release from the

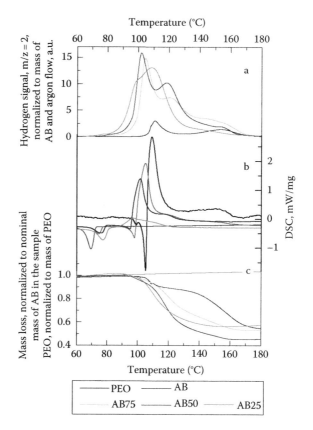

FIGURE 4.9 (a) Combined mass spectrometry hydrogen data (m/z=2), (b) differential scanning calorimetry, and (c) thermogravimetric analysis for thermolysis of neat ammonia borane (AB) and electrospun AB–polyethylene oxide fibers in three compositions by weight: 25%, 50%, and 75% AB. The MS hydrogen signal is normalized against the flow rate of Ar carrier gas and weighted by the mass of AB in the sample. (Reproduced from Nathanson, A.S., Ammonia borane composites for solid state hydrogen storage and calcium-ammonia solutions in graphite, Doctoral thesis, UCL, London, U.K., 2014. With permission.)

thermolyzed PEO-AB fibers at three AB/PEO weight ratios, together with thermogravimetric sanalysis and differential scanning calorimetry data. The temperature of hydrogen release, but also the overall hydrogen capacity in wt%, decreases with increasing PEO content. Hydrogen release occurred at lower temperature than for the neat AB, with suppressed expansion of the material; however, the large mass loss (~50%) suggests that the solids and impurities are released in similar quantities to AB. In the DSC, increasing PEO proportion shifts the PEO melting higher in temperature and decreases the size and temperature of the AB melting endotherm to sub-100°C. At 75% PEO content, the AB melting is no longer clear. As the release of hydrogen goes to lower temperatures, there is an overlap of the exothermic release with the shifted melting point. This suggests that a two-phase system akin to a solid solution of AB in PEO is providing some of the benefits associated with the nanoconfinement of AB. The largest capacity of available hydrogen is in the 75% AB fibers and equates to 9.75 wt% H_2 in the composite material.

4.8 SUMMARY AND PROSPECTS

Hydrogen storage is a critically anticipated technology to allow the rollout of fuel cell systems for low-carbon stationary and portable power, particularly road transport. The field of solid-state hydrogen storage incorporates a large variety of materials, divided loosely between physisorptive materials (with high surface areas, IM, and the need to strengthen weak H_2 binding energies) and chemisorptive systems (high hydrogen content but poor reversibility, slow kinetics and the requirement to destabilize the strong H bonds). The most successful materials are moving toward the middle ground of 20–50 kJ/mol hydrogen binding energy, with doped or catalytic physisorptive structures on one side and nanostructured or catalyzed hydrides approaching from the other.

Electrospinning has the capability to combine a nanostructuring step with scaffolding or matrix embedding of active materials in a single process. It is also an efficient way of generating high-aspect ratio materials where the diffusion distance and the hydriding phase transition energy barriers can be much lower than bulk materials. It is therefore surprising that more work has not been done in this application field to date. The exception is the large number of studies in high-SSA and doped and undoped physisorptive-type materials where the hydrogen storage capacity is typically low. However, for these latter systems there is a good crossover from the generation of nanofiber materials, particularly carbons, for other applications such as biological scaffolds, battery, and capacitor electrodes, and these are relatively easily produced using spinning techniques, from well-known precursors using carbonization and activation routes that have been studied outside of the electrospinning context for many years.

Conversely, some of the high-hydrogen-content complex hydrides are new, and others have only been studied in the hydrogen storage context for a few years. A respected review of solid-state hydrogen storage materials did not list AB as recently as 2004 (Züttel 2004), though it and its derivatives are now among the most promising materials for this application. Furthermore, many of the hydrides are difficult to process by an electrospinning route. They have to be combined with a viscoelastic material in order to spin; they have to be suspended or dissolved in a solution in order to be well dispersed in the resulting fibers, and some are sensitive to air or moisture or are even pyrophoric, making studies even at the lab scale rather difficult. They may also suffer from nonstable or variable spinning behavior. However, some of the complex hydrides, notably AB, have the useful property that they are soluble in common electrospinning solvents including water, and this has made spinning easier. An attempt to formalize the design of solutions in order to spin nonpolymeric components in a core–shell system with an outer protective and permselective layer may have implications for other applications of the technique (Kurban et al. 2010b). As for all composite hydrogen storage materials, the addition of a structural or processing material reduces the overall weight percentage of stored hydrogen, and trying to maximize the benefits of this trade for the loss in specific energy is a large part of the motivation in studies of this type.

It should be clear from the discussions that electrospinning has a significant role to play in the development of nanoscale hydrogen storage materials and particularly in the fine-tuning of parameters to reach an optimized material. This is a nascent field where few investigations have been followed up in detail so far. However, electrospinning has to compete in a crowded space against many alternative synthesis methods that can also derive nanostructured and functionalized hydrogen storage materials. The unique benefits of the technique need to be identified more strongly. One of these is the low-temperature processing (if carbonized or calcined fibers are not required) and another is the ability to create internally complex, multiphase structures via a small number of synthesis steps. On the downside, the constraints of the electrospinning process parameters can make the development work complex and slow, and variability in spinning can lead to a lack of consistency or reproducibility, which has also historically been an issue with hydrogen uptake reporting particularly in physisorptive systems (Zlotea et al. 2009). Another question comes from the ultimate scales of material required in order to supply the world's mobile energy requirements. Great progress has been made with the scaled manufacture of electrospun fiber material (Persano et al. 2013) but the low dimensionality of fibers necessarily makes their bulk generation a time-consuming process. This is particularly the case with coelectrospinning where concentric nozzle delivery is generally vital, rather than the multijet free-surface fiber generation possible with single-phase electrospinning. It is not obvious that electrospinning will be the best route to make the megatons of material required for a billion-plus car. Nevertheless, the field is relatively young and developing rapidly. For hydrogen storage in physisorptive materials, even with pore optimization and enhancement of the H_2 sorption interaction through addition of catalytic and dopant species, no material has been found yet with hydrogen capacity that looks likely to achieve the U.S. DoE targets (5.5 wt% *system* capacity and above in the −40°C to +85°C range). Chemical hydrides, particularly those with a starting hydrogen capacity >10 wt%, look more promising as stores in both volume and specific energy terms. Spinning hydrides directly can be done, although the spinning process parameters can be difficult to optimize, and there is scope for much more development of this approach, following the first reports in the literature. For example, the role of encapsulation in the case of air-sensitive hydrogen storage materials would seem to have promise. It may become possible to generate a core–shell structure spontaneously from a scaled single-phase spinning process, in fibers that combine the hydride scaffold with a protective role. The most valuable role for electrospinning may ultimately be in the generation of lightweight, low-density structures with high SSA to provide nanoconfinement and surface sites for optimization of these hydrides for fast release and, importantly, to retain the dehydrogenated material for regeneration.

REFERENCES

Aardahl, C. L. and S. D. Rassat. 2009. Overview of systems considerations for on-board chemical hydrogen storage. *Int. J. Hydrogen Energy* 34: 6676–6683.
Adelhelm, P., J. B. Gao, M. H. W. Verkuijlen et al. 2010. Comprehensive study of melt infiltration for the synthesis of $NaAlH_4$/C nanocomposites. *Chem. Mater.* 22: 2233–2238.

Ahluwalia, R. K., T. Q. Hua, and J. K. Peng. 2009. Automotive storage of hydrogen in alane. *Int. J. Hydrogen Energy* 34: 7731–7740.

Ahluwalia, R. K., T. Q. Hua, J.-K. Peng et al. 2010. Technical assessment of cryo-compressed hydrogen storage tank systems for automotive applications. *Int. J. Hydrogen Energy* 35: 4171–4184.

Ahluwalia, R. K., T. Q. Hua, and J.-K. Peng. 2012. On-board and off-board performance of hydrogen storage options for light-duty vehicles. *Int. J. Hydrogen Energy* 37: 2891–2910.

Alipour, J., A. M. Shoushtari, and A. Kaflou. 2014. Electrospun PMMA/AB nanofiber composites for hydrogen storage applications. *E-Polymers* 14: 305–311.

Arshad, S. N., M. Naraghi, and I. Chasiotis. 2011. Strong carbon nanofibers from electrospun polyacrylonitrile. *Carbon* 49: 1710–1719.

Balde, C. P., B. P. C. Hereijgers, J. H. Bitter et al. 2006. Facilitated hydrogen storage in NaAlH$_4$ supported on carbon nanofibers. *Cheminform* 37: 33025.

Beaufils, J. P., T. Crowley, T. Rayment et al. 1981. Tunnelling of hydrogen in alkali metal intercalation compounds. *Mol. Phys.* 44: 1257–1269.

Bérubé, V., G. Radtke, M. Dresselhaus et al. 2007. Size effects on the hydrogen storage properties of nanostructured metal hydrides: A review. *Int. J. Energy Res.* 31: 637–663.

Bogdanovic, B. and M. Schwickardi. 1997. Ti-doped alkali metal aluminum hydrides as potential novel reversible hydrogen storage materials. *J. Alloy Compd.* 1–9: 253–254.

Broom, D. P. 2011. *Hydrogen Storage Materials.* London, U.K.: Springer.

Brunauer, S., P. H. Emmett, and E. Teller. 1938. Adsorption of gases in multimolecular layers. *J. Am. Chem. Soc.* 60: 309–319.

Cho, S. J., K. S. Song, J. W. Kim et al. 2002. Hydrogen sorption in HCl-treated polyaniline and polypyrrole: New potential hydrogen storage media. *Fuel Chem. Div. Preprints* 47: 790–791.

Chung, G. S., S. M. Jo, and B. C. Kim. 2005a. Properties of carbon nanofibers prepared from electrospun polyimide. *J. Appl. Polym. Sci.* 97: 165–170.

Chung, H. J., S. M. Jo, D. Y. Kim et al. 2005b. Poly(vinylidene fluoride)-based porous carbon nanofibers. *Trans. Korean Hydrogen New Energy Soc.* 16: 334–342.

Deng, W.-Q., X. Xu, and W. A. Goddard. 2004. New alkali doped pillared carbon materials designed to achieve practical reversible hydrogen storage for transportation. *Phys. Rev. Lett.* 92: 166103.

Dong, Z., S. J. Kennedy, and Y. Wu. 2011. Electrospinning materials for energy-related applications and devices. *J. Power Sources* 196: 4886–4904.

Dresselhaus, M. S. and G. Dresselhaus. 1981. Intercalation compounds of graphite. *Adv. Phys.* 30: 139–326.

Edge, J. S., N. T. Skipper, F. Fernandez-Alonso et al. 2014. Structure and dynamics of molecular hydrogen in the interlayer pores of a swelling 2:1 clay by neutron scattering. *J. Phys. Chem. C* 118: 25740–25747.

Enoki, T., K. Shindo, and N. Sakamoto. 1993. Electronic properties of alkali-metal-hydrogen-graphite intercalation compounds. *Z. Phys. Chem.* 181: 75–82.

Feaver, A., S. Sepehri, P. Shamberger et al. 2007. Coherent carbon cryogel-ammonia borane nanocomposites for H$_2$ storage. *J. Phys. Chem. B* 111: 7469–7472.

Graetz, J., J. Reilly, G. Sandrock et al. 2006. Aluminum hydride, AlH$_3$, as a hydrogen storage compound. Formal Report BNL-77336-2006. Brookhaven National Laboratory, Upton, NY.

Greiner, A. and J. H. Wendorff. 2007. Electrospinning: A fascinating method for the preparation of ultrathin fibers. *Angew. Chem. Int. Ed.* 46: 5670–5703.

Gross, A. F., J. J. Vajo, S. L. Van Atta et al. 2008. Enhanced hydrogen storage kinetics of LiBH$_4$ in nanoporous carbon scaffolds. *J. Phys. Chem. C* 112: 5651–5657.

Gutowska, A., L. Li, Y. Shin et al. 2005. Nanoscaffold mediates hydrogen release and the reactivity of ammonia borane. *Angew. Chem.* 117: 3644–3648.

Hansen, C. M. 2007. *Hansen Solubility Parameters*, 2nd edn. Boca Raton, FL: CRC Press.

He, T., Z. Xiong, G. Wu et al. 2009. Nanosized Co- and Ni-catalyzed ammonia borane for hydrogen storage. *Chem. Mater.* 21: 2315–2318.

Herley, P. J., W. Jones, and B. Vigeholm. 1985. Characterization of the whiskerlike products formed by hydriding magnesium metal powders. *J. Appl. Phys.* 58: 292–296.

Hong, S. E., D.-K. Kim, S. M. Jo et al. 2007. Graphite nanofibers prepared from catalytic graphitization of electrospun poly(vinylidene fluoride) nanofibers and their hydrogen storage capacity. *Catal. Today* 120: 413–419.

Horvath, G. and K. Kawazoe. 1983. Method for calculation of effective pore size distribution in molecular sieve carbon. *J. Chem. Eng. Jpn.* 16: 470.

Hua, T. Q., R. K. Ahluwalia, J.-K. Peng et al. 2011. Technical assessment of compressed hydrogen storage tank systems for automotive applications. *Int. J. Hydrogen Energy* 36: 3037–3049.

Huang, J., S. Virji, B. H. Weiller et al. 2003. Polyaniline nanofibers: Facile synthesis and chemical sensors. *J. Am. Chem. Soc.* 125: 314–315.

Huang, Z.-M., C.-L. He, A. Yang et al. 2006. Encapsulating drugs in biodegradable ultrafine fibers through co-axial electrospinning. *J. Biomed. Mater. Res. A* 77A: 169–179.

Hudson, M. S. L., H. Raghubanshi, D. Pukazhselvan et al. 2012. Carbon nanostructures as catalyst for improving the hydrogen storage behavior of sodium aluminum hydride. *Int. J. Hydrogen Energy* 37: 2750–2755.

Im, J. S., O. Kwon, Y. H. Kim et al. 2008b. The effect of embedded vanadium catalyst on activated electrospun CFs for hydrogen storage. *Micropor. Mesopor. Mater.* 115: 514–521.

Im, J. S., S.-J. Park, T. J. Kim et al. 2008a. The study of controlling pore size on electrospun carbon nanofibers for hydrogen adsorption. *J. Colloid Interface Sci.* 318: 42–49.

Im, J. S., S.-J. Park, T. Kim et al. 2009a. Hydrogen storage evaluation based on investigations of the catalytic properties of metal/metal oxides in electrospun carbon fibers. *Int. J. Hydrogen Energy* 34: 3382–3388.

Im, J. S., S.-J. Park, and Y.-S. Lee. 2007. Preparation and characteristics of electrospun activated carbon materials having meso- and macropores. *J. Colloid Interface Sci.* 314: 32–37.

Im, J. S., S.-J. Park, and Y.-S. Lee. 2009b. The metal–carbon–fluorine system for improving hydrogen storage by using metal and fluorine with different levels of electronegativity. *Int. J. Hydrogen Energy* 34: 1423–1428.

Im, J. S., S.-J. Park, and Y.-S. Lee. 2009c. Superior prospect of chemically activated electrospun carbon fibers for hydrogen storage. *Mater. Res. Bull.* 44: 1871–1878.

Jacobson, M. Z. 2009. Review of solutions to global warming, air pollution, and energy security. *Energy Environ. Sci.* 2: 148.

Jeon, K.-J., H. R. Moon, A. M. Ruminski et al. 2011. Air-stable magnesium nanocomposites provide rapid and high-capacity hydrogen storage without using heavy-metal catalysts. *Nat. Mater.* 10(4): 286–290.

Jiang, H.-L., B. Liu, Y.-Q. Lan et al. 2011. From metal–organic framework to nanoporous carbon: Toward a very high surface area and hydrogen uptake. *J. Am. Chem. Soc.* 133: 11854–11857.

Jurczyk, M., A. Kumar, S. Srinivasan et al. 2007. Polyaniline-based nanocomposite materials for hydrogen storage. *Int. J. Hydrogen Energy* 32: 1010–1015.

Kaburagi, Y., Y. Hishiyama, H. Oka et al. 2001. Growth of iron clusters and change of magnetic property with carbonization of aromatic polyimide film containing iron complex. *Carbon* 39: 593–603.

Kim, B. H., W. G. Hong, S. M. Lee et al. 2010. Enhancement of hydrogen storage capacity in polyaniline-vanadium pentoxide nanocomposites. *Int. J. Hydrogen Energy* 35: 1300–1304.

Kim, D.-K., S. H. Park, B. C. Kim et al. 2005. Electrospun polyacrylonitrile-based carbon nanofibers and their hydrogen storages. *Macromol. Res.* 13: 521–528.

Kim, H., D. Lee, and J. Moon. 2011. Co-electrospun pd-coated porous carbon nanofibers for hydrogen storage applications. *Int. J. Hydrogen Energy* 36: 3566–3573.

Klebanoff, L. E. and J. O. Keller. 2013. 5 Years of hydrogen storage research in the U.S. DOE Metal Hydride Center of Excellence (MHCoE). *Int. J. Hydrogen Energy* 38: 4533–4576.

Kurban, Z. 2011. Electrospun nanostructured composite fibres for hydrogen storage applications. Doctoral thesis, UCL, London, U.K., http://discovery.ucl.ac.uk/1333231/ (accessed on April 29, 2015.)

Kurban, Z., A. Lovell, S. M. Bennington et al. 2010b. A solution selection model for coaxial electrospinning and its application to nanostructured hydrogen storage materials. *J. Phys. Chem. C* 114: 21201–21213.

Kurban, Z., A. Lovell, D. Jenkins et al. 2010a. Turbostratic graphite nanofibres from electrospun solutions of PAN in dimethylsulphoxide. *Eur. Polym. J.* 46: 1194–1202.

Lachawiec, A. J., G. Qi, and R. T. Yang. 2005. Hydrogen storage in nanostructured carbons by spillover: Bridge-building enhancement. *Langmuir* 21: 11418–11424.

Laudenslager, M. J., R. H. Scheffler, and W. M. Sigmund. 2010. Electrospun materials for energy harvesting, conversion, and storage: A review. *Pure Appl. Chem.* 82: 2137–2156.

Li, D. and Y. Xia. 2004. Electrospinning of nanofibers: Reinventing the wheel? *Adv. Mater.* 16: 1151–1169.

Li, S. F., Z. W. Tang, Y. B. Tan et al. 2012. Polyacrylamide blending with ammonia borane: A polymer supported hydrogen storage composite. *J. Phys. Chem. C* 116: 1544–1549.

Lovell, A., F. Fernandez-Alonso, N. T. Skipper et al. 2008. Quantum delocalization of molecular hydrogen in alkali-graphite intercalates. *Phys. Rev. Lett.* 101: 126101.

Lueking, A. and R. T. Yang. 2002. Hydrogen spillover from a metal oxide catalyst onto carbon nanotubes—Implications for hydrogen storage. *J. Catal.* 206: 165–168.

Melaina, M. W., O. Antonia, and M. Penev. 2013. Blending hydrogen into natural gas pipeline. NREL technical report NREL/TP-5600-51995. NREL, Golden, CO.

Moghe, A. K. and B. S. Gupta. 2008. Co-axial electrospinning for nanofiber structures: Preparation and applications. *Polym. Rev.* 48: 353–377.

Moussa, G., S. Bernard, U. B. Demirci et al. 2012. Room-temperature hydrogen release from activated carbon-confined ammonia borane. *Int. J. Hydrogen Energy* 37: 13437–13445.

Nakamori, Y., H. W. Li, M. Matsuo et al. 2008. Development of metal borohydrides for hydrogen storage. *Phys. Chem. Solids* 69: 2292.

Nathanson, A. S. 2014. Ammonia borane composites for solid state hydrogen storage and calcium-ammonia solutions in graphite. Doctoral thesis, UCL, London, U.K.

Nielsen, T. K., F. Besenbacher, and T. R. Jensen. 2011. Nanoconfined hydrides for energy storage. *Nanoscale* 3: 2086.

Panella, B., L. Kossykh, U. Dettlaff-Weglikowska et al. 2005. Volumetric measurement of hydrogen storage in HCl-treated polyaniline and polypyrrole. *Synth. Metals* 151: 208–210.

Park, S. H., S. M. Jo, D. Y. Kim et al. 2005. Effects of iron catalyst on the formation of crystalline domain during carbonization of electrospun acrylic nanofiber. *Synth. Metals* 150: 265–270.

Patchkovskii, S., J. S. Tse, S. N. Yurchenko et al. 2005. Graphene nanostructures as tunable storage media for molecular hydrogen. *Proc. Natl. Acad. Sci. U.S.A.* 102: 10439–10444.

Peng, Y., T. Ben, Y. Jia et al. 2012. Dehydrogenation of ammonia borane confined by low-density porous aromatic framework. *J. Phys. Chem. C* 116: 25694–25700.

Persano, L., A. Camposeo, C. Tekmen et al. 2013. Industrial upscaling of electrospinning and applications of polymer nanofibers: A review. *Macromol. Mater. Eng.* 298: 504–520.

Reardon, H., J. M. Hanlon, R. W. Hughes et al. 2012. Emerging concepts in solid-state hydrogen storage: The role of nanomaterials design. *Energy Environ. Sci.* 5: 5951.

Rose, M., E. Kockrick, I. Senkovska et al. 2010. High surface area carbide-derived carbon fibers produced by electrospinning of polycarbosilane precursors. *Carbon* 48: 403–407.

Rude, L. H., T. K. Nielsen, D. B. Ravnsbæk et al. 2011. Tailoring properties of borohydrides for hydrogen storage: A review. *Phys Status Solidi A* 208: 1754–1773.

Satyapal, S., J. Petrovic, C. Read et al. 2007. The U.S. department of energy's national hydrogen storage project: Progress towards meeting hydrogen-powered vehicle requirements. *Catal. Today* 120: 246–256.

Schlapbach, L. and A. Züttel. 2001. Hydrogen-storage materials for mobile applications. *Nature* 414: 353–358.

Shahgaldi, S., Z. Yaakob, N. M. Jalil et al. 2012b. Synthesis of high-surface-area hexagonal LaNi$_5$ nanofibers via electrospinning. *J. Alloy Compd.* 541: 335–337.

Shahgaldi, S., Z. Yaakob, D. J. Khadem et al. 2012a. Characterization and the hydrogen storage capacity of titania-coated electrospun boron nitride nanofibers. *Int. J. Hydrogen Energy* 37: 11237–11243.

Sirosh, N. 2002. Hydrogen composite tank program. *Proceedings of the 2002 US DoE Hydrogen Program Review* NREL/CP-610-32405. https://www1.eere.energy.gov/hydrogenandfuelcells/pdfs/32405b27.pdf (accessed on April 29, 2015.)

Skowroński, J. M. and J. Urbaniak. 2008. Nickel foam/polyaniline-based carbon/palladium composite electrodes for hydrogen storage. *Energy Convers. Manage.* 49: 2455–2460.

Solin, S. A. and H. Zabel. 1988. The physics of ternary graphite intercalation compounds. *Adv. Phys.* 37: 87–254.

Srinivas, G., J. Ford, W. Zhou et al. 2011. Nanoconfinement and catalytic dehydrogenation of ammonia borane by magnesium-metal-organic-framework-74. *Chemistry* 17: 6043–6047.

Srinivas, G., J. Ford, W. Zhou et al. 2012. Zn-MOF assisted dehydrogenation of ammonia borane: Enhanced kinetics and clean hydrogen generation. *Int. J. Hydrogen Energy* 37: 3633–3638.

Srinivas, G., A. Lovell, C. A. Howard et al. 2010. Structure and phase stability of hydrogenated first-stage alkali- and alkaline-earth metal–graphite intercalation compounds. *Synth. Metals* 160: 1631–1635.

Srinivasan, S. S., R. Ratnadurai, M. U. Niemann et al. 2010. Reversible hydrogen storage in electrospun polyaniline fibers. *Int. J. Hydrogen Energy* 35: 225–230.

Stephens, F. H., V. Pons, and R. T. Baker. 2007. Ammonia borane: The hydrogen source par excellence? *Dalton Trans.* 25: 2613.

Stephens, R. D., A. F. Gross, S. L. Van Atta et al. 2009. The kinetic enhancement of hydrogen cycling in NaAlH(4) by melt infusion into nanoporous carbon aerogel. *Nanotechnology* 20: 204018.

Sun, Z., E. Zussman, A. L. Yarin et al. 2003. Compound core–shell polymer nanofibers by co-electrospinning. *Adv. Mater.* 15: 1929–1932.

Sundaramurthy, J., N. Li, P. S. Kumar et al. 2014. Perspective of electrospun nanofibers in energy and environment. *Biofuel Res. J.* 2: 44–54.

Sutton, A. D., A. K. Burrell, D. A. Dixon et al. 2011. Regeneration of ammonia borane spent fuel by direct reaction with hydrazine and liquid ammonia. *Science* 331: 1426–1429.

Tang, Z., S. Li, Z. Yang et al. 2011. Ammonia borane nanofibers supported by poly(vinyl pyrrolidone) for dehydrogenation. *J. Mater. Chem.* 21: 14616.

Thavasi, V., G. Singh, and S. Ramakrishna. 2008. Electrospun nanofibers in energy and environmental applications. *Energy Environ. Sci.* 1: 205–221.

Wahab, M. A. and J. N. Beltramini. 2014. Catalytic nanoconfinement effect of in-situ synthesized Ni-containing mesoporous carbon scaffold (Ni-MCS) on the hydrogen storage properties of LiAlH$_4$. *Int. J. Hydrogen Energy* 39: 18280–18290.

Wahab, M. A., Y. Jia, D. Yang et al. 2013. Enhanced hydrogen desorption from Mg(BH$_4$)$_2$ by combining nanoconfinement and a Ni catalyst. *J. Mater. Chem. A* 1: 3471.

Wang, Y. and X. Jing. 2008. Formation of polyaniline nanofibers: A morphological study. *J. Phys. Chem. B* 112: 1157–1162.

Watanabe, K., T. Kondow, M. Soma et al. 1973. Molecular-sieve type sorption on alkali graphite intercalation compounds. *Proc. R. Soc. Lond. A* 333: 51–67.

Wolstenholme, D. J., K. T. Traboulsee, Y. Hua et al. 2012. Thermal desorption of hydrogen from ammonia borane: Unexpected role of homopolar B–H···H–B interactions. *Chem. Commun.* 48: 2597.

Xiong, Z., C. K. Yong, G. Wu et al. 2008. High-capacity hydrogen storage in lithium and sodium amidoboranes. *Nat. Mater.* 7: 138–141.

Xu, W., K. Takahasi, Y. Matsuo et al. 2007. Investigation of hydrogen storage capacity of various carbon materials. *Int. J. Hydrogen Energy* 32: 2504–2512.

Yaakob, Z., D. J. Khadem, S, Shahgaldi et al. 2012. The role of Al and Mg in the hydrogen storage of electrospun ZnO nanofibers. *Int. J. Hydrogen Energy* 37: 8388–8394.

Yang, J., A. Sudik, C. Wolverton et al. 2010. High capacity hydrogen storage materials: Attributes for automotive applications and techniques for materials discovery. *Chem. Soc. Rev.* 39: 656–675.

Yang, R. T. and Y. Wang. 2009. Catalyzed hydrogen spillover for hydrogen storage. *J. Am. Chem. Soc.* 131: 4224–4226.

Yao, J., C. Bastiaansen, and T. Peijs. 2014. High strength and high modulus electrospun nanofibers. *Fibers* 2(2): 158–186.

Yasuda, N., S. Sasaki, N. Okinaka et al. 2010. Self-ignition combustion synthesis of LaNi$_5$ utilizing hydrogenation heat of metallic calcium. *Int. J. Hydrogen Energy* 35: 11035–11041.

Yildirim, T. and S. Ciraci. 2005. Titanium-decorated carbon nanotubes as a potential high-capacity hydrogen storage medium. *Phys. Rev. Lett.* 94: 175501.

Zhang, T., X. Yang, S. Yang et al. 2011. Silica hollow nanospheres as new nanoscaffold materials to enhance hydrogen releasing from ammonia borane. *Phys. Chem. Chem. Phys.* 13: 18592.

Zhao, J., J. Shi, X. Zhang et al. 2010. A soft hydrogen storage material: Poly(methyl acrylate)-confined ammonia borane with controllable dehydrogenation. *Adv. Mater.* 22: 394–397.

Zhou, Z., C. Lai, L. Zhang et al. 2009. Development of carbon nanofibers from aligned electrospun polyacrylonitrile nanofiber bundles and characterization of their microstructural, electrical, and mechanical properties. *Polymer* 50: 2999–3006.

Zlotea, C., P. Moretto, and T. Steriotis. 2009. A Round Robin characterisation of the hydrogen sorption properties of a carbon based material. *Int. J. Hydrogen Energy* 34: 3044–3057.

Zussman, E., X. Chen, W. Ding et al. 2005. Mechanical and structural characterization of electrospun PAN-derived carbon nanofibers. *Carbon* 43: 2175–2185.

Züttel, A. 2004. Hydrogen storage methods. *Naturwissenschaften* 91: 157–172.

5 Electrospun Nanofibers for Supercapacitors

Laura Coustan and Frédéric Favier

CONTENTS

5.1 INTRODUCTION

One of the most convenient ways to briefly introduce supercapacitors is certainly to compare their performances toward those of batteries. The obvious discrepancies between these two electrochemical storage devices are firstly evidenced in the Ragone plot depicted in Figure 5.1. Supercapacitor energy densities roughly lie in the range from 0.05 to 10 W h/kg, while their power density is from 1 to 10^5 W/kg.[1] Time constants, which are related to charge/discharge rate, are typically from 0.01 s to 30 min. In terms of energy and power densities and time constants, supercapacitors tend to fill the gap in between conventional electrolytic capacitors and batteries. They do not aim at replacing any of these two types of devices. They should be considered as dedicated power sources for specific uses including electronic tools, automotive start/stop systems, and brake energy harvesting or as a part of an advanced storage system, eventually including other storage and/or conversion devices for renewable energy harvesting and grid regulation, for example.

Therefore, supercapacitors are particularly adapted for applications requiring energy pulses for short periods of time. The energy density (E) of supercapacitors is directly proportional to the voltage (V) and the capacitance (C) (Formula 5.1), and it is then possible to increase it by either increasing the voltage of the device, the capacitance of the electrode materials, or both:

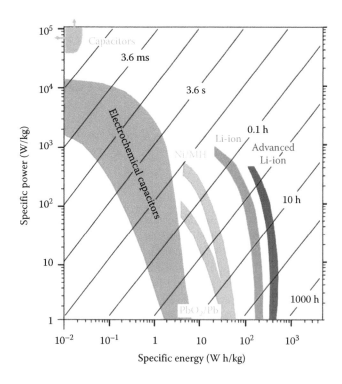

FIGURE 5.1 Ragone plot (specific power against specific energy) for electrochemical storage devices. (Reprinted by permission from Macmillan Publishers Ltd. *Nat. Mater.*, Simon, P. and Gogotsi, Y., Materials for electrochemical capacitors, 7, 845–854, Copyright 2008.)

$$E = \frac{1}{2}CV^2 \qquad (5.1)$$

$$P_{max} = \frac{V^2}{4R} \qquad (5.2)$$

The maximum power (P_{max}) that can be delivered by a supercapacitor is given by Formula 5.2. Again, increasing the operation voltage should contribute to the increase of the power and so should be any design and fabrication options aiming at decreasing the total resistance (R) of the device. Nowadays, R&D activities in the field of supercapacitors are mainly targeted on various strategies for the development of high-capacitance electrode materials with original morphologies and compositions and innovative electrolytes with large potential window, eventually specifically designed for a given electrode material. On the electrode formulation, the use of nano- /microfiber-based material offers attractive perspectives in terms of device energy and power density improvement. The main motivation lies then on an enhancement of the developed surface area and electronic percolation capability of the electrode.

On the other hand, among the various applications for supercapacitors, there are currently many research works on the development of flexible supercapacitors for applications in flexible and wearable technology.[2–6] Wearable electronic is arising as a major technological advance, and the development of wearable energy sources is mandatory to power the foreseen microdevices.[7–9] Beside the basic power and energy requirements, these storage or conversion devices have to fit some extra and specific needs including flexibility, safety, small size, lightweight, low cost, material recyclability, and should eventually be washable, ironable, allergen-free, and bio-compatible. As such, the design of flexible supercapacitors is a great technological breakthrough for applications in wearable technology. As the main characteristic of these devices, flexibility is probably the hardest to achieve.[10] Not only should the individual parts of the device, including current collectors, electrode materials, and the electrolyte, be flexible but should the various interfaces in the assembled device be electrochemically efficient and mechanically robust or adaptive for actual use. A number of recent studies, initiatives, and products have been reported and proposed for the development of flexible energy devices based on various materials assembled in different ways in various designs. Numerous approaches are actually available to build high-performance flexible supercapacitors (Figure 5.2).[6,11,12]

Considering the production chain from yarns to fabrics and from fabrics to clothes, any methods for the fabrication of fibers, either of electroactive materials or electrolytes, can be envisioned.[1,6,13] As such, electrospinning, allowing the elaboration of ultrathin fibers with outstanding properties including large surface area, flexible and stretchable surface functionalities, and good mechanical behavior, has been recently considered as a promising method for the design and fabrication of materials for flexible energy storage devices, including supercapacitors.[14–16] Indeed, fibers can be mixed in fabrics of active materials to assemble in a completely flexible and lightweight supercapacitor. Furthermore, it is a technique offering several opportunities as regards to the shape and composition of the produced nanofibers with targeted architectures such as core–shell materials or with hierarchical porosities.[14] The possibility of playing with the morphology of the electrospun materials, with their composition as well as their multiscale-assembly opportunities, makes it a versatile synthetic approach, which has increased the variety of materials that can be prepared and geometries that can be achieved for integration in supercapacitors. Indeed, the following sections aim at demonstrating that electrospun fibers can be used in supercapacitors as performing electrode materials and current collect or substrate materials. Despite some potential opportunities, the investigation of electrospun fiber-based electrolytes remains fairly limited.[17,18]

5.2 ELECTROSPUN FIBERS AS ELECTRODE MATERIALS

Electrochemical supercapacitors store charges through two different mechanisms. In electrochemical double-layer capacitors (EDLCs), the charge storage mechanism is purely electrostatic. It involves the electrosorption of electrolytic species at the solid/liquid interface at the surface of the electrode material (Figure 5.3).[19,20] As shown in Formula 5.3, in such EDLC electrode, the capacitance is proportional to the surface. Electrode material showing large surface area will be obviously preferred.

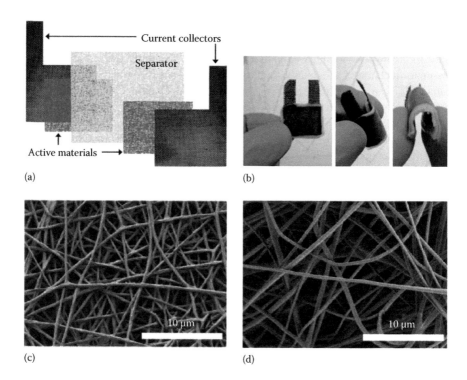

FIGURE 5.2 Assembly of a textile supercapacitor. (a) Schematic representation of the stacked textile layers; (b) photographs of a 3 cm² all-textile flexible supercapacitor; (c) scanning electron microscopy (SEM) image of the active material (PEDOT nanofibers); (d) SEM image of the separator (polyacrylonitrile nanofibers). (Reprinted from *J. Power Sources*, 196, Laforgue, A., All-textile flexible supercapacitors using electrospun poly(3,4-ethylenedioxy-thiophene) nanofibers, 559–564, Copyright 2011, with permission from Elsevier.)

EDLC devices show high-power capabilities, thanks to the fast kinetics of the electrostatic mechanism but an energy density limited by the absence of a charge transfer:

$$C = \frac{\varepsilon_0 \varepsilon S}{d} \tag{5.3}$$

where
  ε_0 is the vacuum permittivity
  ε is the effective dielectric constant of the double layer
  S is the surface in contact with the electrolyte
  d is the double-layer thickness

In contrast, in pseudocapacitors, charge storage involves fast faradic reactions, either at the surface or in the bulk of electroactive oxides, nitrides, and polymers.[21,22] The resulting energy density is greater than with carbon-based electrode

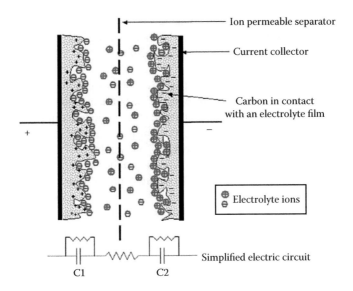

FIGURE 5.3 Scheme of an electrochemical double-layer capacitor. Under polarization, ionic charges absorb at the surface of the carbon electrode of opposite sign. Ion-permeable separator, usually made of silica fibers, prevents electrode shortcuts. Electrode materials are laminated as thick composite films at the surface of aluminum or stainless steel current collectors. (Reprinted from *J. Power Sources*, 157(1), Pandolfo, A.G. and Hollenkamp, A.F., Carbon properties and their role in supercapacitors, 11–27, Copyright 2006, with permission from Elsevier.)

materials, but the low kinetics of involved redox reactions is detrimental to the developed power.

Various combinations of these two electrode material types have been integrated in asymmetric (one electrode is EDLC, the other one, pseudocapacitive) or hybrid (one electrode is EDLC, the other one, battery type) devices.

Electrospinning is used for the development of fibrous electrode nanomaterials for supercapacitor electrodes, either for EDLCs or pseudocapacitors, with the objective to take benefit of this specific morphology either for performance improvement and shaping opportunities.

5.2.1 Electrospun Carbon Nanofibers as Electrode Materials

In the field of supercapacitor, carbon is the most investigated electrode material,[20] either for EDLCs, asymmetric, or hybrid devices. As an electrode material, carbon attractiveness comes from its good electronic conductivity, the plurality of processes to get it as high surface area material (activated carbon, graphene, carbon onions, etc.), its good (electro-) chemical stability in aqueous or organic electrolytes, low cost as raw material, and ease of scalability for the industrial fabrication of electrode materials.[23]

The main performance markers are typically ~100 F/g of electrode material, 2–5 W h/kg and 5–18 kW/kg energy and power densities, respectively, for a complete device. As commercial devices with capacitance from a few millifarads to several kilofarads, they are usually meant to be operated in organic electrolyte (~2.7 V potential window) over hundred thousand or millions of 0.5–5 s charge/discharge cycles and a 20-year calendar life.

As a matter of fact, more than 70% of papers on electrospun fibers for supercapacitors deal with carbon-based materials. Most of carbon nanofibers (CNFs) prepared by electrospinning are from polyacrylonitrile (PAN) solutions in dimethylformamide (DMF). The PAN concentration is an important parameter in this technique, and the polymer is generally dissolved at about 10% in weight.[14,24,25] Resulting electrospun polymer fibers are then carbonized by heating under a neutral atmosphere and, finally, activated by heating under an oxidative atmosphere. These two postprocesses can be done sequentially or at the same time.[26] The activation of CNFs allows the development of the material micro- /mesoporosity leading to a drastic increase of the surface area and consequently of the capacitance of the material. Zhang reported in a recent review, on the preparation and applications of CNFs prepared by electrospinning, carbonization, and activation, and demonstrated the critical role of PAN as precursor.[25] Additionally, he pointed out the effect of Si-containing compounds such as phenylsilane, tetraethyl orthosilicate, and polymethylhydrosiloxane mixed with the PAN precursor on the characteristics of the prepared electrospun nanofibers. During the carbonization step, because of the presence of these Si-based species, micropores are generated at the outer surface of the resulting CNFs for a remarkable development of specific surface area up to 1200 m^2/g.

Kim was the first to report on the electrochemical properties of a CNF-based web prepared by electrospinning of a PAN solution and used as a supercapacitor electrode (Figure 5.4a and b).[26] An increase of the temperature of activation from 700°C to 800°C led to a decrease of the micropore/mesopore ratio and, consequently, of the specific surface area. As a main characteristic of EDLCs, these decreases induced a proportional decrease of the capacitance of the electrode (Figure 5.4c) from 143 to 123 and 75 F/g (measured at 1 mA/cm^2) after activation at 900°C, 800°C, and 700°C, respectively. Critical characteristics such as morphology, diameter, porosity, and crystallinity of the resulting fibers also depend on the nature of the gas used during the heat treatment process, usually nitrogen or argon.

PAN is not the only precursor available for the preparation of CNFs by electrospinning. Indeed, alternative polymeric solutions have been used to prepare electrospun carbon fibers. These polymeric solutions were electrospun the same way as PAN solutions, to produce pure or composite nanofibers. For example, numerous works were done on polybenzimidazole solubilized in dimethylacetamide,[27] polyvinyl alcohol in water,[28] and polyamic acid in tetrahydrofuran/methanol (THF/MeOH).[29] In the latter case, the imidization of the electrospun fibers followed by a carbonization step at 1000°C under nitrogen atmosphere led to a net of carbonized polyimide nanofibers. Pitch providing from the condensation of pyrolyzed fuel oil with Cl_2 was also used as precursor when solubilized in THF.[30]

A special attention should be paid to the variety of porous characteristics of the electrospun CNFs depending on the preparation parameters, mainly the polymer nature and activation conditions. As mentioned above, the porosity of the fibers is a critical parameter on the electrochemical properties.[31–34] PAN-based CNFs are

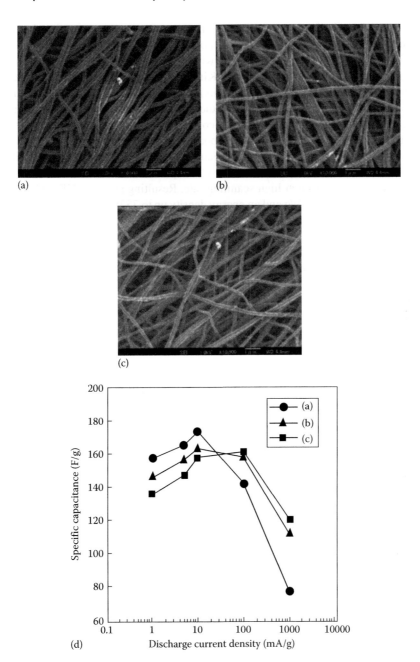

FIGURE 5.4 Scanning electron microscopy images of polyacrylonitrile-based activated carbon nanofiber web as a function of activated temperatures: (a) 700°C, (b) 750°C, and (c) 800°C; (d) corresponding dependence of specific capacitances on the discharge current density. (Reprinted with permission Kim, C. and Yang, K.S., Electrochemical properties of carbon nanofiber web as an electrode for supercapacitor prepared by electrospinning, *Appl. Phys. Lett.*, 83(6), 1216, Copyright 2003, American Institute of Physics.)

characterized by shallower pores with larger average diameter compared to the fibers produced from alternative polymers or mixtures or commercially available.[26] The addition of another polymer as a mixture with PAN influences the pore size distribution in the resulting CNFs. For example, Kim investigated the impact on the electrospinning of PAN nanofibers of the addition of pitch in THF to the DMF precursor solution.[30] With the objective to narrow the pore size distribution, the same author succeeded in the preparation of a skin–core structure by generating ultrapores at the surface of the prepared CNFs. The pore size was found to directly depend on the concentration of PAN/pitch in the precursor solution. Refinement of this critical concentration has allowed to increase the CNF specific surface area for a more efficient charge storage even at high scanning rate. Resulting porous CNFs exhibited a capacitance of about 150 F/g and an energy density up to 75 W h/kg toward the values of 60 F/g and 6 W h/kg, for capacitance and energy density, respectively, measured for raw PAN CNF.[19,25,26]

Using a different approach to obtain porous and flexible CNFs, Gogotsi et al. started from TiC nanofelts.[5,19] These were obtained by carboreduction at 1400°C of TiO_2/C composite fibers, electrospun from a mixture of titanium (IV) butoxide (TiBO, Ti source), furfuryl alcohol (FA, carbon source), polyvinylpyrrolidone (PVP, carrying polymer) in DMF, and acetic acid (for TiBO hydrolysis and FA polymerization). After dry chlorination of the resulting TiC nanofelts at temperatures ranging from 200°C to 1000°C, authors obtained porous carbide-derived carbons nanofibers with a specific capacitance of 135 F/g (Figure 5.5) and attractive

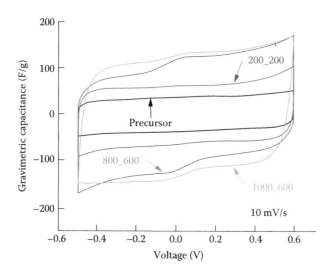

FIGURE 5.5 Cyclic voltammograms of the TiC-CDC nanofelts at 10 mV/s in 1 M H_2SO_4. x–y labels correspond to chlorination (x) and annealing (y) temperatures. (Reprinted from *J. Power Sources*, 201, Gao, Y., Presser, V., Zhang, L., Niu, J. J., McDonough, J., Perez, C., Lin, H., Fong, H., and Gogotsi, Y., High power supercapacitor electrodes based on flexible TiC-CDC nano-felts, 368–375, Copyright 2012, with permission from Elsevier.)

power capabilities since only 10% of the capacitance was lost by increasing the scan rate from 10 to 100 mV/s.

5.2.2 ELECTROSPUN METAL OXIDE NANOFIBERS AS ELECTRODE MATERIALS

Pseudocapacitive metal oxide materials have attractive characteristics with capacitance ranging from 100 to 700 F/g of electrode material. Thanks to their faradic behavior, energy densities are greater than 5 W h/kg. Because of higher densities than carbon-based materials, their volumetric energy density is usually greater than those of EDLCs. In contrast, their power density, typically of 10–100 kW/kg, is limited by the kinetics of the redox reactions involved in the charge storage.

Metal oxides have also been investigated as electrospun fibers for the fabrication of pseudocapacitive electrode materials. As for their carbon counterparts, electrochemical performances are expected to take advantages of the fiber-based net morphology of the prepared electrode on the ion diffusion and transport, on the electronic charge percolation, and on the enhanced surface of electroactive material in contact with the electrolyte. In the case of pseudocapacitive oxides, the micro- /mesoporosity into the fiber is neither mandatory nor critical. Because of their attractive electrochemical performances, the most studied metal oxides for supercapacitive applications are definitively MnO_2 and RuO_2. As a matter of fact, they are also focusing most of the research in the field as electrospun fibers, even if some researches deal with V_2O_5 too.[35] None of these oxides can be directly electrospun as raw material. They are usually obtained as composite fibers by electrospinning a polymer/metal oxide precursor mixture. A thermal treatment allows to decompose the sacrificial polymer to get pure oxide fibers.

5.2.2.1 Manganese Dioxide Electrospun Nanofibers

Manganese dioxide, MnO_2, is a promising material for supercapacitor electrodes. Its low cost, environmental friendliness, and good faradic behavior (thanks to fast redox reactions) while being operated in aqueous-based electrolytes (in contrast with the toxic, dangerous, and/or costly organic electrolytes used with EDLCs) participate in its attractiveness. In a typical synthesis of MnO_2 fibers, a PVP solution containing manganese acetate was electrospun to produce composite nanofibers (Figure 5.6).[36,37] These were calcined at a temperature of 480°C in argon atmosphere. In a 1 M Na_2SO_4 water-based electrolyte, the specific capacitance of the resulting MnO_2 fibers at 340 F/g at a current density of 0.5 mA/g is almost twice that of casted thin film of MnO_2 particles at the same current density. These works also demonstrated electrospun MnO_2 nanofibers to exhibit a higher power density than MnO_2 nanoparticles from conventional synthesis, since 95% of the initial capacitance was retained at high current density (10 mA/g). These attractive performances are mainly assigned to the electrode fabrication method as electrospinning allowed to obtain a net of nanofibers down to 10 nm in diameter.

This architecture enhances the ion diffusion and electronic percolation through the electrode material. Moreover, the polymer decomposition induced a drastic

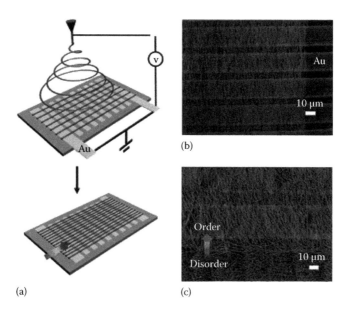

(a)   (b)   (c)

FIGURE 5.6 (a) Schematic illustration of electrospinning setup for producing aligned nanofibers deposited onto the microgold-electrode collector. (b) Scanning electron microscopy (SEM) image of aligned nanofibers on the microgold-electrode-based collector surface. (c) SEM image showing the orientation of fibers deposited at the edges of gold electrodes. (Reprinted from Xue, M., Xie, Z., Zhang, L., Ma, X., Wu, X., Guo, Y., Song, W., Li, Z., and Cao, T., Microfluidic etching for fabrication of flexible and all-solid-state micro supercapacitor based on MnO₂ nanoparticles, *Nanoscale*, 3(7), 2703–2708, Copyright 2011. With permission of The Royal Society of Chemistry.)

increase of the developed specific surface area and porous volume. These increases are strongly related to the calcination temperatures.

5.2.2.2   Ruthenium Dioxide Electrospun Nanofibers

Ruthenium dioxide (RuO_2) presents, as bulk material, a remarkably high specific capacitance (up to 700 F/g, either in H_2SO_4 or KOH 1 M electrolytes), in particular due to its excellent proton and hydroxide conduction characteristics.[38] The synthesis of RuO_2 nanofibers can be achieved by electrospinning a mixture of ruthenium chloride hydrate and polyvinyl acetate in DMF.[39] The resulting fibers were calcined at temperatures ranging from 200°C to 450°C. The resulting RuO_2 nanofiber mat shown in Figure 5.7 exhibits a remarkable capacitive behavior as demonstrated by the rectangular shape of the voltammetry cycle (1 M H_2SO_4 at 10–2000 mV/s). It also showed an attractive metallic conductivity, of about 3×10^2 S/cm, which, however, remains two times lower than the conductivity of RuO_2 bulk at about 10^4 S/cm. Finally, it was possible to enhance the electrochemical performances of the pristine RuO_2 electrospun nanofibers by decorating their surface with electrodeposited RuO_2 nanoparticles. Such RuO_2/RuO_2 nanofibers exhibited a remarkable specific capacitance of 890 F/g at a scan rate of 10 mV/s.[40]

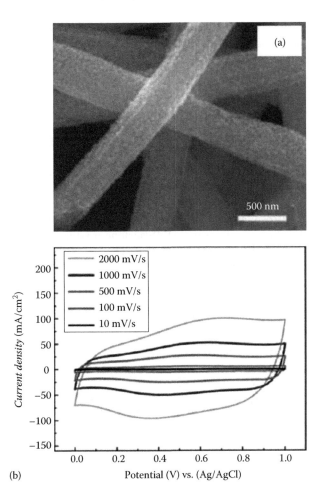

FIGURE 5.7 (a) Scanning electron microscopy image of RuO_2 nanofiber mats calcined at 400°C and (b) cyclic voltammetry of RuO_2/RuO_2 fibers at various scan rates in 1 M H_2SO_4. (Reprinted from Hyun, T.-S., Tuller, H.L., Youn, D.-Y., Kim, H.-G., and Kim, I.-D., Facile synthesis and electrochemical properties of RuO_2 nanofibers with ionically conducting hydrous layer, *J. Mater. Chem.*, 20, 9172–9179, Copyright 2010. With permission of The Royal Society of Chemistry.)

5.3 COMPOSITE ELECTRODE MATERIALS–BASED ELECTROSPUN CARBON NANOFIBERS

The formulation of composite electrode materials is a generalized approach in the field of supercapacitors.[14,41–47] As for any other composites, a synergistic effect is anticipated from the smart mixture of the components. As already mentioned, the key characteristics for supercapacitor electrode materials are the electroactive surface area, ion diffusion and transport, and electronic conductivity. Various approaches have been developed for the formulation of composite electrodes based on electrospun CNFs. The next sections focus on composites associating the textural and shaping potential as well as electronic behavior of the electrospun fibers with the high capacitance of carbons, polymers, and oxides for the fabrication of EDLC and pseudocapacitor electrode materials.

5.3.1 ACTIVATED CARBON/POLYMER FIBER COMPOSITE MATERIALS

Lust et al. recently developed an original and efficient two-step preparation method for the fabrication of EDLC half-cells based on activated carbon supported on electrospun polyvinylidene fluoride (PVDF) fibers.[17] At first, the nanofiber polymer separator layer has been prepared by electrospinning a PVDF solution in a mixture of DMF and acetone. In a second step, the electrode material layer has been directly electrospun onto the separator from a suspension of activated carbon in a PVDF solution in the same DMF/acetone mixture. With a high specific capacitance (up to 120 F/g at 1 mV/s in 1 M $(C_2H_5)_3CH_3NBF_4$ acetonitrile electrolyte), low equivalent series resistance, short time constant (0.043 s), wide operation temperature range (from $-30°C$ to $24°C$),[18] and high specific energy (22 W h/kg) and specific power values (245 kW/kg), Lust results are a rare demonstration of performing half-cells and a step forward to the fabrication of fully electrospun supercapacitor devices.

5.3.2 CARBON/CARBON COMPOSITE MATERIALS

With the objective to enhance the electrochemical properties of electrospun CNFs, carbon nanoobjects can be added in the electrospinning polymer solution. Most of the research works in this field have been targeted on the effects of multiwalled carbon nanotube (CNT) addition.[41–47] CNTs, discovered by Iijima, have been extensively investigated because of their remarkable physical properties.[48] Despite a moderate developed surface area, their high aspect ratio, high tensile modulus, and electronic properties make them suitable for supercapacitor applications.[23] CNTs have often been added to precursor solutions to provide or improve the mechanical behaviors of the resulting electrospun fibers.[49–51] Furthermore, the addition of CNTs into PAN has firstly been studied by Vaisman and has been shown to favor the orientation of PAN chains during heating, leading to an enhancement of the crystallinity and mechanical strength of the resulting composite.[50] It has also been observed that the diameter of electrospun CNTs-embedded CNFs decreases when the loading of CNTs is increased in the PAN-DMF suspension. The higher conductivity of the polymer mixture induced by the presence of CNTs is suggested to increase the electrostatic forces during the formation of the electrospinning jet, leading to thinner fibers.[50] Prilutsky has shown that CNTs embedded in electrospun PAN can act as nucleation centers for the growth of a graphitic structure during the carbonization of the fibers.[46] In a more recent study, the same author showed by transmission electronic microscopy that the orientation of the graphitic layers depends more on the temperature of the carbonization step than on its duration.[51] Furthermore, whatever the shape, orientation, and assemble morphology of CNTs, their presence in the electrospun fibers induces the formation of graphitic structures at temperature way lower than the

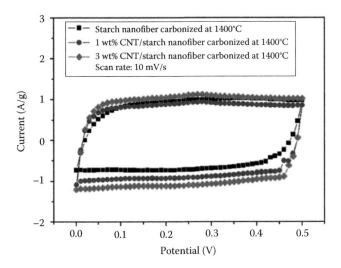

FIGURE 5.8 Cyclic voltammetry of composite carbon nanotube (CNT)/carbon nanofiber (CNF) with various loadings of CNT and CNF (10 mV/s scan rate in 1 M H_2SO_4 electrolyte). (Reprinted from Kim, Y. K., Cha, S.II, Hong, S.H., and Jeong, Y.J., A new hybrid architecture consisting of highly mesoporous CNT/carbon nanofibers from starch, *J. Mater. Chem.*, 22, 20554, Copyright 2012. With permission of The Royal Society of Chemistry.)

2000°C usually needed for the graphitization of some raw polymers.[42,51,52] As shown by Kim, the loading of CNTs in the electrospun fibers allows an enhancement of the capacitance.[42] At a 1–3 wt% loading of CNTs in the electrospun fibers, the resulting capacitance increased from 144 to 193 F/g at a scan rate of 10 mV/s in 1 M H_2SO_4 electrolyte (Figure 5.8).

In a similar approach, graphene was added to the polymer solution to produce composite electrospun CNFs. With a high conductivity, a better chemical stability than porous carbons, and a large specific surface area (usually about 2000 m²/g),[53] graphene has been foreseen as the perfect electrode material for EDLC supercapacitor. It indeed showed some attractive electrochemical performances.[54] The addition of graphene oxide (GO) nanosheets (GO is obtained by exfoliation of graphite in a strong acidic and oxidative medium. For more details on the so-called Hummer's modified method, refer on the work of Marcano on the synthesis of GO.[55]) in a PAN/DMF precursor solution led, after electrospinning and carbonization, to materials with specific capacitances reaching up 260 F/g at a discharge current density of 100 mA/g.[56] More generally, oxidized graphene nanosheets do not alter the flexible and stretchable behavior of electrospun CNFs. Despite an increase in the average diameter of electrospun embedded-graphene CNFs in comparison to that of the corresponding graphene-free CNFs, the specific surface area of the composite material is greater than that of raw GO nanosheets.[56,57] Together with the good electronic conductivity provided by the CNFs, this surface area enhancement accounts for the attractive electrochemical performances of the composite. In an alternative approach, Hsu prepared composite graphene/CNF fibers, by using graphene-like

FIGURE 5.9 HRSEM images of carbon nanofiber (CNF) (a) and graphene-like carbon nanowall at carbon nanotube coated on the CNF surface (b). (Reprinted from *Diam. Relat. Mater.*, 25, Hsu, H.-C., Wang, C.-H., Nataraj, S. K., Huang, H.-C., Du, H.-Y., Chang, S.-T., Chen, L.-C., and Chen, K.-H., Stand-up structure of graphene-like carbon nanowalls on CNT directly grown on polyacrylonitrile-based carbon fiber paper as supercapacitor, 176–179, Copyright 2012, with permission from Elsevier.)

carbon nanowall (GNW) grown at the surface of CNTs.[54] By deposition of these GNW at CNT by microwave plasma-enhanced chemical vapor deposition at the surface of electrospun CNFs (Figure 5.9), a strong improvement of the capacitance was observed at a current density of 0.5 mA/g, from about 55 F/g for raw CNTs to 176 F/g for as prepared composite GNW at CNT.

5.3.3 Conducting Polymer/Carbon Composite Materials

In the previous section, CNFs are introduced as coated with graphene nanosheets and a CNTs/graphene composite. These approaches aim at increasing the developed specific surface area and improve the electronic percolation. A strong effect is observed on the measured capacitance due to the enhancement of the electrostatic charge storage capabilities of carbon electrodes. It is also attractive to combine CNFs with pseudocapacitive materials such as conducting polymers. The charge storage mechanism in conducting polymer electrodes lies on structural defects induced by polymer reduction or oxidation upon charge/discharge. Depending on the reduction/oxidation state, the polymer is n- or p-doped. As a matter of fact, there are several literature reports on the polymer coating of the surface of electrospun CNFs. The most widely used conducting polymers in supercapacitor applications are polyaniline (PANI),[19,58–60] polypyrrole (PPy),[61] poly(3,4-ethylenedioxythiophene),[14] and poly(4-fluorophenyl-3-thiophene).[14] Because of their limited electroactive surface in contact with the electrolyte, these materials, however, provide limited electrochemical performances in terms of energy and power densities and cyclability. Despite these limitations, many research groups have investigated the coating of electrospun CNFs by these polymers.

PANI can be obtain via a simple synthesis and presents a high specific capacitance and good (electro-)chemical stability in comparison with the alternative

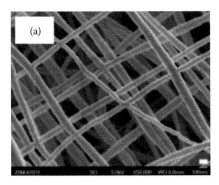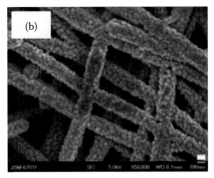

FIGURE 5.10 Scanning electron microscopy pictures of raw electrospun (a) carbon nanofiber and decorated with (b) polyaniline. (Reprinted from Yan, X., Tai, Z., Chen, J., and Xue, Q., Fabrication of carbon nanofiber-polyaniline composite flexible paper for supercapacitor, *Nanoscale*, 3, 212–216, Copyright 2011. With permission of The Royal Society of Chemistry.)

polymers.[14,58–60] In 2011, Yan et al. prepared a composite electrode material from electrospun PAN-based CNFs coated by PANI.[58] The authors processed by the immersion of a web of electrospun, stabilized, and activated CNFs in a solution of aniline and ammonium peroxydisulfate. The resulting paper-like electrode was fully flexible and mechanically robust (Figure 5.10).

In comparison with CNF-free PANI electrode, its capacitance was doubled, going from 317 to 638 F/g at 2 A/g, while its equivalent series resistance decreased from 62.7 to 21.6 Ω. The observed performance enhancement is assigned to the synergistic effect of the net structure formed by the CNF assemble and the specific surface area developed by the nanoparticles of PANI at the CNF surface. With the same objective of improving the electrochemical behavior of such PANI/carbon composites, He proposed an alternative approach for the decoration of graphitized electrospun CNFs with PANI.[61] An acidic treatment of the CNFs prior to the immersion in the aniline solution was shown to induce a rapid polymerization of PANI resulting in the growth of PANI nanowires at the surface of the CNFs. The nanowire length depends on the aniline concentration in the coating solution. Here again, the presence of PANI at the surface of CNFs led to a large surface area in contact with the electrolyte and allowed a drastic enhancement of the electrochemical properties.

PPy has also been used for the coating of electrospun PAN-based CNFs (Figure 5.11).[20,43,62] The PPy coating is made of nanosized particles, building conductive bridges favoring either the ion diffusion through the material or the overall electrical conductivity of the electrode. Furthermore, it increased the specific surface area and the average pore diameter, allowing more electrochemically active sites to be in contact with the electrolyte and a better electronic percolation through both electrode material and current collector. As a matter of fact, PPy particles at the surface of electrospun CNFs improved the electrochemical performances of the material thanks to the enhanced charge transfer in the material. Electrospun PPy/CNFs developed a specific capacitance of about 250 F/g at a current density of 1 mA/cm^2 in 1 M H_2SO_4, almost twice the CNF-free PPy capacitance. The addition of PPy

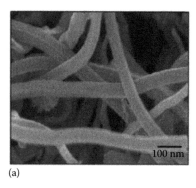 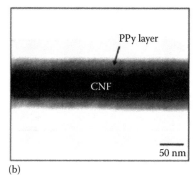

(a)   (b)

FIGURE 5.11   (a) Scanning electron microscopy and (b) TEM images of polypyrrole-coated carbon nanofibers. (From *Sens. Actuators B*, 122, Jang, J. and Bae, J., Carbon nanofiber/poly-pyrrole nanocable as toxic gas sensor, 7–13, Copyright 2007, with permission from Elsevier.)

also contributed to the enhancement of the power since the loss of capacitance was limited to 20% for the PPy/CNF electrode when the current was increased from 1 to 10 mA/cm^2. In comparison, it decreased by 26% for CNF-free PPy when operated in similar conditions.[43]

5.3.4   METAL OXIDE/CARBON COMPOSITE MATERIALS

As already mentioned earlier, selected faradic metal oxides are used in superca-pacitors. In terms of energy density improvement, they are the most promising elec-trode materials for industrial development. They are usually easy to synthesize and can be inexpensive as raw materials. Their main limitation is on the power density because of the kinetics of the involved redox reactions. The decoration of carbon nanoobjects with metal oxides has proven to increase the power density by improving the (sometime) limited intrinsic electronic conductivity of the metal oxides and by providing an enhanced electric percolation to the electrode material.[63,64] Moreover, it also induces a favorable increase of the developed specific surface area. Electrospun CNFs are chosen backbones to address this composite approach. Most of work in the field has been done on the coating of electrospun CNFs with MnO_2,[64–67] Fe_3O_4,[68,69] and RuO_2.[70,71]

MnO_2 is certainly one of the most attractive pseudocapacitive materials. Its capacitance can theoretically reach about 1300 F/g for a single e$^-$ exchange in the Mn^{3+}/Mn^{4+} redox system. Nevertheless, it is characterized by a low electrical con-ductivity, which limits its power density.[72] In the aim of increasing the power perfor-mances of MnO_2-based electrodes, MnO_2 coatings were conducted by immersion of electrospun CNFs in a $KMnO_4$ solution. The carboreduction of Mn^{7+} to Mn^{4+} results in the decoration of CNFs by MnO_2 nanoparticles.[65] The presence of nanoflakes, characteristic of the birnessite form of MnO_2 (δ -MnO_2) was clearly visible on high-magnification scanning electron microscopy (SEM) images (Figure 5.12).

The average diameter of the coated fibers roughly remained as the coating thick-ness was very thin, about 40 nm in this case. Several studies confirm that the coating

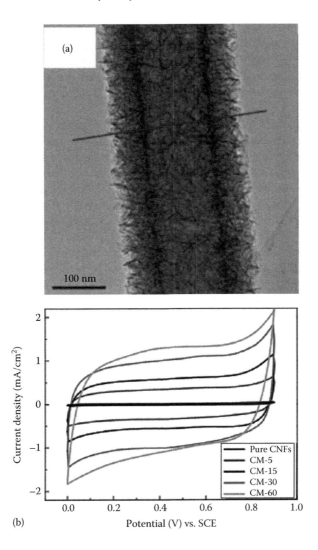

FIGURE 5.12 (a) TEM image and EDS spectra of carbon/MnO$_2$ nanocomposite treated at 60°C (CM-60). (b) Cyclic voltammograms of pure carbon nanofibers and carbon nanofibers/MnO$_2$ nanocomposites (CM) at a scan rate of 2 mV/s in 0.1 M Na$_2$SO$_4$ electrolyte. (Reprinted from *Electrochim. Acta*, 56, Wang, J.-G., Yang, Y., Huang, Z.-H., and Kang, F., Coaxial carbon nanofibers/MnO$_2$ nanocomposites as freestanding electrodes for high-performance electrochemical capacitors, 9240–9247, Copyright 2011, with permission from Elsevier.)

of electrospun CNFs by MnO$_2$ is a rather simple and inexpensive method to fabricate MnO$_2$-based electrodes showing high energy and power densities as a result of the enhancement of the ionic conductivity and specific capacitance of CNFs coupled to that of the electronic conductivity.

Fe$_3$O$_4$ can also be coated at the surface of electrospun CNFs. As for Mn in MnO$_2$, Fe in Fe$_3$O$_4$ can accommodate different valence states allowing a faradic charge storage. Moreover, just like MnO_2, iron oxides suffer from a poor electronic conductivity. As a composite with CNFs, an improvement of the electrochemical performances is expected.[67] Thanks to their magnetic properties, Fe_3O_4 nanoparticles are easily deposited on chosen surfaces, but again, their limited electronic conductivity leads to a poor adhesion at the carbon surface. Consequently, because of a low redox activity, a limited capacitive response was observed.[67,71,73] In contrast to MnO_2 coating, the deposition of iron oxide particles at the surface of electrospun CNFs was done through a quite complex route. PAN solution was firstly electrospun, stabilized, and carbonized. Prepared nanofibers were mixed and stirred in an iron salt solution. Finally, Fe_3O_4 synthesis took place in an autoclave for 18 h at 160°C. Resulting Fe_3O_4/CNF composites showed remarkable capacitive performances thanks to the electronic conductivity provided by the CNFs. The high specific surface area, developed by either electrospun CNFs or Fe_3O_4 nanosheets, facilitates the ion transfer and enhances the electrochemical behavior of both components. Indeed, Mu reports on a greater specific capacitance for the Fe_3O_4 at CNF composite at 135 versus 83 F/g for raw Fe_3O_4. Moreover, as shown in Figure 5.13, a prepared composite exhibited a greater cycle life along with 91% of the initial capacitance retained after 1000 cycles at a current density of 420 mA/g.[67]

Some works have also been performed by Wu on a composite material based of RuO_2 and NiO particles and electrospun CNFs.[69] In contrast with the coatings discussed earlier, NiO/RuO_2/CNFs composite were prepared in a single-step synthesis. Ruthenium and nickel salts were dissolved in the PAN solution. After complete

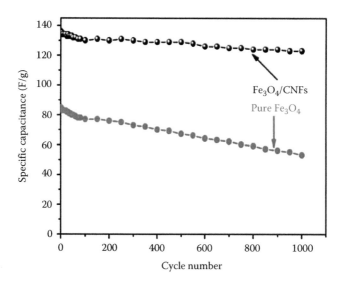

FIGURE 5.13 Comparison of Fe_3O_4 and Fe_3O_4 at carbon nanofiber performances at a current density of 420 mA/g. (Reprinted from Mu, J., Chen, B., Guo, Z., Zhang, M., Zhang, Z., Zhang, P., Shao, C., and Liu, Y., Highly dispersed Fe_3O_4 nanosheets on one-dimensional carbon nanofibers: Synthesis, formation mechanism, and electrochemical performance as supercapacitor electrode materials, *Nanoscale*, 3, 5034, Copyright 2011. With permission of The Royal Society of Chemistry.)

mixing, the mixture was electrospun. Resulting electrospun nanofibers were stabilized and carbonized. After heat treatment, the specific surface area was considerably decreased down to 40 m²/g because of the densities of the oxides. Such materials are quite attractive, especially regarding the ease of the one step synthesis, but the measured capacitance at about 60 F/g limits its actual use as supercapacitor electrodes.

5.4 CONCLUSION

For either electrochemical performance improvements or design of flexible electrode materials, electrospinning appears as a versatile route for the fabrication of fiber-like electrodes for supercapacitors. Both types of supercapacitor electrode materials, EDLC and pseudocapacitor, can be prepared with attractive performances in terms of energy but, especially, power densities. Supercapacitive fibers are usually prepared through multistep syntheses including the electrospinning of a chosen polymer as nanofibers, which are converted as CNFs before being activated (EDLC) or decorated (EDLC and pseudocapacitor) with a capacitive material. The electrochemical performances are assigned to the large developed surface area and the enhanced electronic conductivity provided by the carbon nanofibers.

In the field of supercapacitors, most of the research works on the electrospun nanofibers are devoted to the formulation of electrode materials. Although these are successful with regard to the resulting electrochemical performances, quite surprisingly, the efforts on the use of electrospun nanofibers as solid electrolyte or electrolyte support, for the mechanical reinforcement of gel electrolyte, for example, are currently very limited.

REFERENCES

1. Simon, P., Gogotsi, Y. Materials for electrochemical capacitors. *Nat. Mater.* **7**, 845–854 (2008).
2. Chen, T., Dai, L. Flexible supercapacitors based on carbon nanomaterials. *J. Mater. Chem. A* **2**, 10756 (2014).
3. Cai, W., Lai, T., Dai, W., Ye, J. A facile approach to fabricate flexible all-solid-state supercapacitors based on MnFe₂O₄/graphene hybrids. *J. Power Sources* **255**, 170–178 (2014).
4. Jost, K., Stenger, D., Perez C. R., McDonough, J. K., Lian, K., Gogotsi, Y., Dion, G. Knitted and screen printed carbon-fiber supercapacitors for applications in wearable electronics. *Energy Environ. Sci.* **6**, 2698 (2013).
5. Gao, Y., Presser, V., Zhang, L., Niu, J. J., McDonough, J. K., Perez, C. R., Lin, H., Fong, H., Gogotsi, Y. High power supercapacitor electrodes based on flexible TiC-CDC nanofelts. *J. Power Sources* **201**, 368–375 (2012).
6. Laforgue, A. All-textile flexible supercapacitors using electrospun poly(3,4-ethylenedioxythiophene) nanofibers. *J. Power Sources* **196**, 559–564 (2011).
7. Chou, S.-L., Wang, J.-Z., Chew, S.-Y., Liu, H.-K., Dou, S.-X. Electrodeposition of MnO₂ nanowires on carbon nanotube paper as free-standing, flexible electrode for supercapacitors. *Electrochem. Commun.* **10**, 1724–1727 (2008).
8. Hu, L., Pasta, M., La Mantia, F., Cui L. F., Jeong, S., Deshazer, H. D., Choi, J. W., Han, S. M., Cui, Y. Stretchable, porous, and conductive energy textiles. *Nano Lett.* **10**, 708–714 (2010).

9. Gniotek, K., Krucinska, I. The basic problems of textronics. *Fibers Text. Eastern Europe* **12**(1), 13–16 (2004).

10. Yu, G., Hu, L., Vosgueritchian, M., Wang, H., Xie, X., McDonough J. R., Cui, X., Bao, Z. Solution-processed graphene/MnO₂ nanostructured textiles for high-performance electrochemical capacitors. *Nano Lett.* **11**, 2905–2911 (2011).

11. Wang, Y., Shi, Y., Zhao, C. X., Wong, J. I., Sun, X. W., Yang, H. Y. Printed all-solid flexible microsupercapacitors: Towards the general route for high energy storage devices. *Nanotechnology* **25**, 094010 (2014).

12. Kaempgen, M., Chan, C. K., Ma, J., Cui, Y., Gruner, G. Printable thin film supercapacitors using single-walled carbon nanotubes. *Nano Lett.* **9**, 1872–1876 (2009).

13. Wu, Q., Xu, Y., Yao, Z., Liu, A., Shi, G. Supercapacitors based on flexible graphene/polyaniline nanofiber composite films. *ACS Nano* **4**, 1963–1970 (2010).

14. Cavaliere, S., Subianto, S., Savych, I., Jones, D. J., Rozière, J. Electrospinning: Designed architectures for energy conversion and storage devices. *Energy Environ. Sci.* **4**, 4761 (2011).

15. Dong, Z., Kennedy, S. J., Wu, Y. Electrospinning materials for energy-related applications and devices. *J. Power Sources* **196**, 4886–4904 (2011).

16. Miao, J., Miyauchi, M., Simmons, T. J., Dordick, J. S., Linhardt, R. J. Electrospinning of nanomaterials and applications in electronic components and devices. *J. Nanosci. Nanotechnol.* **10**, 5507–5519 (2010).

17. Tõnurist, K., Vaas, I., Thomberg, T., Jänes, A., Kurig, H., Romann, T., Lust, E. Application of multistep electrospinning method for preparation of electrical double-layer capacitor half-cells. *Electrochim. Acta* **119**, 72–77 (2014).

18. Liivand, K., Vaas, I., Thomberg, T., Jänes, A., Lust, E. Low temperature performance of electrochemical double-layer capacitor based on electrospun half-cells. *J. Electrochem. Soc.* **162**(5), A5031–A5036 (2015).

19. Presser, V., Zhang, L., Niu, J. J., McDonough, J., Perez, C., Fong, H., Gogotsi, Y. Flexible nano-felts of carbide-derived carbon with ultra-high power handling capability. *Adv. Energy Mater.* **1**(3), 423–430 (2011).

20. Pandolfo, A. G., Hollenkamp, A. F. Carbon properties and their role in supercapacitors. *J. Power Sources* **157**(1), 11–27 (2006).

21. Ko, R., Carlen, M. Principles and applications of electrochemical capacitors. *Electrochim. Acta* **45**, 2483–2498 (2000).

22. Winter, M., Brodd, R. J. What are batteries, fuel cells, and supercapacitors. *Chem. Rev.* **104**, 4245–4269 (2004).

23. Frackowiak, E., Béguin, F. Carbon materials for the electrochemical storage of energy in capacitors. *Carbon N. Y.* **39**, 937–950 (2001).

24. Feng, L., Xie, N., Zhong, J. Carbon nanofibers and their composites: A review of synthesizing, properties and applications. *Materials (Basel)* **7**, 3919–3945 (2014).

25. Zhang, L., Aboagye, A., Kelkar, A., Lai, C., Fong, H. A review: Carbon nanofibers from electrospun polyacrylonitrile and their applications. *J. Mater. Sci.* **49**, 463–480 (2013).

26. Kim, C., Yang, K. S. Electrochemical properties of carbon nanofiber web as an electrode for supercapacitor prepared by electrospinning. *Appl. Phys. Lett.* **83**(6), 1216 (2003).

27. Kim, C., Park, S.-H., Lee, W.-J., Yang, K.-S. Characteristics of supercapacitor electrodes of PBI-based carbon nanofiber web prepared by electrospinning. *Electrochim. Acta* **50**, 877–881 (2004).

28. Yu, H., Wu, J., Fan, L., Xu, K., Zhong, X., Lin, Y., Lin, J. Improvement of the performance for quasi-solid-state supercapacitor by using PVA–KOH–KI polymer gel electrolyte. *Electrochim. Acta* **56**, 6881–6886 (2011).

29. Kim, C., Choi, Y.-O., Lee, W.-J., Yang, K.-S. Supercapacitor performances of activated carbon fiber webs prepared by electrospinning of PMDA-ODA poly(amic acid) solutions. *Electrochim. Acta* **50**, 883–887 (2004).

30. Kim, B.-H., Yang, K. S., Kim, Y. A., Kim, Y. J., An, B., Oshida, K. Solvent-induced porosity control of carbon nanofiber webs for supercapacitor. *J. Power Sources* **196**, 10496–10501 (2011).

31. Lee, P.-C., Han, T.-H., Hwang, T., Oh, J.-S., Kim, S.-J., Kim, B.-W., Lee, Y., Choe, H. R., Jeoung, S. K., Yoo, S. E., Nam, J.-D. Electrochemical double layer capacitor performance of electrospun polymer fiber-electrolyte membrane fabricated by solvent-assisted and thermally induced compression molding processes. *J. Membr. Sci.* **409–410**, 365–370 (2012).

32. Kim, B.-H., Yang, K. S. Enhanced electrical capacitance of porous carbon nanofibers derived from polyacrylonitrile and boron trioxide. *Electrochim. Acta* **88**, 597–603 (2013).

33. Kim, B.-H., Yang, K. S., Ferraris, J. P. Highly conductive, mesoporous carbon nanofiber web as electrode material for high-performance supercapacitors. *Electrochim. Acta* **75**, 325–331 (2012).

34. Kim, S. Y., Kim, B.-H., Yang, K. S., Oshida, K. Supercapacitive properties of porous carbon nanofibers via the electrospinning of metal alkoxide-graphene in polyacrylonitrile. *Mater. Lett.* **87**, 157–161 (2012).

35. Wee, G., Soh, H. Z., Cheah, Y. L., Mhaisalkar, S. G., Srinivasan, M. Synthesis and electrochemical properties of electrospun V_2O_5 nanofibers as supercapacitor electrodes. *J. Mater. Chem.* **20**, 6720–6725 (2010).

36. Li, X., Wang, G., Wang, X., Li, X., Ji, J. Flexible supercapacitor based on MnO_2 nanoparticles via electrospinning. *J. Mater. Chem. A* **1**, 10103 (2013).

37. Xue, M., Xie, Z., Zhang, L., Ma, X., Wu, X., Guo, Y., Song, W., Li, Z., Cao, T. Microfluidic etching for fabrication of flexible and all-solid-state micro supercapacitor based on MnO_2 nanoparticles. *Nanoscale* **3**(7), 2703–2708 (2011).

38. Conway, B. Transition from "Supercapacitor" to "Battery" behavior in electrochemical energy storage. *J. Electrochem. Soc.* **138**, 1539–1548 (1991).

39. Conway, B. E., Birss, V., Wojtowicz, J. The role and utilization of pseudocapacitance for energy storage by supercapacitors. *J. Power Sources* **66**, 1–14 (1997).

40. Hyun, T.-S., Tuller, H. L., Youn, D.-Y., Kim, H.-G., Kim, I.-D. Facile synthesis and electrochemical properties of RuO_2 nanofibers with ionically conducting hydrous layer. *J. Mater. Chem.* **20**, 9172–9179 (2010).

41. Prilutsky, S., Zussman, E., Cohen, Y. The effect of embedded carbon nanotubes on the morphological evolution during the carbonization of poly(acrylonitrile) nanofibers. *Nanotechnology* **19**, 165603 (2008).

42. Kim, Y. K., Cha, S. II., Hong, S. H., Jeong, Y. J. A new hybrid architecture consisting of highly mesoporous CNT/carbon nanofibers from starch. *J. Mater. Chem.* **22**, 20554 (2012).

43. Ju, Y.-W., Choi, G.-R., Jung, H.-R., Lee, W.-J. Electrochemical properties of electrospun PAN/MWCNT carbon nanofibers electrodes coated with polypyrrole. *Electrochim. Acta* **53**, 5796–5803 (2008).

44. Ervin, M. H., Miller, B. S., Hanrahan, B., Mailly, B., Palacios, T. A comparison of single-wall carbon nanotube electrochemical capacitor electrode fabrication methods. *Electrochim. Acta* **65**, 37–43 (2012).

45. Pilehrood, M. K., Heikkilä, P., Harlin, A. Preparation of carbon nanotube embedded in polyacrylonitrile (pan) nanofibre composites by electrospinning process. *Autex Res. J.* **12**, 1–6 (2012).

46. Prilutsky, S., Zussman, E., Cohen, Y. Carbonization of electrospun poly (acrylonitrile) nanofibers containing multiwalled carbon nanotubes observed by transmission electron microscope with in situ heating. *J. Polym. Sci.* **48**, 2121–2128 (2010).

47. Hou, H., Ge, J. J., Zeng, J., Li, Q., Reneker, D. H., Greiner, A., Cheng, S. Z. D. Electrospun polyacrylonitrile nanofibers containing a high concentration of well-aligned multiwall carbon nanotubes. *Chem. Mater.* **81**, 967–973 (2005).

48. IIjima, S. Helical microtubules of graphitic carbon. *Nature* **354**, 56–58 (1991).
49. Chae, H. G., Minus, M. L., Kumar, S. Oriented and exfoliated single wall carbon nanotubes in polyacrylonitrile. *Polymer* **47**, 3494–3504 (2006).
50. Vaisman, L., Larin, B., Davidi, I., Watchel, E., Marom, G., Wagner, H. D. Processing and characterization of extruded drawn MWNT-PAN composite filaments. *Compos. Part A Appl. Sci. Manuf.* **38**, 1354–1362 (2007).
51. Guo, Q., Zhou, X., Li, X., Chen, S., Seema, A., Greiner, A., Hou, H. Supercapacitors based on hybrid carbon nanofibers containing multiwalled carbon nanotubes. *J. Mater. Chem.* **19**, 2810 (2009).
52. Niu, H., Zhang, J., Xie, Z., Wang, X., Lin, T. Preparation, structure and supercapacitance of bonded carbon nanofiber electrode materials. *Carbon N. Y.* **49**, 2380–2388 (2011).
53. Chakrabarti, M. H., Low, C. T. J., Brandon, N. P., Yufit, V., Hashim, M. A., Irfan, M. F., Akhtar J., Ruiz-Trejo, E., Hussain, M. A. Progress in the electrochemical modification of graphene-based materials and their applications. *Electrochim. Acta* **107**, 425–440 (2013).
54. Hsu, H.-C., Wang, C.-H., Nataraj, S. K., Huang, H.-C., Du, H.-Y., Chang, S.-T., Chen, L.-C., Chen, K.-H. Stand-up structure of graphene-like carbon nanowalls on CNT directly grown on polyacrylonitrile-based carbon fiber paper as supercapacitor. *Diam. Relat. Mater.* **25**, 176–179 (2012).
55. Marcano, D. C., Kosynkin, D. V., Berlin, J. M., Sinitskii, A., Sun, Z., Slesarev, A., Alemany, L. B., Lu, W., Tour, J. M. Improved synthesis of graphene oxide. *ACS Nano* **8**, 4806–4814 (2010).
56. Zhou, Z., Wu, X.-F. Graphene-beaded carbon nanofibers for use in supercapacitor electrodes: Synthesis and electrochemical characterization. *J. Power Sources* **222**, 410–416 (2013).
57. Tai, Z., Yan, X., Lang, J., Xue, Q. Enhancement of capacitance performance of flexible carbon nanofiber paper by adding graphene nanosheets. *J. Power Sources* **199**, 373–378 (2012).
58. Yan, X., Tai, Z., Chen, J., Xue, Q. Fabrication of carbon nanofiber-polyaniline composite flexible paper for supercapacitor. *Nanoscale* **3**, 212–216 (2011).
59. Giray, D., Balkan, T., Dietzel, B., Sezai Sarac, A. Electrochemical impedance study on nanofibers of poly(m-anthranilic acid)/polyacrylonitrile blends. *Eur. Polym. J.* **49**, 2645–2653 (2013).
60. Sun, B., Long, Y.-Z., Chen, Z.-J., Liu, S.-L., Zhang, H.-D., Zhang, J.-C., Han, W.-P. Recent advances in flexible and stretchable electronic devices via electrospinning. *J. Mater. Chem. C* **2**, 1209 (2014).
61. He, S., Hu, X., Chen, S., Hu, H., Hanif, M., Hou, H. Needle-like polyaniline nanowires on graphite nanofibers: Hierarchical micro/nano-architecture for high performance supercapacitors. *J. Mater. Chem.* **22**, 5114 (2012).
62. Wang, J.-G., Yang, Y., Huang, Z.-H., Kang, F. Rational synthesis of MnO_2/conducting polypyrrole@carbon nanofiber triaxial nano-cables for high-performance supercapacitors. *J. Mater. Chem.* **22**, 16943 (2012).
63. Wu, Z. S. Graphene/metal oxide composite electrode materials for energy storage. *Nano Energy* **1**(1), 107–131 (2012).
64. Zhi, M., Manivannan, A., Meng, F., Wu, N. Highly conductive electrospun carbon nanofiber/MnO_2 coaxial nano-cables for high energy and power density supercapacitors. *J. Power Sources* **208**, 345–353 (2012).
65. Wang, J.-G., Yang, Y., Huang, Z.-H., Kang, F. Coaxial carbon nanofibers/MnO_2 nanocomposites as freestanding electrodes for high-performance electrochemical capacitors. *Electrochim. Acta* **56**, 9240–9247 (2011).

66. Jang, J., Bae, J. Carbon nanofiber/polypyrrole nanocable as toxic gas sensor. *Sens. Actuators B* **122**, 7–13 (2007).

67. Mu, J., Chen, B., Guo, Z., Zhang, M., Zhang, Z., Zhang, P., Shao, C., Liu, Y. Highly dispersed Fe_3O_4 nanosheets on one-dimensional carbon nanofibers: Synthesis, formation mechanism, and electrochemical performance as supercapacitor electrode materials. *Nanoscale* **3**, 5034 (2011).

68. Ren, T., Si, Y., Yang, J., Ding, B., Yang, X., Hong, F., Yu, J. Polyacrylonitrile/ polybenzoxazine-based Fe_3O_4@carbon nanofibers: Hierarchical porous structure and magnetic adsorption property. *J. Mater. Chem.* **22**, 15919 (2012).

69. Wu, Y., Balakrishna, R., Reddy, M. V., Nair, A. S., Chowdari, B. V. R., Ramakrishna, S. Functional properties of electrospun NiO/RuO_2 composite carbon nanofibers. *J. Alloy Compd.* **517**, 69–74 (2012).

70. Ju, Y.-W., Choi, G.-R., Jung, H.-R., Kim, C., Yang, K.-S., Lee, W.-J. A hydrous ruthenium oxide-carbon nanofibers composite electrodes prepared by electrospinning. *J. Electrochem. Soc.* **154**, A192–A197 (2013).

71. Du, X., Wang, C., Chen, M., Jiao, Y., Wang, J. Electrochemical performances of nanoparticle Fe_3O_4/activated carbon supercapacitor using KOH electrolyte solution. *J. Phys. Chem. C* **113**, 2643–2646 (2009).

72. Ghodbane, O., Pascal, J. L., Favier, F. Microstructural effects on charge-storage properties in MnO_2-based electrochemical supercapacitors. *ACS Appl. Mater. Interfaces* **1**(5), 1130–1139 (2009).

73. Shi, W., Zhu, J., Sim, D. H., Tay, Y. Y., Lu, Z., Zhang, X., Sharma, Y., Srinivasan, M., Zhang, H., Hng, H. H., Yan, Q. Achieving high specific charge capacitances in Fe_3O_4/ reduced graphene oxide nanocomposites. *J. Mater. Chem.* **21**, 3422 (2011).

6 Electrospinning for the Development of Improved Lithium-Ion Battery Materials

Marcus Fehse and Lorenzo Stievano

CONTENTS

6.1 INTRODUCTION

Energy storage plays a key role in the context of future technological requirements and increasing world population. On the one hand, a rapid growth of renewable sources is envisioned in the next years in order to cover increasing fractions of the final energy generation portfolio, thus achieving a sustainable energy production. However, the penetration of renewable energy sources is slowed down by their characteristic intermittency and fluctuating behavior. These issues can be addressed only via the implementation of improved storage technologies in the electricity networks. On the other hand, efficient and affordable energy storage is needed to power the myriad portable consumer electronic devices (cell phones, PDAs, laptops, or for implantable medical applications, such as pace makers), as well as future hybrid and/or electric vehicles, which are central to the reduction of CO_2 emissions arising from transportation. In these scenarios, the development of clean and efficient energy storage systems with improved performance, availability, durability, safety, and reduced cost becomes crucial.

Among the different available technologies, electrochemical energy storage plays a major role and, in the large family of electrochemistry-based systems, lithium-ion batteries (Li-ion battery) take a sublime position (Scrosati and Garche, 2010). To achieve high energy, high-power density, and efficiency in such systems, designed functional materials are needed that associate the required electrochemical properties for improved performance, with chemical and electrochemical stability, while leaving a minimally low environmental footprint (Armand and Tarascon, 2008; Tarascon and Armand, 2001). The synthesis of tailored electrolyte and electrode structures and architectures is thus fundamental for improving device performances, such as energy density, power, cycling life, charge/discharge rate, safety, and cost. Among these materials, nanosized and nanostructured materials occupy a key position, since their size and/or structure correspond to the length scale over which many of the elementary steps such as charge transfer or ion insertion occur. Moreover their large surface-to-volume ratio provides a high-volume fraction of interfaces, favoring increased reaction rates while reducing the stress related to volume changes often occurring upon electrochemical cycling (Liu et al., 2010; Mukherjee et al., 2012; Wang et al., 2010b).

Several different techniques are nowadays available to synthesize nanostructured materials of controlled size and architecture for application in Li-based systems. Among these techniques, electrospinning was introduced quite recently and has rapidly taken an outstanding importance (Aravindan et al., 2015; Cavaliere et al., 2011; Dai et al., 2011; Dong et al., 2011). By December 2014, about 660 scientific papers had been published (Figure 6.1), with a very rapid increase of the number of publications observed in only a few years.

In fact, electrospinning remains a simple, versatile, and cost-effective technique allowing the production of 1D nanostructures such as nanotubes and nanofibers in a continuous process and at long-length scales with controllable and reproducible diameter, particle size and shape, and phase composition (Dai et al., 2011; Reneker et al., 2007; Sawicka and Gouma, 2006).

This method can thus be used to create mesoporous patterns with 1D structures allowing enlarged contact of electrode and electrolyte, facilitating simultaneously electron percolation toward the current collectors and the diffusion of the ionic species in the electrolyte toward and from the electrode materials. This is a promising approach to overcome intrinsic kinetic drawbacks of battery materials, thus enhancing the electrochemical properties of the cell (Aravindan et al., 2015; Cavaliere et al., 2011; Dai et al., 2011; Dong et al., 2011).

In this chapter, after a schematic introduction of the configuration of a Li-ion cell and of its components, the use of electrospinning for producing different parts of the cell, allowing in several cases significant improvements of the cycling performances, will be reviewed.

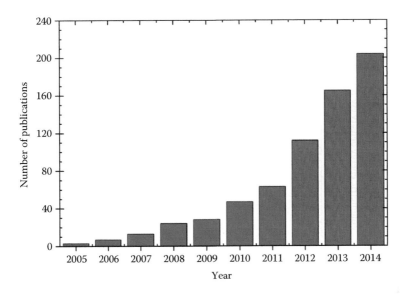

FIGURE 6.1 Number of indexed publications in the field of Li-based accumulators featuring electrospinning up to December 2014. (Data from *Web of Science*, Thomson Reuters.)

6.2 LI-ION BATTERIES AND THEIR COMPONENTS

Almost 30 years ago, Yoshino et al. presented a new secondary battery system $C/LiCoO_2$ based on the preceding work of Goodenough and Whittingham (Mizushima et al., 1981; Thackeray et al., 1983; Whittingham, 1976) on the "rocking-chair system" (a metaphor is used to picture the back and forth movement of lithium ions between negative and positive electrode during charge and discharge). This system was adapted and introduced commercially in 1991 by SONY and Asahi Kasei/Toshiba (Yoshino, 2012). This battery system, which became worldwide known as Li-ion battery, was one of the biggest milestone in the development of electrochemical energy storage since Alessandro Volta presented the first voltaic pile in 1800.

Originally conceptualized to feed small portable electronic devices, they spread quickly, due to their superior storage capacity (Figure 6.2), long cycle life, and high charging currents to become indispensable for modern high-technology society (Buchmann, 2014). Furthermore, Li-ion battery do not show a memory effect, feature low self-discharge (Nishi, 2001), and can operate in a wide temperature range of typically −20°C to 60°C, which further contribute to their superior role. Today, Li-ion battery are at the dawn of conquering new markets such as automotive application, backbone for smart power grids, and are even implemented in the aviation industry (Mukherjee et al., 2012). This triumph is enabled by a consistent progress in research and development of the Li-ion battery components, which has led to doubling its capacity (≥200 W h/kg) while at the same time mass production cut cost by over 80% to less than 400$/kW h within the last two decades (Buchmann, 2014; Mukherjee et al., 2012; Sankey et al., 2010; Thackeray et al., 2012).

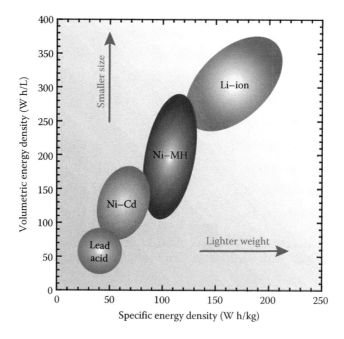

FIGURE 6.2 Comparison of gravimetric and volumetric energy density of various secondary battery systems. (Reproduced from Landi, B.J., Ganter, M.J., Cress, C.D., DiLeo, R.A., and Raffaelle, R.P., Carbon nanotubes for lithium-ion batteries, *Energy Environ. Sci.*, 2(6), 638–654, 2009. With permission of The Royal Society of Chemistry.)

Similar to all electrochemical cells, also Li-ion cells have a mutual assembly, which comprises two electrodes (the positive and the negative), an electrolyte, and other ancillary constructional components. In Figure 6.3, the typical setup of an Li-ion battery is schemed. In such systems, electrical energy is generated by conversion of chemical energy via redox reactions at the electrodes. In the "rocking-chair" Li-ion battery, lithium cations are reversibly inserted/extracted into/from the host electrode matrix, and this process is accompanied by an electron flow through an external circuit. Upon charging, Li^+ are released from the positive electrode and inserted in the negative one, while upon discharging the mechanism is reversed, and Li^+ are extracted from the negative electrode and inserted into the positive one.

6.2.1 ELECTRODE MATERIALS

Differently from other battery technologies, Li-ion batteries have the characteristic of allowing a very large number of possible host materials for lithium ions, both at the positive and at the negative electrode. These materials feature very different properties in terms of working potential, capacity, reaction properties versus Li^+, etc. The most studied electrode materials are resumed in Figure 6.4. At the negative electrode, graphitic carbon has remained since its introduction the most used material, even though other materials such as silicon allow much larger theoretical capacities.

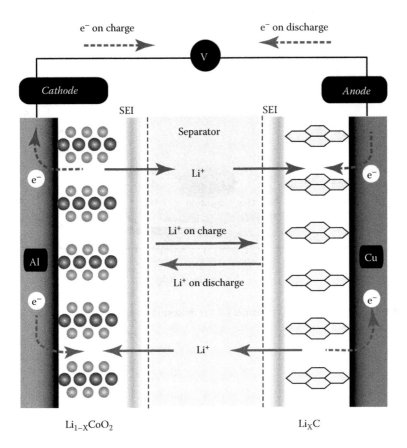

$$Li_{1-x}CoO_2 \qquad\qquad Li_xC$$

FIGURE 6.3 Schematic assembly of standard lithium-ion battery with graphite as the negative and $LiCoO_2$ as the positive electrode material. (Reproduced from Landi, B.J., Ganter, M.J., Cress, C.D., DiLeo, R.A., and Raffaelle, R.P., Carbon nanotubes for lithium-ion batteries, *Energy Environ. Sci.*, 2(6), 638–654, 2009. With permission of The Royal Society of Chemistry.)

Graphite allows the intercalation of one lithium ion per six carbon atoms in the interlayer spaces between the graphene layers, with a theoretical capacity of 372 mA h/g. At the positive electrode, several materials are commercially used. Most of them can be categorized in three main families, namely, layered lithium metal oxides (e.g., $LiCoO_2$), polyanionic compounds of transition metal (e.g., $LiFePO_4$), or spinels (e.g., $LiMn_2O_4$). These materials can accommodate up to one Li atom per transition metal center, with capacities very seldom exceeding 200 mA h/g.

Depending on their working mechanism, three main groups of Li-ion battery electrode materials can be identified, as shown in Figure 6.5 (Palacín, 2009).

6.2.1.1 Intercalation Materials

This family of materials is currently dominating the Li-ion battery market. The storage in this type of material is based on the topotactic incorporation of Li^+ into an

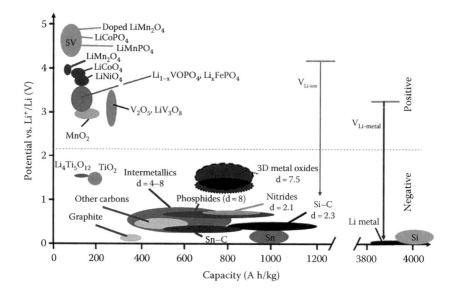

FIGURE 6.4   Capacity versus potential for the most studied electrode materials for lithium-ion cell systems.

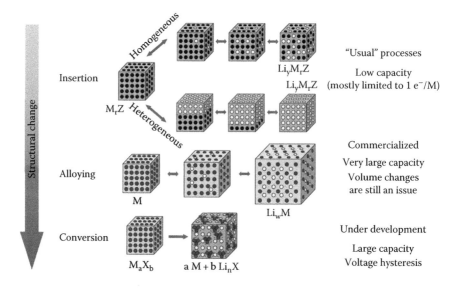

FIGURE 6.5   Schematic presentation of reaction mechanism and properties of three main electrode material types for use in lithium-ion batteries. Black circles, voids in the crystal structure; gray circles, metal atoms; white circles, lithium ions. (Reproduced from Palacín, M.R., Recent advances in rechargeable battery materials: A chemist's perspective, *Chem. Soc. Rev.*, 38(9), 2565–2575, 2009. With permission of The Royal Society of Chemistry.)

open host whereupon none or only minor structural modifications occur. In order to compensate the charge, a reduction of the host lattice occurs upon cation intercalation. These reaction characteristics result in relatively small theoretical capacity (Wu and Cui, 2012) but, at the same time, provide intrinsic advantages such as high reversibility, good cycling stability, and long cycle life, which make them the materials of choice for most of today's application. Virtually all positive electrode materials are based on this mechanism, as well as the most used negative electrode ones (graphite and titanates).

6.2.1.2   Alloy Materials

These materials can deliver very large theoretic capacities (e.g., 4200 and 992 mA h/g, for Si and Sn, respectively) by multistep alloying with Li^+, yielding compositions as high as $Li_{4.4}M$ (M = Si, Ge, Sn). However, the main drawback of these reactions, which hampers their application in commercial batteries, is the very large volumetric expansion observed upon lithiation (Scrosati and Garche, 2010). Even after decades of research, leading to elaborated nanostructures and the implication of sophisticated additives and binders to buffer these volume changes, they still fail to meet life span, security, and cycle stability requirements for commercial applications. All alloy materials react at sufficiently low potentials (below 1 V) and can thus be used as the negative electrode.

6.2.1.3   Conversion Materials

A conversion reaction, which was first verified for transition metal oxides, occurs between lithium and a binary compound containing a transition metal (M = Ti, Mn, Fe, Co, Ni, etc.) and a p-block element (X = O, P, Sb, Sn, etc.), and lithium according to the following general equation:

$$M_aX_b + (b \cdot n)Li \rightleftarrows aM + bLi_nX$$

Materials undergoing the conversion reaction, called conversion materials, allow reversible capacities as high as 1500 mA h/g through the full reduction of the transition metal to the metallic state enabling a large lithium uptake by the formation of a Li_nX matrix embedding the transition metal nanoparticles. Similarly to alloy materials, conversion materials also undergo very large volume expansion upon reaction with Li^+. Moreover, they often reveal strong hysteresis in electrochemical cycling curve, which reduces the energy efficiency (Cabana et al., 2010). As in the case of alloy materials, almost all conversion materials (with the exception of fluorides) can be used at the negative electrode due to their low reaction potential versus Li.

In summary, while negative electrode materials can generate quite large capacities, those of positive electrode materials, based on the insertion mechanism, are relatively small. In order to maximize the specific energy and the energy density of a cell, it is thus important to increase both capacity and the cell voltage. While the first goal is obtained by maximizing the capacity of the negative electrode, the second one can be attained by using positive electrode materials at the highest possible working potential. In addition, both electrode materials must assure a high reversibility of the electrochemical reactions in order to maintain the specific charge for thousands of electrochemical charge–discharge cycles.

6.2.2 Electrolytes and Separators

One of the key components of a battery is the electrolyte, which allows ionic diffusion of the Li$^+$ ions between the two electrodes. Three main families of electrolytes can be identified: liquid electrolytes (which must be used together with a porous separator), solid electrolytes (dry polymer membranes or ceramics), and gelled polymer membranes:

Liquid electrolytes: The most common commercial systems are based on liquid electrolytes, usually a mixture of organic carbonates such as propylene carbonate, ethylene carbonate, and/or diethyl carbonate in which a lithium salt (e.g., lithium hexafluorophosphate, LiPF$_6$) is dissolved. The separator used with these electrolytes is usually a porous polymer or ceramic membrane, which assures the mechanical separation of the two electrodes avoiding possible short circuits. Commercial battery systems are mainly based on polymers such as polyethylene (PE) or polypropylene (PP) and can have complex architectures to maximize the porosity (and thus the diffusion of the electrolyte) while maintaining very good mechanical properties. In addition to the good ionic conductivity, these membranes must assure the so-called shut-down effect, that is, the clogging of the porosity at high temperatures due to the fusion of the polymer, an intrinsic security system that is based on the drastic increase of the resistivity when the working temperature of the battery increases too much (above 100°C). Typical complex separators with high-security properties are those created by Celgard®, consisting of multilayered PE–PP mesoporous membranes: in these systems, the PE collapses at relatively low temperatures (below 130°C), while the PP maintains its rigidity up to 150°C, thus assuring sufficiently good mechanical properties even when the PE layer is already melted and thus preventing possible short circuits.

Polymer electrolytes: Polymer membranes are usually based on salts (or acids) dissolved in a polymer matrix. In this way, the membrane works both as an ionic conductor and as an electronic insulator. Such systems, mainly based on LiClO$_4$ dissolved in polyoxyethylene (PEO) or polyoxypropylene, have been known for 40 years (Wright, 1975). The advantages of this technology are the absence of volatile and flammable compounds, their simple shaping and application, and their low cost compared to liquid electrolytes. The main drawback that is hampering the large-scale application of such membranes in batteries is their low conductivity at ambient temperature, which is of the order of 10^{-6} S/cm. This low conductivity is mainly due to the semicrystalline structure of the polymer, with the crystalline domains blocking the migration of the Li$^+$ cations at temperatures below 50°C.

Gelled polymer electrolytes: Gelled polymer electrolytes are obtained by adding a large amount of solvent to a polymeric membrane. The addition of the solvent strongly increases the ionic conductivity by enhancing the mobility of the polymer chains. Conductivities as high as 10^{-3} S/cm can be obtained by using different polymers swollen with liquid electrolytes, such as PEO, polyacrylonitrile (PAN), or copolymers of polyvinylidene fluoride (PVDF) and poly(hexafluoropropylene) (HFP). Despite their advantages, polymer gel membranes also have drawbacks due to poor mechanical properties and sometimes unsatisfactory thermal stability. Moreover, elaboration of free-standing films gelled membranes is quite complicated because they must be carried out under controlled atmosphere. In some cases, their mechanical properties can be improved by, for example, using highly crystalline PVDF homopolymer; however, the crystallinity of the polymer hinders the migration of lithium ions and as such is detrimental to the ionic conductivity of the membrane and thus to the performance of the battery. An interesting alternative to bypass the burden of assembling the cell in a moisture-free environment was developed by Tarascon et al. (1996) at Bellcore, by using a Li salt-free plasticizer, which is substituted by the liquid electrolyte at the last stage of the cell processing through an extraction/activation step.

As already mentioned in Section 6.1, Li-ion batteries have enabled the development of small and portable electronic devices, but future challenges concerning larger and more demanding applications such as electric or hybrid electric vehicles necessitate significant advances in battery technology as they require fast charging and discharging at high-power rates. Thus, further improvements are needed in various aspects to achieve satisfactory performance for such demanding applications. One of the strategies used to improve the efficiency and the durability of Li-ion batteries is to maintain a low weight, volume, and cost by designing and fabricating nanomaterials and nanostructured materials for use as electrodes and electrolytes. In this regard, the versatility of electrospinning makes them highly suitable to prepare the different battery components with the benefits of smaller, nanostructured geometries and architectures and reduced manufacturing costs. In the next sections of this chapter, the main advances in the areas of electrospun electrodes and separator membranes for Li-ion batteries are described.

6.3   ELECTROSPINNING IN THE PREPARATION OF ELECTRODES FOR LI-ION BATTERIES

Nanostructuring has been extensively explored for the preparation of high-performance Li-ion battery electrodes and electrode materials, since in many cases they can provide a variety of notable advantages (Aricò et al., 2005; Bruce et al., 2008; Liu et al., 2011; Manthiram et al., 2008; Mukherjee et al., 2012; Pitchai et al., 2011; Scrosati and Garche, 2010; Wang and Cao, 2008; Wang et al., 2010b; Wu and Cui, 2012; Yang et al., 2009b; Zu and Manthiram, 2013). Primarily, particles of electrode material with a nanometer size have a very large surface-to-bulk ratio, leading to a high electrode/electrolyte contact area and a short diffusion length for lithium-ion transport within the particles. These properties usually produce a significant increase of the rate of Li$^+$ insertion/extraction, enhancing the rate capability and the power density of the material. The same is also true for electron transport, which is also usually enhanced in nanomaterials. Moreover, in the case of alloy or conversion reactions leading to large volume changes, the composition range of solid solutions is often more extensive for nanoparticles than for bulk materials, and the strain associated with insertion/extraction reactions is often better accommodated, avoiding destruction of the material structure (electrochemical grinding) and improving the cycle life of the cell.

In spite of these benefits, the use of nanomaterials in Li-ion batteries also introduces new challenges. The large surface area might enhance possible parasitic reactions between the electrode and the electrolyte, leading to self-discharge of the cell and to the formation of thicker solid-electrolyte interphases (SEI), with a global decrease of the cycle life due to both irreversible electrolyte consumption and increase of the impedance. In addition, compared to dense micrometric bulk materials, the low packing density of nanostructured materials results in lower volumetric energy density (tap density). Further issues could be caused by the complexity of the synthetic methods and the difficulty of accurately controlling the size of nanomaterials, which may result in greater processing and manufacturing costs. Moreover, the manipulation of nanomaterials requires the implication of restrictive regulations. Taking these challenges into account, nanowires show great potential in this area among the wide range of morphologies because of their improved percolation behavior compared to particles (Reddy et al., 2010). Soft chemistry and template syntheses are usually employed to synthesize nanostructured electrodes, and after sintering treatment, they often result in large grain size and aggregation, losing the high specific surface area and its associated advantages (Hosono et al., 2009). Compared to other synthetic methods, electrospinning remains therefore a simple and versatile alternative approach for preparing nanostructured positive and negative electrodes for Li-ion batteries (Cavaliere et al., 2011; Kim et al., 2010).

Such versatility in processing battery materials was impressively demonstrated by Jayaraman et al., who built an all 1D full cell (Figure 6.6) (Jayaraman et al., 2014). In this full cell, all components were derived from scalable electrospinning technique. Through adequately adjusting the synthesis parameters, the fiber properties such as morphology, length, porosity, and crystallinity of the different electrospun fibers were optimized according to their specific requirements. For the positive electrode, hollow and porous LiMn$_2$O$_4$ (LMO) nanofibers with average diameter of ≈500 nm and thickness of ≈75 nm were used (Jayaraman et al., 2013). The membrane consisted of continuous, interconnected bead-free fibers, poly(vinylidene fluoride-co-hexafluoropropylene) (PVDF-HFP) fibers with a diameter of ≈650 nm. For the negative electrode, interconnected TiNb$_2$O$_7$ fibers with average diameter ≈200 nm were used. Such fully electrospun electrochemical cell (LiMn$_2$O$_4$|PVDF-HFP|TiNb$_2$O$_7$) stored 116 mA h/g at 150 mA/g reversibly over 200 cycles with an excellent average coulombic efficiency of 99.895%. In combination with operating voltage of approximately 2.4 V, this yielded to an energy density of 278 W h/kg, corresponding to state-of-the-art high-energy lithium batteries.

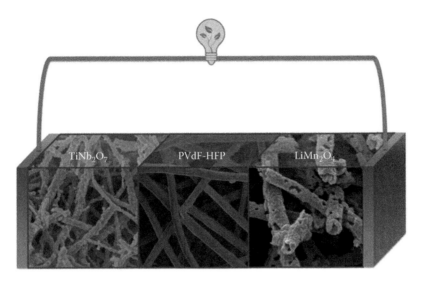

FIGURE 6.6   Schematic representation of a typical Li-ion battery comprising a 1D component. (Reprinted with permission from Jayaraman, S., Aravindan, V., Suresh Kumar, P., Chui Ling, W., Ramakrishna, S., and Madhavi, S., Exceptional performance of $TiNb_2O_7$ anode in all one-dimensional architecture by electrospinning, *ACS Appl. Mater. Interfaces,* 6(11), 8660–8666. Copyright 2014 American Chemical Society.)

Even though this full 1D cell remains a very specific case, many other publications have dealt with different positive and negative electrospun electrode materials. A comprehensive selection of these researches on positive and negative electrode materials is summarized in Tables 6.1 and 6.2, respectively, and shortly presented in the following sections. A selection of the works on the use of electrospinning for manufacturing separators and gelled electrolytes is summarized in Table 6.3.

6.3.1   NANOSTRUCTURED POSITIVE ELECTRODE MATERIALS

A growing number of studies concerning electrospun positive electrode materials of both transition metal oxides with different crystal structures (layered materials, spinel structure, etc.) and polyanionic compound (phosphates, sulfates, etc.) have been carried out. The modification of the electrospinning configuration and different post-treatments made available a wide variety of morphologies and architectures, such as nanowires (Mai et al., 2010), nanobelts (Ban et al., 2009), nanonuggets (Le Viet et al., 2011), coaxial (Gu et al., 2007) and triaxial (Hosono et al., 2010) core–sheath, and hollow (Gu and Jian, 2008) fibers.

Among layered oxides, the most widely commercially used positive electrode material remains $LiCoO_2$, which provides a reasonable theoretical capacity of 140 mA h/g and very good cycle life (Mizushima et al., 1981). The preparation by electrospinning of nanostructured $LiCoO_2$ fibers was performed by several groups to achieve a fast solid-state diffusion by decreasing the diffusion distance of Li$^+$ cations (Chen et al., 2011). Gu et al. obtained $LiCoO_2$ fibers with large surface area,

TABLE 6.1

Electrospun Positive Electrode Materials for Li-Ion Batteries

Material	Fiber Diameter (nm)/Structure	Crystallite Size (nm)	BET Surface (m²/g)	Discharge Capacity First Cycle (mA h/g) [Rate (C)]	Discharge Capacity nth Cycle (mA h/g) [Rate (C)]	References
$LiCoO_2$	500–2000	20–35	18	182 [0.14]	n=2, 173; n=20, 123 [0.14]	Gu et al. (2005)
$LiCoO_2$/MgO	1000–2000/shell 10–100	<10	20	181 [0.14]	n=40, 163 [0.14]	Gu et al. (2007)
LiPON-coated $LiCoO_2$	60–80/3D architecture	NA	NA	129 [NA]	n=2, 128; n=100, 120 [NA]	Lu et al. (2008)
$LiNi_{0.8}Co_{0.1}Mn_{0.1}O_2$/ MgO	1000–2000; 30–60 shell 300–500 wall/hollow coaxial	NA	30	195 [0.14]	n=50, 174 [0.14]	Gu and Jian (2008)
$Li(Ni_{1/3}Co_{1/3}Mn_{1/3})$ O_2/$Li(Ni_{1/2}Mn_{1/2})O_2$	<5000/shell 300	20–50	NA	192 [0.1]	n=2, 171	Yang et al. (2010a)
$LiMn_2O_4$	Nanofibers	NA	NA	118 [1.1]		Jayaraman et al. (2014)
$LiMn_2O_4$	Hollow NF 500, thick 60	56	NA	119 [1]; 56 [16]	n=1250, 103 [1]	Jayaraman et al. (2013)
Ta–Nb_2O_5	≈100–200, NFs/ nanonuggets	NA	3–6	≈500–800 [1], ≈240–150 [1.5]	n=40, ≈350–150 [1]; 50, ≈170–100 [1.5]	Le Viet et al. (2011)
$LiFePO_4$/C	100–300	18–28	NA	166 [0.05], 140 [0.1], 125 [0.2], 98 [0.5], 71 [1], 37 [2]	n=50, ≈140 [0.1]; ≈120 [0.2]	Toprakci et al. (2011)
$LiFePO_4$/C + Fe_2P	Carbon-coated NF 130			155 [0.1], 84 [1]	n=50, ≈80 [1]	Qiu et al. (2013)
$LiMg_{0.05}Mn_{0.95}PO_4$	Carbon-coated NF 80	NA	NA	135 [0.1]	n=200, 120 [2]	Lu et al. (2014)

(Continued)

TABLE 6.1 (*Continued*)
Electrospun Positive Electrode Materials for Li-Ion Batteries

Material	Fiber Diameter (nm)/Structure	Crystallite Size (nm)	BET Surface (m²/g)	Discharge Capacity First Cycle (mA h/g) [Rate (C)]	Discharge Capacity nth Cycle (mA h/g) [Rate (C)]	References
$LiMn_xFe_{1-x}PO_4$/VGCF	NW 100 around VGCF core, bundle 500	NA	NA	x=0: 140 [0.05]; x=0.2: 130 [0.05]	n=30, 140 [0.05]	Kagesawa et al. (2014)
$C/LiFePO_4/C$	≈400–1000/triaxial	NA	NA	160 [0.06], 130 [0.6], 80 [6]	n=20, ≈160 [0.06] n=≈130 [0.6] n=≈80 [6]	Hosono et al. (2010)
$LiFePO_4$/CNT/C	NF 168	37	NA	169 [0.05]; 121 [2]	n=50, ≈169 [0.05]	Toprakci et al. (2012)
$LiFePO_4$/C	100/shell 2–5	NA	NA	169 [0.1], 162 [0.5], 150 [1], 114 [5], 93 [10]	n=100, 146 [1]	Zhu et al. (2011)
V_2O_5	40–70 × 10–20/nanobelts	NA	NA	350 [NA]	n=25, 270 [NA]	Ban et al. (2009)
V_2O_5 $(+V_xO_2)$	100–200	50 × 100 NRs	NA	275–390 [NA]	n=50, 187–201 [NA]	Mai et al. (2010)
V_2O_5	300–800	70–80	NA	300 [0.1], 160 [1.0]	n=20, 200 [0.1]; 112 [1.0] n=50, 150 [0.1]	Cheah et al. (2011)

Notes: NA, data not available; NR, nanorod; NS, nanosheet; NF, nanofibers; NT, nanotubes; NP, nanoparticles; VGCF, vapor grown carbon fibers.

TABLE 6.2
Electrospun Negative Electrode Materials for Li-Ion Batteries

Material	Fiber/NP Diameter (nm)/Structure	Crystallite/Pore Size (nm)	BET Surface (m²/g)	Discharge Capacity First Cycle (mA h/g) [Rate (C)]	Discharge Capacity nth Cycle (mA h/g) [Rate (C)]	References
CNFs	200–300	0.5–5	NA	1000 [0.08]	n=2, 500 [0.08]	Kim et al. (2006)
CNFs	≈150–300	NA	NA	621 [0.13]	n=20, 479 [0.13] n=50, 454 [0.13] n=2, 440 [0.27] n=2, 345 [0.54]	Ji et al. (2010)
CNFs + Si NPs	NA	NA	NA	1710 [0.1] 870 [0.33]	n=5, 1150 [0.1] n=50, 1100 [0.1] n=30, ≈700 [0.33]	Wang et al. (2010a)
CNFs	≈200–300	NA	NA	233 [0.13] 82 [5.38]	n=1, 173 [0.53] n=200, 169 [0.53]	Chen et al. (2014b)
HCNF	HCNF-800 1080	1.2/6.52	20.35	1022 [0.14] 436 [0.54]	n=10, 390 [0.14]	Lee et al. (2012b)
CNFs, coaxial	CNF-800 200 CNF-1250 200	6.7 13.8	186.6 14.6	≈550 [0.054] 271 [0.054]	n=100, ≈380 [0.27] n=100, ≈210 [0.27]	Jin et al. (2014)
C sphere at NT, triaxial	Spheres 15 NT 200	NA	NA	1597 [0.14]	n=100, 965 [0.14], n=650, 330 [10]	Chen et al. (2012)
CNFs (600, 800, 1000)	310 210 200	1.05 1.88 2.16	750 260 230	≈520 [0.27] ≈810 [0.27] ≈215 [0.27]	n=550, ≈215 [0.27] n=550, ≈425 [0.27] n=550, ≈220 [0.27]	Wu et al. (2013)
CNFs + Fe₂O₃ NC	Fiber 220–600	23	46	820 [0.2] 262 [5]	n=100, 820 [0.2]	Zhang et al. (2014)

(Continued)

TABLE 6.2 (Continued)
Electrospun Negative Electrode Materials for Li-Ion Batteries

Material	Fiber/NP Diameter (nm)/Structure	Crystallite/Pore Size (nm)	BET Surface (m²/g)	Discharge Capacity First Cycle (mA h/g) [Rate (C)]	Discharge Capacity nth Cycle (mA h/g) [Rate (C)]	References
CNFs+ Fe$_2$O$_3$ NC	Fibers 100 Particles 150	18.3	61.3	≈1000 [0.27] ≈350 [13.44]	n=40, 830 [0.13]	Wu et al. (2014c)
Fe$_2$O$_3$ (15%) NPs in CNFs	NPs 20 CNFs 487	NA	NA	422 [0.27] 288 [1.34]	n=2, 519, [0.13] n=75, 488 [0.13]	Ji et al. (2012)
C/Fe$_3$O$_4$	Porous microbelts 4000	0.11 cm³/g	174.6	1751 [0.2]	n=50, 710 [0.2]	Lang and Xu (2013)
Sb/CNFs	Sb 15–20 CNFs 200	15–20	NA	631 [0.07] 337 [5]	n=400, ≈450 [0.33]	Wu et al. (2014a)
SnSb/CNFs	CNFs 200	NA/27	89.3	1088 [0.07] 354 [2.16]	n=150, 637 [0.07]	Xue et al. (2013)
Si–C–TiO$_2$	Si-C micropellets 300 Tio2 NFs 300–80	Pellets 300, Fibers 300–800	NA	610 [0.15]	n=55, 677 [0.08]	Wu et al. (2014b)
Si NPs at CNTs	SiNP at CNTs	150 nm	NA	1000 [0.8] 700 [8]	n=200, 810 [1]	Wu et al. (2012)
Si NPs–CNTs/C	C encapsulated 50 SiNP + CNTs (1%) 200	NA	NA	1760 [0.1]	n=1, 1350 [1] n=100, 1000 [1]	Hieu et al. (2014)
Si NPs at C	Si NP CNTs 1000	NA	NA	1381 [0.1] 721 [12]	n=20, 446 [4.91] n=1500, 366 [4.91]	Hwang et al. (2012)
Si NPs at CNFs	Si core 930 C shell 1287	NA	NA	967 [0.07]	n=5, 596 [0.07] n=50, 590 [0.07]	Lee et al. (2012a)
Si NPs at HCNF	440, wall 40	NA	NA	2691 [0.023] 300 [4.54]	n=50, 1601 [0.023]	Kong et al. (2013)

(Continued)

TABLE 6.2 (Continued)
Electrospun Negative Electrode Materials for Li-Ion Batteries

Material	Fiber/NP Diameter (nm)/Structure	Crystallite/Pore Size (nm)	BET Surface (m²/g)	Discharge Capacity First Cycle (mA h/g) [Rate (C)]	Discharge Capacity nth Cycle (mA h/g) [Rate (C)]	References
N-doped TiO$_2$/C	NFs 250	NA/10	191	389 [0.1] 150 [5]	n=100, 264 [0.1]	Ryu et al. (2013)
C at SiNW	Si NWs 250	NA	NA	1900 [0.1]	n=90, 900 [0.1]	Yoo et al. (2012)
CNFs + Si NPs	CNFs 300–500 Si NPs 50–80	NA	NA	1060–780 [0.01]	n=50, 410–380 [0.01]	Choi et al. (2010)
CNFs + Sn NPs	CNFs 100–1000 Sn NPs 1	NA	NA	816 [NA]	n=20, 400 [NA]	Zou et al. (2010)
CNFs + Sn/SnO$_x$ NPs	CNFs ≈50–500 Sn NPs 20–40	NA	352	1192 [0.03]	n=40, ≈520 [0.03]	Zou et al. (2011)
C/SnO$_2$/C	C-core 200 SnO$_2$-shell 50 C-skin 15	NA	NA	1050 [0.1] 180 [10]	n=200, 837 [0.1]	Kong et al. (2012)
CNFs + Sn NPs	CNFs 150–350 Sn NPs 30–40	NA	NA	1211.7 [0.025]	n=2, ≈800 [0.025]	Yu et al. (2010)
CNFs + Sn at CNPs	C hollow NFs 150 wall thickness 30 Sn NP 100 at 10	NA	NA	1156 [0.1]	n=100, ≈780 [0.1] n=100, ≈450 [5]	Kong et al. (2012)
CNFs + Ni NPs	CNFs ≈ 150–200 Ni NPs 20	NA	NA	795 [0.13] ≈570 [0.8] ≈450 [1.34] ≈400 [2.7]	n=50, ≈500 [0.13] n=50, ≈400 [0.8] n=50, ≈350 [1.34] n=50, ≈250 [2.7]	Ji et al. (2009)

(Continued)

TABLE 6.2 (*Continued*)
Electrospun Negative Electrode Materials for Li-Ion Batteries

Material	Fiber/NP Diameter (nm)/Structure	Crystallite/Pore Size (nm)	BET Surface (m²/g)	Discharge Capacity First Cycle (mA h/g) [Rate (C)]	Discharge Capacity nth Cycle (mA h/g) [Rate (C)]	References
Ni/C	NF 400	NA	37.3	1698 [NA] 135 [NA]	n=50, 457 [NA]	Wang et al. (2013)
CNFs + MnO_x NPs	CNFs ≈ 150–200 MnO_x NPs 21–42	NA	NA	≈650 [NA]	n=2, 558 [NA] n=50, ≈500 [NA]	Lin et al. (2010)
CNFs + Fe_3O_4 NPs	CNFs ≈ 350 Fe_3O_4 NPs 8–52	NA	NA	1551 [0.22] 623 [1.73]	n=7, 763 [0.22] n=80, 1007 [0.22]	Wang et al. (2008)
Sb NP at C	NF 400	NA	NA	422 [0.16] 88 [10]	n=300, 350 [0.16]	Zhu et al. (2013)
SnO_2	≈50–80	10	NA	1650 [0.1]	n=50, ≈480 [0.13]	Yang et al. (2010b)
SnO_2	NTs 220 wall thickness 25	Pores 11	28	1650 [0.23]	n=2, ≈1100 [0.23] n=50, 807 [0.23] n=40, ≈350 [2.4] n=30, ≈150 [10]	Li et al. (2010)
SnO_2/TiO_2 composite	Homogeneous SnO_2/TiO_2 NFs Heterogeneous SnO_2/TiO_2 NFs SnO_2 NPs/TiO_2 NFs	1500–2000 Nano/250–500 <100/800–1200	NA	780 [0.16] 560 [0.19] 500 [0.2] 140 [2.62] 130 [2.99] 160 [3.19]	n=75, 60 [0.2] n=75, 20 [0.16] n=75, 40 [0.19]	Tran et al. (2014)
TiO_2	60	10	54	175 [0.45]	n=50, ≈90 [0.45]	Reddy et al. (2010)
TiO_2–graphene NF	150	6	191	260 [0.18] 71 [9.7]	n=214; 300, 131 [1.23]	Zhang et al. (2012)

(Continued)

TABLE 6.2 (Continued)
Electrospun Negative Electrode Materials for Li-Ion Batteries

Material	Fiber/NP Diameter (nm)/Structure	Crystallite/Pore Size (nm)	BET Surface (m²/g)	Discharge Capacity First Cycle (mA h/g) [Rate (C)]	Discharge Capacity nth Cycle (mA h/g) [Rate (C)]	References
TiO_2 rice grains	20	20	34.1	207 [0.45]; 89 [4.5]	n=800, 140 [0.45]	Zhu et al. (2012)
$Nb_{0.05}Ti_{0.95}O_2$	NTs 50	7	66	250 [0.05]	n=20, 140 [0.05]	Fehse et al. (2013)
$TiNb_2O_7$	NFs 100–300	57	NA	271 [1.92]	n=100, 257 [1.92]	Jayaraman et al. (2014)
$Li_4Ti_5O_{12}$	100–200/3D cross-bar NFs	<100	NA	192 [0.5]; 170 [1.5]	n=30, 140 [0.5]; n=30, 87 [1.5]	Lu et al. (2007)
$Li_4Ti_5O_1$	NFs <300 nm	NA	NA	162 [0.1]; 138 [10]	n=50, 159 [0.1]	Jo et al. (2012)
$Li_4Ti_5O_1$	NFs 200–300	NA	NA	154 [0.07]; 125 [1]	n≈60, 150 [0.07]	Sandhya et al. (2013)
Co_3O_4	200	<50	NA	1336 [0.5]	n=40, 604 [0.5]	Ding et al. (2008)
Co_3O_4/C	200–300	50–100	NA	1146 [0.11]	n=20, 861 [0.11]	Zhang et al. (2011)
Co_3O_4 at TiO_2	NSs on NFs 200–500	NA	NA	632 [0.22]	n=480, 603 [0.22]	Wang et al. (2012)
α-Fe_2O_3	NRs 150	46	NA	1515 [0.05]; ≈750 [2.5]	n=50, 1095 [0.05]	Cherian et al. (2012)
α-Fe_2O_3	Hollow NFs 180	21	25	1795 [0.06]	n=40, 1293 [0.06]	Chaudhari and Srinivasan (2012)
Mn_3O_4	NFs 100–300	20–25	NA	2200 [NA]	n=5, ≈650 [NA]; n=50, ≈450 [NA]	Fan and Whittingham (2007)
CuO	NFs 200	30	NA	1071 [1]; 167 [2.22]	n=100, 426 [1]	Sahay et al. (2012)
$Zn_{1-x}Mn_xFe_2O_4$	Porous NFs 140–200	NA	NA	1350 [NA] (x=0.7)	n=50, 612 [NA] (x=0.7)	Teh et al. (2013)

(Continued)

TABLE 6.2 (Continued)
Electrospun Negative Electrode Materials for Li-Ion Batteries

Material	Fiber/NP Diameter (nm)/Structure	Crystallite/Pore Size (nm)	BET Surface (m²/g)	Discharge Capacity First Cycle (mA h/g) [Rate (C)]	Discharge Capacity nth Cycle (mA h/g) [Rate (C)]	References
$ZnCo_2O_4$ NT	NTs 200–300 crystallite size 30	3	NA	1710 [0.11]	n=30, 1454 [0.11] n=30, 794 [2.21]	Luo et al. (2012)
NiO	NFs 1000	100	NA	675 [0.3] 204 [11.14]	n=100, 583 [0.11]	Aravindan et al. (2013b)
$NiFe_2O_4$	NFs 200	20	50	1450 [1]	n=100, 1000 [1]	Cherian et al. (2013)
$NiFe_2O_4$	NFs 218	60	NA	1398 [0.2]	n=100, 660 [0.2]	Lee et al. (2013b)
Ag or Au/TiO_2	NFs 50 NPs 5–10	6/3.5	25–53	180–300 [1]	n=2, 160–170 [1]	Nam et al. (2010)
NaS	NFs 150	NA	NA	250 [0.1] 72 [6]	n=500, 150 [1]	Hwang et al. (2013)
$Li_4Ti_5O_{12}$ + graphene	<1000	NA	169	164 [0.2] 110 [22]	n=1200, 100 [22]	Zhu et al. (2010a)

Notes: NA, data not available; NR, nanorod; NS, nanosheet; NF, nanofibers; NT, nanotubes; NP, nanoparticles.

TABLE 6.3

Electrospun Separators and Polymer Gelled Electrolytes for Li-Ion Batteries

Material	Average Fiber Diameter (nm)	Anode/Cathode Materials	Conductivity $(10^{-3}$ S/cm)	Discharge Capacity First Cycle (mA h/g) [Rate (C)]	Discharge Capacity after (n) Cycles (mA h/g)	References
PVDF	250		1.7	NA	NA	Choi et al. (2003)
PVDF	450		>1	NA	NA	Kim et al. (2004)
PVDF	100–800		1.6–2.0	NA	NA	Choi et al. (2004)
PVDF-HFP	500		>1	NA	NA	Kim et al. (2005)
PAN	330		>1	145 [0.5]	136 (n = 150)	Choi et al. (2005)
PVDF, PVDF-HFP	420	MCMB/LiCoO$_2$	1.04	141 [1]	NA	Lee et al. (2006)
PVDF-HFP	1 μm		4.7	125 [1]	117 (n = 100)	Li et al. (2007)
PVDF-HFP/IL	~700		2.3	149 [0.1]	152 (n = 24)	Cheruvally et al. (2007)
PVDF-HFP/SiO$_2$	1–2 μm		8.06	170 [0.1]	NA	Raghavan et al. (2008a)
PVDF-HFP/NP	1–5 μm		7.2	164 [0.1]	156 (n = 30)	Raghavan et al. (2008b)
PI	300–625	MCMB/LiNi$_{0.8}$Co$_{0.2}$O$_2$	NA	~320 [0.1]	~305 (n = 21)	Bansal et al. (2008)
PAN	380/250		NA	124.5 [0.5]	102.7 (n = 250)	Cho et al. (2008)
PVDF-HFP/SiO$_2$	2 μm		4.3	170 [0.1]	160.5 (n = 80)	Kim et al. (2008)
PAN/SiO$_2$	270		11	139 [0.5]	127 (n = 150)	Jung et al. (2009)
PVDF	~125		N/A	177 [0.1]	150 (n = 10)	Yang et al. (2009a)
PVDF-HFP/PAN	320–490		3.9–6.5	145 [0.1]	136 (n = 50)	Raghavan et al. (2010b)
PVDF-HFP	1 μm		4.32	140 [0.1]	128 (n = 50)	Raghavan et al. (2010a)
PVDF-CTFE/Al$_2$O$_3$	2–4 μm	MCMB/LiCoO$_2$	0.49	143.5 [0.1]	131.8 (n = 200)	Lee et al. (2010)
PVDF/SiO$_2$	490		4.7	NA	NA	Kim et al. (2011)
PVDF-CTFE	230		>2	120 [0.33]	114 (n = 10)	Croce et al. (2011)
PAN/LaTO	230		1.95	162 [0.2]	156 (n = 50)	Liang et al. (2011a)
PVDF	500–900	Li/LiMn$_2$O$_4$	NA	123 [0.5]	120 (n = 50)	Gao et al. (2006)

(Continued)

TABLE 6.3 (Continued)

Electrospun Separators and Polymer Gelled Electrolytes for Li-Ion Batteries

Material	Average Fiber Diameter (nm)	Anode/Cathode Materials	Conductivity (10^{-3} S/cm)	Discharge Capacity First Cycle (mA h/g) [Rate (C)]	Discharge Capacity after (n) Cycles (mA h/g)	References
PAN	900	C_{GR}/LiCoO$_2$	0.017	113 [0.05]	95 (n=30)	Carol et al. (2011)
LATP/PAN	200–250	Li/LiFePO$_4$	3.6	160 [0.2]	152 (n=50)	Liang et al. (2011b)
PAN	350	Li/LiFePO$_4$	2.14	145 [0.1]	130 (n=50)	Raghavan et al. (2011)
PVDF-CTFE and PVDF	120–200		NA	NA	NA	Alcoutlabi et al. (2012)
PAN/PMMA (IL)	450–600	Li/LiFePO$_4$	3.6	134 [0.2]	125 (n=50)	Rao et al. (2012)
PVDF-PMMA/TiO$_2$	330		2.95	NA	NA	Cui et al. (2013)
PVDF-HFP	1000	Li/LiMn$_2$O$_4$	3.2	109 [1]	100 (n=800)	Aravindan et al. (2013a)
PVDF/SiO$_2$	300–380	Li/LiFePO$_4$	2.6	159 [0.2]	163 (n=50)	Yanilmaz et al. (2013)
PVDF-CTFE/PVDF-HFP	130–190		NA	NA	NA	Lee et al. (2013a)
PVDF	100–300	Li/LiFePO$_4$	1.3–1.8	163 [0.2]	150 (n=50)	Lee et al. (2013a)
PI	200	Li/Li$_4$Ti$_5$O$_{12}$	NA	160 [0.2]	160 (n=100)	Miao et al. (2013)
PET	300–420	Li/LiFePO$_4$	2.27	147 [0.1]	140 (n=50)	Hao et al. (2013)
PVDF-HFP/nanoclay	1–2.5 μm	Li/LiFePO$_4$	5.5–2.9	160 [0.1]	147 (n=50)	Shubha et al. (2013)
PI/Al$_2$O$_3$	300	C_{GR}/NMC-LiMn$_2$O$_4$	0.36	136 [0.2] / 125 [1]	132 (n=200) / 120 (n=200)	Lee et al. (2014)
PDA at PVDF	400–500	Li/LiMn$_2$O$_4$	0.96	105 [0.5]	100 (n=100)	Cao et al. (2014)
PVDF/SiO$_2$	300–1000	Li/LiFePO$_4$	2.1–2.6	162 [0.2] / 140 [1]	160 (n=50) / 132 (n=100)	Yanilmaz et al. (2014)
PVDF/Al$_2$O$_3$ on PE	200–800	Li/LiCoO$_2$	1.25	138 [0.5]	108 (n=100)	An et al. (2014)

Notes: NA, data not available; PAN, polyacrylonitrile; PI, polyimide; PVDF, polyvinylidene fluoride; PVDF-HFP, poly(vinylidene fluoride-co-hexafluoropropylene); PVDF-CTFE, poly(vinylidene fluoride-co-chlorotrifluoroethylene); PET, polyethylene terephthalate; PE, polyethylene; PDA, polydopamine; MCMB, mesocarbon microbeads; IL, room temperature ionic liquids; LaTO, lanthanum titanate oxide; LATP, lithium aluminum titanium phosphate.

providing a high initial discharge capacity of 182 mA h/g, that is, slightly larger than that of conventional powder and film electrodes. However, a rapid capacity fading was observed for these materials, mostly provoked by the partial dissolution of the active material and the formation of a SEI layer containing precipitated Li_2CO_3 and CoF_2 (Gu et al., 2007).

A possible way of reducing these parasite effects and to improve cycling performances is by coating. For instance, Gu et al. produced electrodes from MgO-coated Li_2CO_3 nanofibers obtained by coaxial spinning (Gu et al., 2007). Such electrodes exhibited an excellent reversibility and a smaller impedance growth, with obvious improvement of cyclability compared to bare $LiCoO_2$ fiber electrodes (cf. Table 6.1). Coaxial electrospinning was employed with success in several cases, as resumed in the recent review of Qu et al. (2013). To remain in the field of layered material, one can cite the example of the nickel–manganese–cobalt-layered material $LiNi_{0.8}Co_{0.1}Mn_{0.1}O_2$, which showed a substantial improvement of reversibility and cyclability upon MgO coating (Gu and Jian, 2008). Similar results were also obtained for other core–shell architectures prepared by different methods allowing the protection of the layered material core by a shell of a more stable compound, such as $Li(Ni_{1/3}Co_{1/3}Mn_{1/3})O_2/Li(Ni_{1/2}Mn_{1/2})O_2$ obtained by electrospinning assisted by a sol–gel posttreatment (Yang et al., 2010a) or $LiCoO_2/LiPON$ in which the LiPON layer was deposited on the electrospun fibers by radio-frequency sputtering (Lu et al., 2008). The particular 3D electrode structure obtained in this way, with highly accessible surface area and continuous networks, is very promising for the design of improved batteries maximizing both power and energy density.

Among the transition metal oxide with a spinel crystal structure, very good results were obtained also in the case of the $LiMn_2O_4$ (LMO). For instance, Jayaraman et al. reported excellent lithium storage performances for electrospun hollow LMO nanofibers, which retained 87% of their initial reversible capacity after 1250 cycles at the 1 C rate (Jayaraman et al., 2013), in the range of the values that were achieved for the full 1D cell cited earlier (119 mA h/g at 150 mA/g) (Jayaraman et al., 2014). Such values are slightly higher than those of the LMO nanofibers of Zhou et al. (2014) and Liu et al. (2012) achieving 112 and 80 mA h/g, respectively. Mixed metal spinels such as $LiNi_{0.5}Mn_{1.5}O_4$ were also obtained by electrospinning, yielding to very interesting stable capacities of ~110 mA h/g for more than 40 cycles (Arun et al., 2014).

Electrospun transition metal phosphates with olivine structure are probably the most studied polyanionic positive electrode materials. The most important compound of this family is $LiFePO_4$ (LFP), which is characterized by a large specific capacity (170 mA h/g), a relatively high discharge potential, very good thermal stability and safety, low toxicity, and especially low cost. The main drawback of LFP is for sure the very low electronic conductivity, which is the main cause of its low rate capability in lithium batteries. In order to dodge this limitation, the used strategies are to reduce the particle size down to the mesoscale or even to the nanoscale and to coat the particles with a conducting carbon layer in order to improve electron percolation (Zhang, 2011). Such coatings have also the asset of protecting the surface of the active material from undesired side reactions leading to its partial decomposition. Electrospinning is therefore well adapted to obtain in a single-step mesostructured $LiFePO_4/C$ composites (Hosono et al., 2010; Toprakci et al., 2011, 2012; Zhu et al., 2011).

As an example, Zhu et al. prepared very thin single-crystal carbon-coated $LiFePO_4$ nanowires (Zhu et al., 2011). The carbon coating and the 3D connectivity of the network ensured high electron conductivity and helped the Li^+ diffusion by optimizing the diffusion of the liquid electrolyte, leading to very good rate performances and cycling capability. For instance, at 0.1 C discharge rate, the capacity was close to the theoretical value (169 mA h/g), and after 100 cycles at 1 C, it remained at 86% of initial one. Postmortem investigation of the deeply cycled material showed that the morphology of the carbon-coated $LiFePO_4$ fibers was largely preserved. Similar results were obtained by other research groups also for other olivine phosphates containing different transition metals, for instance, carbon-coated Mg-modified $LiMnPO_4$ showed a very good and stable performance of 135 mA h/g at a C/10 rate (Lu et al., 2014). Carbon-coated mixed-metal Fe–Mn lithium phosphates with variable Fe/Mn ratios were also obtained by this method, leading to excellent rate capabilities and cycling lives (Kagesawa et al., 2014).

Also vanadium pentoxide V_2O_5 has several drawbacks in spite of its high theoretical capacity (~400 mA h/g), for instance, in addition to its intrinsic structural instability upon charge/discharge cycling, it also exhibits strong kinetic limitations due to low electron conductivity. Electrospinning was thus also used to produce materials overcoming such limitations (Ban et al., 2009; Cheah et al., 2011; Mai et al., 2010). For instance, Cheah et al. prepared single phase electrospun V_2O_5 fibers providing an initial discharge capacity of 316 mA h/g in the voltage range of 1.75 and 4.0 V and a coulombic efficiency close to 100% throughout 50 charge–discharge cycles (Cheah et al., 2011). This study highlights once again the importance of the network morphology of electrospun fibers to access the maximum amount of electroactive material, thus improving the accessibility of electrolyte to the cathode material and the overall performances of the battery.

6.3.2 Nanostructured Negative Electrode Materials

Most commercialized batteries are nowadays based on graphitic carbon, which, due to security issues caused by the growth of metal dendrites during the charge process, substituted lithium metal electrodes in early batteries, marking the transition from lithium to Li-ion battery. In addition to safety, graphite has several advantages, such as low cost and good cycle life. The main limitations of graphite concern its relatively low rate capability and its low capacity; in fact, graphite is able to store up to one Li per six carbon atoms, yielding a theoretical capacity of 372 mA h/g (compared to the 3860 mA h/g for Li metal). Moreover, the negative electrode has usually a major impact on the cyclability compared to the positive one. In fact, the low working potential of the negative electrode is often at the limits of electrochemical stability window of the electrolyte and causes the precipitation, at the electrode surface, of more or less stable SEI layers, strongly influencing the cycle life. A performing negative electrode in terms of security, capacity, rate capability, and cycle life remains thus a major challenge, and a great deal of the use of electrospinning for Li-based batteries has dealt with the preparation of innovating negative electrode materials.

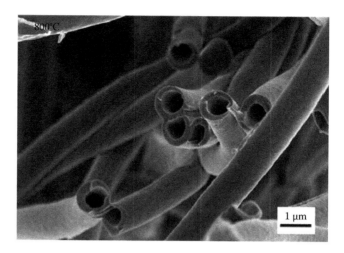

FIGURE 6.7   FE-SEM micrograph hollow carbon nanofibers after carbonization at 800°C. (Reprinted from *J. Power Sources*, 199, Lee, B.-S., Son, S.-B., Park, K.-M., Yu, W.-R., Oh, K.-H., and Lee, S.-H., Anodic properties of hollow carbon nanofibers for Li-ion battery, 53–60, Copyright 2012, with permission from Elsevier.)

Carbon nanofibers (CNFs) with different diameters (Wu et al., 2013) and shapes (Chen et al., 2012) as well as hollow (Figure 6.7) (Lee et al., 2012b) and porous structures can be obtained by electrospinning. Jin et al. (2014) have been produced via electrospinning of polymer blends followed by an adapted thermal treatment. All these efforts aim primarily at improving the electrolyte–electrode contact as well as reducing the length of lithium diffusion paths in order to enhance overall capacity but more importantly to enable cycling at elevated currents (Ji et al., 2010). For instance, CNFs prepared starting from PAN and heat treated at 1000°C produced reversible capacities very close to the theoretical ones at relatively high rates (350 mA h/g at a charge current of 100 mA/g) and largely exceeding it at low rates (450 mA h/g at a charge current of 10 mA/g). Such good performances are due to the specific nanometric texture obtained by electrospinning, including a large accessible surface area and a highly reduced lithium-ion diffusion path within the active material (derived from the nanometer-sized fiber diameter), high-carbon purity, and increased electrical conductivity (Kim et al., 2006). The main drawback of such a material remains its very high irreversible first discharge capacity of about 500 mA h/g, which was originally attributed to electrolyte decomposition and formation of an SEI at ca. 0.8 V, but which has been more recently related to irreversible Li trapping in highly dispersed carbonaceous materials (Memarzadeh Lotfabad et al., 2014). Such extra capacity exceeding the theoretic limit of graphite has been reported by other groups working with nanostructured carbon systems (Chen et al., 2012; Jin et al., 2014). For instance, Chen et al. obtained a very high reversible specific capacity of 969 mA h/g (Chen et al., 2012) with amorphous carbon nanotubes (CNTs) decorated with hollow graphitic carbon nanospheres prepared by using a novel triple-coaxial electrospinning method (Figure 6.8). Such materials displayed a remarkable capacity for more

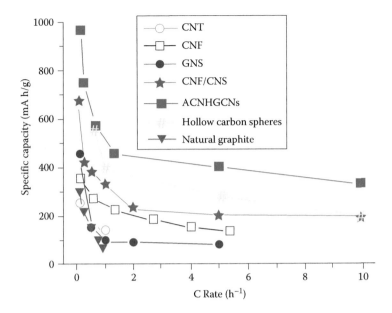

FIGURE 6.8 Comparison of rate capabilities of carbon nanotubes (CNTs), carbon nanofibers (CNF), graphitized nanosheets (GNS), CNF/GNS, amorphous CNTs hollow graphitic carbon nanospheres, hollow carbon spheres, and natural graphite. (Reproduced from Chen, Y., Lu, Z., Zhou, L., Mai, Y.-W., and Huang, H., Triple-coaxial electrospun amorphous carbon nanotubes with hollow graphitic carbon nanospheres for high-performance Li ion batteries, *Energy Environ. Sci.*, 5(7), 7898–7920, 2012. With permission of The Royal Society of Chemistry.)

than 600 cycles and a rate capability of 400 mA h/g at 5 C but also a very high first cycle irreversible capacity exceeding 600 mA h/g. Enhanced rate capability was also obtained by Nan et al. (2014; 473 mA h/g at 2.69 C) for nitrogen-doped porous CNF networks prepared from melamine and PAN precursors via electrospinning followed by carbonization and NH_3 treatment. In this case, nitrogen doping has the advantage of significantly enhancing electronic conductivity, improving electron percolation toward the current collector.

Examples of the application of electrospun carbon-based electrodes concerned also the insertion of sodium for the development of possible Na-ion batteries (*vide infra*), leading to generally lower capacities of about 250 mA h/g (Chen et al., 2014b; Jin et al., 2014). However, the case of sodium is different from the lithium one, since graphite does not easily insert sodium, and only hard-carbon structures (also known as nongraphitizable carbon, usually obtained by calcination of sugars) are known to be sufficiently active for Na^+ insertion, leading to capacities of the order of 300 mA h/g at best (Dahbi et al., 2014).

Another interesting group of negative electrode materials prepared by electrospinning is that of carbon fibers loaded with metal or metal oxide nanoparticles. In this case, the active phase is represented by both the carbon network, which also acts as electron wiring network and the nanosized additional metal or oxide species. Most examples concern alloy materials such as tin (Kong et al., 2012; Ni et al., 2014; Yu et al., 2010; Zhu et al., 2010b; Zou et al., 2010, 2011) or silicon (Choi et al., 2010; Hieu et al., 2014; Hwang et al., 2012; Kong et al., 2013; Lee et al., 2012a; Wang et al., 2010a; Wu et al., 2012, 2014b; Yoo et al., 2012), but also conversion materials such as iron oxide (Ji et al., 2012; Lang and Xu, 2013; Wu et al., 2014c; Zhang et al., 2014) or manganese oxide (Lin et al., 2010). As already mentioned in Section 6.1, alloy materials can produce higher theoretical capacities than carbon, but suffer from poor cycling performances due to enormous volumetric expansion (up to 300%) and contraction during alloying and dealloying with lithium, respectively. This causes the electrochemical grinding of the active material along with formation of SEI on freshly exposed surface. This resembles a continuous, irreversible consumption of lithium and electrolyte and progressively builds up a thick diffusion barrier. A possible strategy to overcome this intrinsic drawback is by dispersing the active material in the form of nanoparticles in a conducting porous matrix, which is able to buffer the volume expansion, stabilizing the particles of active material while assuring simultaneously the electron percolation network. Good results were obtained, for instance, using mesoporous carbon to support tin or tin oxide nanoparticles (Elia et al., 2014; Hassoun et al., 2008; Jahel et al., 2014), but also with CNF networks, which in addition also eliminate the need for binding or conducting additives.

A good example of the advantages of electrospinning for the preparation of Si-based negative electrodes is the work of Choi et al. (2010), who studied the effects of various surrounding confinements of Si nanoparticles on the electrochemical performance of Si-based negative electrodes. Three different types of surrounding confinements were compared: Si nanoparticles–embedded CNFs (prepared via electrospinning), carbon nitride–encapsulated Si nanoparticles with core/shell structure, and binder-enriched Si nanoparticle-based anode. They found that the electrospun material, having conducting hard surrounding confinements, strongly improved the cycling performances via the formation of nanofibrillar networks, which enhance the electronic and the ionic transport to the confined active material nanoparticles.

In other studies (see Table 6.2), excellent capacity retention and outstanding cycle life were obtained by improving the protection of the active material surface to reduce recurrent electrolyte degradation: for instance, Hieu et al. (2014) obtained a stable capacity of 1000 mA h/g for 100 cycles at 1 C together with excellent coulombic efficiency exceeding 99% with core–shell structured nanofibers consisting of silicon nanoparticles and CNTs encased in carbon shells. Such materials were fabricated by coaxial electrospinning using precursor solutions containing a blend of silicon nanoparticles, CNTs, and polyvinylpyrrolidone for the core, and PAN for the shell, and final carbonization at 1000°C for 1 h under nitrogen.

In a similar example, Kong et al. obtained a stable capacity of ≈850 mA h/g at 0.1 C after 200 cycles using carbon/SnO₂/carbon core/shell/shell hybrid nanofibrous mats prepared via single-spinneret electrospinning followed by carbonization and further hydrothermal treatment. In this case, the enhanced cycle life is mainly due to the morphological stability and reduced diffusion resistance induced by both the carbon core and the deposited carbon skin.

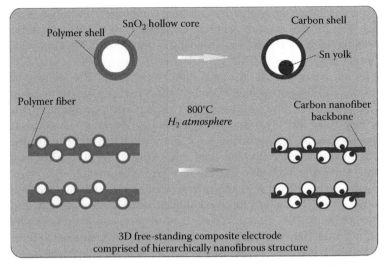

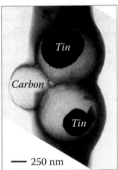

FIGURE 6.9 Sn/C yolk–shell free standing films. (Reproduced from Ni, W., Cheng, J., Shi, L., Li, X., Wang, B., Guan, Q. et al., Integration of Sn/C yolk–shell nanostructures into free-standing conductive networks as hierarchical composite 3D electrodes and the Li-ion insertion/extraction properties in a gel-type lithium-ion battery thereof, *J. Mater. Chem. A*, 2(45), 19122–19130, 2014. With permission of The Royal Society of Chemistry.)

Another possible strategy for preparing protected active material nanoparticles is that of yolk–shell structures integrated into nanofibrous 3D electric conducting structures, such as that proposed by Ni et al. (2014) (Figure 6.9). In the latter case, Sn/C composites were prepared by electrospinning of mixtures of tin oxide nanoparticles in PAN followed by heat treatment at 800°C in hydrogen or nitrogen. The obtained interconnected yolk–shell design forms a flexible conductive network, which inhibits the aggregation of tin nanoparticles, buffers the occurring volume strain, and prevents continuous electrolyte degradation by protecting the active materials surface, resulting in very good capacity retention not only in the classical half-cell configuration but also in full cells versus LiCoO$_2$ positive electrodes.

Fibrous carbon anodes have also been loaded with transition metal oxide nanoparticles. Such conversion-type materials suffer from similar drawbacks as alloy materials, hence the approach of forming composite with carbon is a versatile approach also for this class of material. Composites of electrospun carbon fibers have been reported for the most prominent transition metals Mn (Lin et al., 2010), Fe (Ji et al., 2012; Lang and Xu, 2013; Wu et al., 2014c; Zhang et al., 2014), Co (Zhang et al., 2011), and Ni (Ji et al., 2009; Wang et al., 2013). These materials have several advantages, such as high theoretical capacity, safety, durability, non-toxicity, and relatively low cost, but suffer from a poor electronic conductivity and usually bad cyclability. Two possible synergetic strategies to improve such poor performances are the application of a specific formulation strategy and the down-sizing of the active material to the nanoscale (Aricò et al., 2005). Electrospinning is again a promising method to address such problems, since it allows one to produce composite nanofibers with nanometric metal oxides embedded in electrospun conductive CNFs. The most prominent studied materials are actually iron oxide/carbon composites. For instance, Zhang et al. reported an impressive capacity of 820 mA h/g after 100 cycles at 0.2 C with electrospun fibers prepared from a solution of $Fe(acac)_3$ and polyacrylonitrile (PAN) in N,N-dimethylformamide (DMF) and further heat treatment at 500°C in Ar flow (Zhang et al., 2014).

Another interesting example is that of MnO_x/CNFs composite prepared by electrodepositing MnO_x nanoparticles directly onto electrospun CNFs and used directly as the negative electrode in lithium half cells. A remarkable capacity retention of 410 mA h/g after 100 cycles was observed, together with a very good morphological integrity after cycling (Figure 6.10).

The last family of electrospun negative electrode material is that of metal oxides working through lithium intercalation. Most publications deal with TiO_2 and titanates, such as $Li_4Ti_5O_{12}$ (LTO) or $TiNb_2O_7$. The latter was employed by Jaryaraman

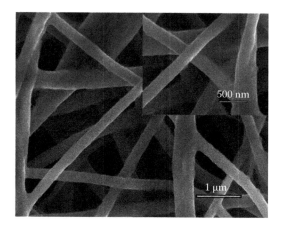

FIGURE 6.10   SEM images of MnO_x/carbon nanofibers after 100 charge/discharge cycles at a constant current density of 50 mA/g. Deposition time: 10 h. The inset shows MnO_x/carbon nanofibers with higher magnification. (Reprinted from *J. Power Sources*, 195(15), Lin, Z., Ji, L., Woodroof, M.D., and Zhang, X., Electrodeposited MnO_x/carbon nanofiber composites for use as anode materials in rechargeable lithium-ion batteries, 5025–5031, Copyright 2010, with permission from Elsevier.)

et al. (2014) in the full 1D cell presented earlier in this chapter, achieving a reversible capacity of 271 mA h/g at 150 mA/g in half-cell configuration versus Li metal . This value can be compared to those typically found for other titanium-based electrospun electrodes at comparable rates such as LTO by Sandhya et al. (2013; 125 mA h/g) and Jo et al. (2012; 155 mA h/g) or for TiO_2 by Reddy et al. (2010; 92 mA h/g) and Fehse et al. (2013; 85 mA h/g for Nb-doped samples). Its noticeable superiority is well in line with the elevated theoretical capacity of $TiNbO_7$ (388 mA h/g) compared to those of LTO and TiO_2 (150 and 336 mA h/g, respectively). Such materials work at relatively high voltage, are chemically very stable, and are well adapted to high cycle rates. The main drawback of these materials is their relatively low electronic and ionic conductivity, which makes their nanoscaling necessary in order to obtain a complete lithiation. Therefore, the use of 1D electrospun nanofibers is an appropriate strategy to improve both electron transport and lithium-ion diffusion during cycling, improving both columbic efficiency and rate capability of the electrochemical cell. Furthermore, such properties yield to long cycle life and good capacity retention, which is demonstrated by Jayaraman et al. (2014) achieving 257 mA h/g after 100 cycles.

Several approaches were followed to further improve the conductivity of such materials. For instance, Nam et al. prepared TiO_2 nanofibers embedding Au or Ag metal nanoparticles (between 5 and 10 nm in diameter) via a one-step electrospinning process followed by calcination at 450°C (Nam et al., 2010). The addition of metal nanoparticles to the network of titania nanofibers positively promoted lithium-ion diffusion and charge transfer, with an increase of the specific capacity of ≈20%, and an almost twofold improvement of the rate performances.

Doping with aliovalent ions is also a facile strategy to modify the electronic properties of electrospun titanium oxide. For instance, Wang et al. show that the conductivity, which is the crucial parameter for high-performance cycling, can be raised by two orders of magnitude by doping mesoporous TiO_2 with Nb (Wang et al., 2010c). Following this example, Fehse et al. prepared Nb-doped TiO_2 nanofibers by simple electrospinning of a titanium isopropoxide solution containing 10% of Nb precursor (Fehse et al., 2013). Such fibers showed very similar capacities at low rates, but a largely improved rate capability at high rates, with a twofold increase capacity at 3 C for the Nb-doped samples. In this case, electrospinning proved to be a very efficient method for producing homogeneously doped nanofibers with improved cycling performances.

6.3.3 NANOFIBER-BASED SEPARATORS AND ELECTROLYTES

Electrolytic separator membranes to be used with liquid electrolytes are among the most highly engineered and critical components of lithium battery systems, since they have to provide a barrier between the anode and the cathode while providing an optimized interconnectivity for ionic conduction within the membrane. Their porosity and soaking properties are thus very important and need to be optimized. To obtain such controlled membranes, typical methods such as phase inversion or casting of a polymer gel are usually employed. However, such methods do not allow a precise control of porosity. Moreover, residual solvent normally remains in the casted polymer and is cumbersome to remove completely. Such drawbacks may thus affect electrochemical properties or stability of the final separator, and thus of the whole cell. Electrospinning, on the contrary, is well adapted to produce polymer membranes, and it is not surprising that such application has been first applied to separator engineering (Choi et al., 2003, 2004; Kim et al., 2004). Electrospun membranes offer an interesting and scalable alternative, assuring high electrolyte uptake in their nanoporous matrix and very good homogeneity with some degree of control over the morphology. A list of the main publications in this field is reported in Table 6.3.

Since the first studies, most publications concern the synthesis of separator membranes based on either PVDF or some of its copolymers, such as PVDF-HFP or poly(vinylidene fluoride-co-chlorotrifluoroethylene) (PVDF-CTFE). Many publications also concern PAN- and polymethyl methacrylate (PMMA)-based gelled electrolytes, or a combination of the two of them, whereas only few papers introduce other types of polymers such as polyimide, PEO, or polyethylene terephthalate (see Table 6.3). Some of these works also concern the preparation of composite structure, either by combining electrospun fibers with inorganic oxides or by supporting nonwoven electrospun bead on top of porous polymer films.

The combination of PVDF and HFP probably provides the best performances because of the very good stability at low potentials (due to the very stable electron-withdrawing fluorinated carbon groups) and to the very good dissociation properties (due to the high dielectric constant, $\varepsilon = 8.4$), which improve the conductivity by increasing the number of charge carriers. Furthermore, the presence of HFP units helps improving the mechanical stability of the membranes. PAN and PMMA also provide very good conductivity at room temperature, but on the other hand, their relatively poor mechanical properties and the syneresis of solvent molecules hamper their use in commercial cells. In general, electrospun separators offer better performances than gelled polymer electrolytes, since the large porous volume provided by their nonwoven structure allows for a high uptake of electrolyte, thus enhancing the transport properties within the membrane structure.

The performances of electrospun separators can also be compared to those of membranes made by classical methods. Raghavan et al. (2010a), for instance, showed that the much higher porous volume and the more uniform structure of PVDF-HFP electrospun membranes led to significantly higher ionic conductivity and thus to much better cell performances than membranes obtained by phase inversion. Such tendency was confirmed also for PAN membranes, which showed improved cycle life and thermal stability compared to a conventional polyolefin microporous separator (Cho et al., 2008).

As already mentioned earlier, several studies have also concerned composite membranes in which such polymers are coupled to inorganic oxides such as silica, alumina, lithium lanthanum titanate (LLTO), titanium oxide, or clays. In these studies, it is clearly shown that the addition of inorganic oxides has a positive effect on ionic conductivity. In some studies on PAN membranes, for instance, it was shown that the addition of up to 12% SiO_2 (Jung et al., 2009) or of 15% LLTO (Liang et al., 2011a) induces a decrease of the diameter of the fibers due to the electrostatic repulsion of the inorganic particles during electrospinning, leading to membranes with greater porosity and electrolyte uptake. On PVDF-HFP, the opposite effect on

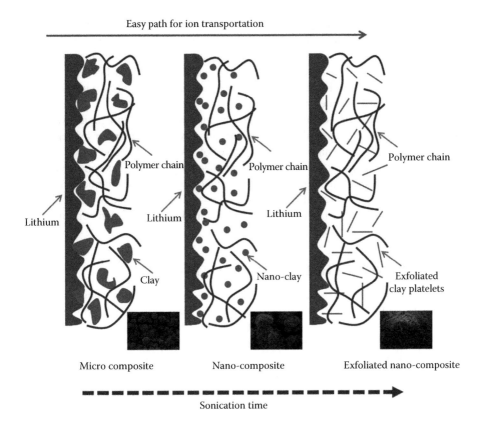

FIGURE 6.11 Schematic representation of the effect of sonication time on composite morphology and ion conduction channel formation in composite polymer electrolyte with nanoclay. (Reprinted from *Mater. Res. Bull.*, 48(2), Shubha, N., Prasanth, R., Hoon, H.H., and Srinivasan, M., Dual phase polymer gel electrolyte based on non-woven poly(vinylidenefluoride-co-hexafluoropropylene)-layered clay nanocomposite fibrous membranes for lithium ion batteries, 526–537, Copyright 2013, with permission from Elsevier.)

the diameter of the fibers was observed with the addition of SiO_2 (Kim et al., 2011; Raghavan et al., 2008a). In this case, the observed improvement of the ionic conductivity and of the electrolyte uptake was attributed to a decrease in PVDF crystallinity induced by the presence of the inorganic oxide. Finally, significantly enhanced ionic conductivity and compatibility with lithium electrodes were obtained by the incorporation of exfoliated layered nanoclays in electrospun PVDF-HFP membranes (Figure 6.11) (Shubha et al., 2013). In this case, the addition of clay was found not only to improve the electrolyte uptake and the conductivity, but also to significantly reduce the interfacial resistance by stabilizing the passivation layer, thus leading to a better compatibility with the lithium electrode.

The last family of separators is that of multilayered composite membrane containing at least one electrospun layer. For instance, Raghavan et al. fabricated a trilayer membrane composed of an internal PVDF-HFP layer sandwiched between two PAN ones via sequential electrospinning of the two polymer solutions and compared it to a membrane obtained from a blend of the two polymers. The performances of the multilayered composite were not as good as those of the mixed membrane, indicating that the polymer blend mixture induces beneficial interactions between the two cospun polymers, resulting in a greater porosity and electrolyte uptake in spite of the larger average fiber diameter.

More complex architectures could be obtained by adding inorganic oxide particles. For instance, Lee et al. (2010) prepared a composite separator by depositing a mixed polymer–inorganic electrospun layer of PVDF-CTFE and Al_2O_3 on both sides of a PE membrane. Such system provided better performances compared to the starting PE film due to a higher electrolyte uptake and improved confinement within the membrane (Lee et al., 2010). Similar results were also obtained by An et al., who also reported improved thermal resistance to shrinkage at high temperatures, and thus of the safety (An et al., 2014).

6.4 TECHNICAL CHALLENGES AND PERSPECTIVES

As it was shown in the previous sections, a large part of the most common battery materials could be easily and successfully manufactured by electrospinning, from high-voltage positive electrodes to metal-based negative ones, to fully functional separators. Such materials provided in several cases improved performances thanks to the favorable and easily controlled 1D morphology of nonwoven fibers, paving the way to the potential application of this technique to the fabrication of high-performance Li-ion battery. Several challenges, however, remain to be overcome before electrospinning can be widely applied. In particular, there are still limitations on the types of materials that can be fabricated by electrospinning, and it may be difficult to fully eliminate defects such as beads and obtain completely homogeneous nanofibers. Supplementary developments are also required to further reduce the diameters of the fibers, especially for applications such as nanofibrous separators.

Nonetheless, these promising results have led several research groups to apply electrospinning to the fabrication of materials for new emerging post-Li systems, that is, lithium–sulfur and sodium-ion cells. A short introduction of these systems together with a few representative results obtained with electrospun materials are presented in the following sections.

6.4.1 ELECTROSPINNING FOR NEXT-GENERATION SODIUM-ION BATTERIES

The actual and future widespread expansion of lithium-based batteries has recently raised the question about the long-term availability of lithium resources. In fact, even though lithium is not really a rare element, lithium-rich reserves are unevenly spread within the Earth's crust, and in the long term, it is expected to become more and more rare and expensive (Tarascon, 2010). Unlike Li, Na is one of the most abundant elements on Earth's crust. Moreover, batteries employing Na instead of Li ions are safer and more environmentally benign. For this reason, sodium-based batteries have very recently regained interest after a first period of active research in the past decades (Ellis and Nazar, 2012; Kim et al., 2012; Melot and Tarascon, 2013;

Palomares et al., 2012; Slater et al., 2012; Yabuuchi et al., 2014). In particular, they have been suggested as a viable and cheap alternative for large-scale storage systems, which do not require as high energy densities as those obtained with lithium-based systems.

Electrospinning was thus recently applied for both positive and negative electrode materials. Interesting examples of positive electrode materials are those of $Na_3V_2(PO_4)_3$ (Li et al., 2015) and of the mixed-metal layered oxide $Li_{1+x}(Mn_{1/3}Ni_{1/3}Fe_{1/3})O_2$ (Kalluri et al., 2014), which exhibited very good performance ascribable to the specific 1D morphology. In the case of negative electrode materials, most examples are related to Sb-based alloy materials (Kim and Kim, 2014; Lv et al., 2015; Wu et al., 2014a; Zhu et al., 2013), whereas only very few concern carbon- (Chen et al., 2014b; Jin et al., 2014) or titanate-based (Liu et al., 2013) negative electrodes. The case of antimony is particularly interesting, due to the surprising and unexpected good cycling behavior of this metal, in spite of the very large volume expansion upon reaction with sodium (Darwiche et al., 2013). Electrospun Sb–carbon materials showed a noticeable improvement of the rate capability and interesting cycling stabilities compared to pure antimony. Such improved electrochemical performances were attributed to the uniform distribution of Sb nanoparticles in the CNFs, providing a conductive and buffering matrix for effective release of the mechanical stress caused by sodium insertion/extraction.

6.4.2 ELECTROSPINNING FOR LITHIUM–SULFUR BATTERIES

The large-scale development of full electric vehicles is strongly related to the development of high-energy batteries, exceeding the actual performances of commercial Li-ion battery. Two promising technologies are lithium–air (Li/O_2) and lithium–sulfur (Li/S) systems. While the Li–air technology is still far from the market, the Li/S batteries are widely considered as one of the next-generation energy storage system (Bruce et al., 2011; Chen et al., 2014a). Li/S system is based on the electrochemical reaction of sulfur with lithium to form Li_2S. A particularity of this system mostly lies in its working mechanism, which is very complex, and has been largely studied in the last few years (Ji and Nazar, 2010; Manthiram et al., 2014; Xu et al., 2014; Yin et al., 2013). The overall reaction between element sulfur and lithium to form Li_2S is not a direct intercalation reaction. In fact, the reduction of solid sulfur is accompanied by the formation lithium polysulfides (Li_2S_x, $2 \leq x \leq 8$), which are usually soluble in the organic electrolyte.

The main limitations of the lithium–sulfur cell are related on the one hand to the bad electronic conductivity of elemental sulfur and Li_2S, which requires the addition of high fractions (up to 50%) of a conducting carbon additive, and on the other hand the diffusion of soluble polysulfides between the two electrodes giving rise to a loss of active material both at the positive and at the negative electrode, and to the well-known "shuttle mechanism," which strongly affects the cell performance. Several methods have been suggested to tackle these limitations, such as confining the sulfur inside a porous carbon structure to improve its accessibility, coating carbon/sulfur composites with polymers to reduce polysulfide migration, or adding small fraction of porous silica or titania to the carbon/sulfur composite.

These additives act as polysulfide reservoirs, decreasing their concentration in the electrolyte and leading to a higher utilization of sulfur and increased capacities (Evers and Nazar, 2013).

Ji et al. (2011) made use of electrospinning to obtain porous CNFs encapsulating sulfur particles. These nanocomposites showed reversible capacities as high as 1400 mA h/g at 0.05 C, good discharge capacity retention, and enhanced rate capability in rechargeable Li/S cells. These promising performance were attributed to the high electrical conductivity and the extremely high surface area of the CNFs that homogeneously disperse and immobilize S on their porous structures, alleviating the polysulfide shuttle phenomenon. Similar results were obtained on sodium–sulfur composites at room temperature (Hwang et al., 2013). These materials showed very stable cycling capacities for ~500 cycles and very high rate performances.

More recently, Yao et al. (2014) used electrospun CNFs with tin-doped indium oxide nanoparticles decorating the surface as hybrid 3D electrodes to maximize the number of deposition sites for polysulfides, obtaining outstanding cycling performances using a catholyte containing Li_2S_8. The nonwoven structure of the fibers allowed enough porous volume to accommodate the electrolyte and to assist the precipitation of sulfur on the surface of the indium oxide additive. A similar strategy was used also by Waluś et al. (2013) who showed that high-sulfur loadings can be efficiently deposited on a commercial electrospun porous carbon current collector. The electrode obtained in this way showed not only improved electrochemical performances compared to classical electrodes supported on aluminum films in terms of reduced polarization but also provided additional porous volume to accommodate sufficient amounts of electrolyte.

6.5   CHAPTER SUMMARY

In this chapter, the application of electrospinning to the development of new materials for Li-ion batteries was presented as follows:

- In the introduction, the growing importance of electrospinning for the development of Li-ion battery materials is presented, as impressively reflected by the ever rising number of scientific publications in this field.
- In the following section, a short review of the constitution and of the working mechanism of Li-ion battery is given, with specific details on different battery parts and on their properties.
- The following section is dedicated to the review of the application of electrospinning to the different battery components. A first subsection on positive electrode materials is followed by a second one on the negative electrode ones and finally by a third section on electrospun polymer electrolytes and separators.
- In the final section, the new challenges and perspectives for the application of electrospinning to battery materials is presented. A particular attention is dedicated to electrospun electrode materials for novel post-lithium chemistries such as lithium–sulfur and sodium-ion cells.

REFERENCES

Alcoutlabi, M., Lee, H., Watson, J. V., and Zhang, X. (2012). Preparation and properties of nanofiber-coated composite membranes as battery separators via electrospinning. *J. Mater. Sci.*, *48*(6), 2690–2700. doi:10.1007/s10853-012-7064-0.

An, M.-Y., Kim, H.-T., and Chang, D.-R. (2014). Multilayered separator based on porous polyethylene layer, Al_2O_3 layer, and electro-spun PVdF nanofiber layer for lithium batteries. *J. Solid State Electrochem.*, *18*(7), 1807–1814. doi:10.1007/s10008-014-2412-4.

Aravindan, V., Sundaramurthy, J., Kumar, P. S., Lee, Y.-S., Ramakrishna, S., and Madhavi, S. (2015). Electrospun nanofibers: A prospective electro-active material for constructing high performance Li-ion batteries. *Chem. Commun.*, *51*(12), 2225–2234. doi:10.1039/c4cc07824a.

Aravindan, V., Sundaramurthy, J., Kumar, P. S., Shubha, N., Ling, W. C., Ramakrishna, S., and Madhavi, S. (2013a). A novel strategy to construct high performance lithium-ion cells using one dimensional electrospun nanofibers, electrodes and separators. *Nanoscale*, *5*(21), 10636–10645. doi:10.1039/c3nr04486f.

Aravindan, V., Suresh Kumar, P., Sundaramurthy, J., Ling, W. C., Ramakrishna, S., and Madhavi, S. (2013b). Electrospun NiO nanofibers as high performance anode material for Li-ion batteries. *J. Power Sources*, *227*, 284–290. doi:10.1016/j.jpowsour.2012.11.050.

Aricò, A. S., Bruce, P. G., Scrosati, B., Tarascon, J.-M., and van Schalkwijk, W. (2005). Nanostructured materials for advanced energy conversion and storage devices. *Nat. Mater.*, *4*(5), 366–377. doi:10.1038/nmat1368.

Armand, M. and Tarascon, J.-M. (2008). Building better batteries. *Nature*, *451*, 652–657.

Arun, N., Aravindan, V., Jayaraman, S., Shubha, N., Ling, W. C., Ramakrishna, S., and Madhavi, S. (2014). Exceptional performance of a high voltage spinel $LiNi_{0.5}Mn_{1.5}O_4$ cathode in all one dimensional architectures with an anatase TiO_2 anode by electrospinning. *Nanoscale*, *6*(15), 8926–8934. doi:10.1039/c4nr01892c.

Ban, C., Chernova, N. A., and Whittingham, M. S. (2009). Electrospun nano-vanadium pentoxide cathode. *Electrochem. Commun.*, *11*(3), 522–525. doi:10.1016/j.elecom.2008.11.051.

Bansal, D., Meyer, B., and Salomon, M. (2008). Gelled membranes for Li and Li-ion batteries prepared by electrospinning. *J. Power Sources*, *178*(2), 848–851. doi:10.1016/j.jpowsour.2007.07.070.

Bruce, P. G., Freunberger, S. A., Hardwick, L. J., and Tarascon, J.-M. (2011). $Li–O_2$ and Li–S batteries with high energy storage. *Nat. Mater.*, *11*(1), 19–29. doi:10.1038/nmat3191.

Bruce, P. G., Scrosati, B., and Tarascon, J.-M. (2008). Nanomaterials for rechargeable lithium batteries. *Angew. Chem. Int. Ed.*, *47*(16), 2930–2946. doi:10.1002/anie.200702505.

Buchmann, I. (2014). Battery University. Retrieved from http://www.batteryuniversity.com/. Accessed on 2014.

Cabana, J., Monconduit, L., Larcher, D., and Palacín, M. R. (2010). Beyond intercalation-based Li-ion batteries: The state of the art and challenges of electrode materials reacting through conversion reactions. *Adv. Mater.*, *22*, E170–E192.

Cao, C., Tan, L., Liu, W., Ma, J., and Li, L. (2014). Polydopamine coated electrospun poly(vinyldiene fluoride) nanofibrous membrane as separator for lithium-ion batteries. *J. Power Sources*, *248*, 224–229. doi:10.1016/j.jpowsour.2013.09.027.

Carol, P., Ramakrishnan, P., John, B., and Cheruvally, G. (2011). Preparation and characterization of electrospun poly(acrylonitrile) fibrous membrane based gel polymer electrolytes for lithium-ion batteries. *J. Power Sources*, *196*(23), 10156–10162. doi:10.1016/j.jpowsour.2011.08.037.

Cavaliere, S., Subianto, S., Savych, I., Jones, D. J., and Rozière, J. (2011). Electrospinning: Designed architectures for energy conversion and storage devices. *Energy Environ. Sci.*, *4*(12), 4761–4785. doi:10.1039/c1ee02201f.

Chaudhari, S. and Srinivasan, M. (2012). 1D hollow α-Fe$_2$O$_3$ electrospun nanofibers as high performance anode material for lithium ion batteries. *J. Mater. Chem.*, *22*(43), 23049–23056. doi:10.1039/c2jm32989a.

Cheah, Y. L., Gupta, N., Pramana, S. S., Aravindan, V., Wee, G., and Srinivasan, M. (2011). Morphology, structure and electrochemical properties of single phase electrospun vanadium pentoxide nanofibers for lithium ion batteries. *J. Power Sources*, *196*(15), 6465–6472. doi:10.1016/j.jpowsour.2011.03.039.

Chen, L., Liao, J., Chuang, Y., Hsu, K., Chiang, Y., and Fu, Y. (2011). Synthesis and characterization of PVP/LiCoO$_2$ nanofibers by electrospinning route. *J. Appl. Polym. Sci*, *121*(1), 154–160. doi:10.1002/app.33499.

Chen, R., Zhao, T., and Wu, F. (2014a). From a historic review to horizons beyond: Lithium–sulphur batteries run on the wheels. *Chem. Commun.*, *51*(1), 18–33. doi:10.1039/c4cc05109b.

Chen, T., Liu, Y., Pan, L., Lu, T., Yao, Y., Sun, Z. et al. (2014b). Electrospun carbon nanofibers as anode materials for sodium ion batteries with excellent cycle performance. *J. Mater. Chem. A*, *2*(12), 4117–4121. doi:10.1039/c3ta14806h.

Chen, Y., Lu, Z., Zhou, L., Mai, Y.-W., and Huang, H. (2012). Triple-coaxial electrospun amorphous carbon nanotubes with hollow graphitic carbon nanospheres for high-performance Li ion batteries. *Energy Environ. Sci.*, *5*(7), 7898–7920. doi:10.1039/c2ee22085g.

Cherian, C. T., Sundaramurthy, J., Kalaivani, M., Ragupathy, P., Kumar, P. S., Thavasi, V. et al. (2012). Electrospun α-Fe$_2$O$_3$ nanorods as a stable, high capacity anode material for Li-ion batteries. *J. Mater. Chem.*, *22*(24), 12198–12204. doi:10.1039/c2jm31053h.

Cherian, C. T., Sundaramurthy, J., Reddy, M. V., Suresh Kumar, P., Mani, K., Pliszka, D. et al. (2013). Morphologically robust NiFe$_2$O$_4$ nanofibers as high capacity Li-ion battery anode material. *ACS Appl. Mater. Interfaces*, *5*(20), 9957–9963. doi:10.1021/am401779p.

Cheruvally, G., Kim, J.-K., Choi, J.-W., Ahn, J.-H., Shin, Y.-J., Manuel, J. et al. (2007). Electrospun polymer membrane activated with room temperature ionic liquid: Novel polymer electrolytes for lithium batteries. *J. Power Sources*, *172*(2), 863–869. doi:10.1016/j.jpowsour.2007.07.057.

Cho, T.-H., Tanaka, M., Onishi, H., Kondo, Y., Nakamura, T., Yamazaki, H. et al. (2008). Battery performances and thermal stability of polyacrylonitrile nano-fiber-based nonwoven separators for Li-ion battery. *J. Power Sources*, *181*(1), 155–160. doi:10.1016/j.jpowsour.2008.03.010.

Choi, H. S., Lee, J. G., Lee, H. Y., Kim, S. W., and Park, C. R. (2010). Effects of surrounding confinements of Si nanoparticles on Si-based anode performance for lithium ion batteries. *Electrochim. Acta*, *56*(2), 790–796. doi:10.1016/j.electacta.2010.09.101.

Choi, S.-S., Lee, Y. S., Joo, C. W., Lee, S. G., Park, J. K., and Han, K.-S. (2004). Electrospun PVDF nanofiber web as polymer electrolyte or separator. *Electrochim. Acta*, *50*(2–3), 339–343. doi:10.1016/j.electacta.2004.03.057.

Choi, S. W., Jo, S. M., Lee, W. S., and Kim, Y.-R. (2003). An electrospun poly(vinylidene fluoride) nanofibrous membrane and its battery applications. *Adv. Mater.*, *15*(23), 2027–2032. doi:10.1002/adma.200304617.

Choi, S. W., Kim, J. R., Jo, S. M., Lee, W. S., and Kim, Y.-R. (2005). Electrochemical and spectroscopic properties of electrospun PAN-based fibrous polymer electrolytes. *J. Electrochem. Soc.*, *152*(5), A989–A995. doi:10.1149/1.1887166.

Croce, F., Focarete, M. L., Hassoun, J., Meschini, I., and Scrosati, B. (2011). A safe, high-rate and high-energy polymer lithium-ion battery based on gelled membranes prepared by electrospinning. *Energy Environ. Sci.*, *4*(3), 921. doi:10.1039/c0ee00348d.

Cui, W.-W., Tang, D.-Y., and Gong, Z.-L. (2013). Electrospun poly(vinylidene fluoride)/poly(methyl methacrylate) grafted TiO$_2$ composite nanofibrous membrane as polymer electrolyte for lithium-ion batteries. *J. Power Sources*, *223*, 206–213. doi:10.1016/j.jpowsour.2012.09.049.

Dahbi, M., Yabuuchi, N., Kubota, K., Tokiwa, K., and Komaba, S. (2014). Negative electrodes for Na-ion batteries. *Phys. Chem. Chem. Phys.*, *16*(29), 15007–15028. doi:10.1039/c4cp00826j.

Dai, Y., Liu, W., Formo, E., Sun, Y., and Xia, Y. (2011). Ceramic nanofibers fabricated by electrospinning and their applications in catalysis, environmental science, and energy technology. *Polym. Adv. Technol.*, *22*(3), 326–338. doi:10.1002/pat.1839.

Darwiche, A., Marino, C., Sougrati, M. T., Fraisse, B., Stievano, L., and Monconduit, L. (2013). Better cycling performances of bulk Sb in Na-ion batteries compared to Li-ion systems: An unexpected electrochemical mechanism. *J. Am. Chem. Soc.*, *135*(27), 10179. doi:10.1021/ja4056195.

Ding, Y., Zhang, P., Long, Z., Jiang, Y., Huang, J., Yan, W., and Liu, G. (2008). Synthesis and electrochemical properties of Co_3O_4 nanofibers as anode materials for lithium-ion batteries. *Mater. Lett.*, *62*(19), 3410–3412. doi:10.1016/j.matlet.2008.03.033.

Dong, Z., Kennedy, S. J., and Wu, Y. (2011). Electrospinning materials for energy-related applications and devices. *J. Power Sources*, *196*(11), 4886–4904.

Elia, G. A., Wang, J., Bresser, D., Li, J., Scrosati, B., Passerini, S., and Hassoun, J. (2014). A new, high energy Sn-C/Li[$Li_{0.2}Ni_{0.4/3}Co_{0.4/3}Mn_{1.6/3}$]$O_2$ lithium-ion battery. *ACS Appl. Mater. Interfaces*, *6*(15), 12956–12961. doi:10.1021/am502884y.

Ellis, B. L. and Nazar, L. F. (2012). Sodium and sodium-ion energy storage batteries. *Curr. Opin. Solid State Mater. Sci.*, *16*(4), 168–177. doi:10.1016/j.cossms.2012.04.002.

Evers, S. and Nazar, L. F. (2013). New approaches for high energy density lithium-sulfur battery cathodes. *Acc. Chem. Res.*, *46*(5), 1135–1143. doi:10.1021/ar3001348.

Fan, Q. and Whittingham, M. S. (2007). Electrospun manganese oxide nanofibers as anodes for lithium-ion batteries. *Electrochem. Solid State Lett.*, *10*(3), A48–A51. doi:10.1149/1.2422749.

Fehse, M., Cavaliere, S., Lippens, P.-E., Savych, I., Iadecola, A., Monconduit, L. et al. (2013). Nb-doped TiO_2 nanofibers for lithium ion batteries. *J. Phys. Chem. C*, *117*(27), 13827–13835. doi:10.1021/jp402498p.

Gao, K., Hu, X., Dai, C., and Yi, T. (2006). Crystal structures of electrospun PVDF membranes and its separator application for rechargeable lithium metal cells. *Mater. Sci. Eng. B*, *131*(1–3), 100–105. doi:10.1016/j.mseb.2006.03.035.

Gu, Y., Chen, D., and Jiao, X. (2005). Synthesis and electrochemical properties of nanostructured $LiCoO_2$ fibers as cathode materials for lithium-ion batteries. *J. Phys. Chem. B*, *109*(38), 17901–17906. doi:10.1021/jp0521813.

Gu, Y., Chen, D., Jiao, X., and Liu, F. (2007). $LiCoO_2$-MgO coaxial fibers: Co-electrospun fabrication, characterization and electrochemical properties. *J. Mater. Chem.*, *17*(18), 1769–1776. doi:10.1039/b614205b.

Gu, Y. and Jian, F. (2008). Hollow $LiNi_{0.8}Co_{0.1}Mn_{0.1}O_2$–MgO coaxial fibers: Sol–gel method combined with co-electrospun preparation and electrochemical properties. *J. Phys. Chem. C*, *112*(51), 20176–20180. doi:10.1021/jp808468x.

Hao, J., Lei, G., Li, Z., Wu, L., Xiao, Q., and Wang, L. (2013). A novel polyethylene terephthalate nonwoven separator based on electrospinning technique for lithium ion battery. *J. Membr. Sci.*, *428*, 11–16. doi:10.1016/j.memsci.2012.09.058.

Hassoun, J., Derrien, G., Panero, S., and Scrosati, B. (2008). A nanostructured Sn–C composite lithium battery electrode with unique stability and high electrochemical performance. *Adv. Mater.*, *20*(16), 3169–3175.

Hieu, N. T., Suk, J., Kim, D. W., Park, J. S., and Kang, Y. (2014). Electrospun nanofibers with a core–shell structure of silicon nanoparticles and carbon nanotubes in carbon for use as lithium-ion battery anodes. *J. Mater. Chem. A*, *2*(36), 15094–15101. doi:10.1039/C4TA02348J.

Hosono, E., Kudo, T., Honma, I., Matsuda, H., and Zhou, H. (2009). Synthesis of single crystalline spinel $LiMn_2O_4$ nanowires for a lithium ion battery with high power density. *Nano Lett.*, *9*(3), 1045–1051. doi:10.1021/nl803394v.

Hosono, E., Wang, Y., Kida, N., Enomoto, M., Kojima, N., Okubo, M. et al. (2010). Synthesis of triaxial LiFePO$_4$ nanowire with a VGCF core column and a carbon shell through the electrospinning method. *ACS Appl. Mater. Interfaces*, *2*(1), 212–218. doi:10.1021/am900656y.

Hwang, T. H., Jung, D. S., Kim, J.-S., Kim, B. G., and Choi, J. W. (2013). One-dimensional carbon-sulfur composite fibers for Na-S rechargeable batteries operating at room temperature. *Nano Lett.*, *13*(9), 4532–4538. doi:10.1021/nl402513x.

Hwang, T. H., Lee, Y. M., Kong, B.-S., Seo, J.-S., and Choi, J. W. (2012). Electrospun core-shell fibers for robust silicon nanoparticle-based lithium ion battery anodes. *Nano Lett.*, *12*(2), 802–807. doi:10.1021/nl203817r.

Jahel, A., Ghimbeu, C. M., Monconduit, L., and Vix-Guterl, C. (2014). Confined ultrasmall SnO$_2$ particles in micro/mesoporous carbon as an extremely long cycle-life anode material for Li-ion batteries. *Adv. Energy Mater.*, *4*(11), 1–7, 1400025. doi:10.1002/aenm.201400025.

Jayaraman, S., Aravindan, V., Suresh Kumar, P., Ling, W. C., Ramakrishna, S., and Madhavi, S. (2013). Synthesis of porous LiMn$_2$O$_4$ hollow nanofibers by electrospinning with extraordinary lithium storage properties. *Chem. Commun.*, *49*(59), 6677–6679. doi:10.1039/c3cc43874k.

Jayaraman, S., Aravindan, V., Suresh Kumar, P., Ling, W. C., Ramakrishna, S., and Madhavi, S. (2014). Exceptional performance of TiNb$_2$O$_7$ anode in all one-dimensional architecture by electrospinning. *ACS Appl. Mater. Interfaces*, *6*(11), 8660–8666. doi:10.1021/am501464d.

Ji, L., Lin, Z., Medford, A. J., and Zhang, X. (2009). In-situ encapsulation of nickel particles in electrospun carbon nanofibers and the resultant electrochemical performance. *Chem. Eur. J.*, *15*(41), 10718–10722. doi:10.1002/chem.200902012.

Ji, L., Rao, M., Aloni, S., Wang, L., Cairns, E. J., and Zhang, Y. (2011). Porous carbon nanofiber–sulfur composite electrodes for lithium/sulfur cells. *Energy Environ. Sci.*, *4*(12), 5053–5059. doi:10.1039/c1ee02256c.

Ji, L., Toprakci, O., Alcoutlabi, M., Yao, Y., Li, Y., Zhang, S. et al. (2012). α-Fe(2)O(3) nanoparticle-loaded carbon nanofibers as stable and high-capacity anodes for rechargeable lithium-ion batteries. *ACS Appl. Mater. Interfaces*, *4*(5), 2672–2679. doi:10.1021/am300333s.

Ji, L., Yao, Y., Toprakci, O., Lin, Z., Liang, Y., Shi, Q. et al. (2010). Fabrication of carbon nanofiber-driven electrodes from electrospun polyacrylonitrile/polypyrrole bicomponents for high-performance rechargeable lithium-ion batteries. *J. Power Sources*, *195*(7), 2050–2056. doi:10.1016/j.jpowsour.2009.10.021.

Ji, X. and Nazar, L. F. (2010). Advances in Li–S batteries. *J. Mater. Chem.*, *20*(44), 9821–9826.

Jin, J., Shi, Z., and Wang, C. (2014). Electrochemical performance of electrospun carbon nanofibers as free-standing and binder-free anodes for sodium-ion and lithium-ion batteries. *Electrochim. Acta*, *141*, 302–310. doi:10.1016/j.electacta.2014.07.079.

Jo, M. R., Jung, Y. S., and Kang, Y.-M. (2012). Tailored Li$_4$Ti$_5$O$_{12}$ nanofibers with outstanding kinetics for lithium rechargeable batteries. *Nanoscale*, *4*(21), 6870–6875. doi:10.1039/c2nr31675g.

Jung, H.-R., Ju, D.-H., Lee, W.-J., Zhang, X., and Kotek, R. (2009). Electrospun hydrophilic fumed silica/polyacrylonitrile nanofiber-based composite electrolyte membranes. *Electrochim. Acta*, *54*(13), 3630–3637. doi:10.1016/j.electacta.2009.01.039.

Kagesawa, K., Hosono, E., Okubo, M., Nishio-Hamane, D., Kudo, T., and Zhou, H. (2014). Electrochemical properties of LiMn$_x$Fe$_{1-x}$PO$_4$ (x = 0, 0.2, 0.4, 0.6, 0.8 and 1.0)/vapor grown carbon fiber core–sheath composite nanowire synthesized by electrospinning method. *J. Power Sources*, *248*, 615–620. doi:10.1016/j.jpowsour.2013.09.133.

Kalluri, S., Pang, W. K., Seng, K. H., Chen, Z., Guo, Z., Liu, H. K., and Dou, S. X. (2014). One-dimensional nanostructured design of $Li_{1+x}(Mn_{1/3}Ni_{1/3}Fe_{1/3})O_2$ as a dual cathode for lithium-ion and sodium-ion batteries. *J. Mater. Chem. A*, *3*(1), 250–257. doi:10.1039/C4TA04271A.

Kim, C., Yang, K. S., Kojima, M., Yoshida, K., Kim, Y. J., Kim, Y. A., and Endo, M. (2006). Fabrication of electrospinning-derived carbon nanofiber webs for the anode material of lithium-ion secondary batteries. *Adv. Funct. Mater.*, *16*(18), 2393–2397. doi:10.1002/adfm.200500911.

Kim, J.-C. and Kim, D.-W. (2014). Synthesis of multiphase SnSb nanoparticles-on-SnO_2/Sn/C nanofibers for use in Li and Na ion battery electrodes. *Electrochem. Commun.*, *46*, 124–127. doi:10.1016/j.elecom.2014.07.005.

Kim, J.-K., Cheruvally, G., Li, X., Ahn, J.-H., Kim, K.-W., and Ahn, H.-J. (2008). Preparation and electrochemical characterization of electrospun, microporous membrane-based composite polymer electrolytes for lithium batteries. *J. Power Sources*, *178*(2), 815–820. doi:10.1016/j.jpowsour.2007.08.063.

Kim, J. M., Joh, H.-I., Jo, S. M., Ahn, D. J., Ha, H. Y., Hong, S.-A., and Kim, S.-K. (2010). Preparation and characterization of Pt nanowire by electrospinning method for methanol oxidation. *Electrochim. Acta*, *55*(16), 4827–4835. doi:10.1016/j.electacta.2010.03.036.

Kim, J. R., Choi, S. W., Jo, S. M., Lee, W. S., and Kim, B. C. (2004). Electrospun PVdF-based fibrous polymer electrolytes for lithium ion polymer batteries. *Electrochim. Acta*, *50*(1), 69–75. doi:10.1016/j.electacta.2004.07.014.

Kim, J. R., Choi, S. W., Jo, S. M., Lee, W. S., and Kim, B. C. (2005). Characterization and properties of P(VdF-HFP)-based fibrous polymer electrolyte membrane prepared by electrospinning. *J. Electrochem. Soc.*, *152*(2), A295–A300. doi:10.1149/1.1839531.

Kim, S.-W., Seo, D.-H., Ma, X., Ceder, G., and Kang, K. (2012). Electrode materials for rechargeable sodium-ion batteries: Potential alternatives to current lithium-ion batteries. *Adv. Energy Mater.*, *2*(7), 710–721. doi:10.1002/aenm.201200026.

Kim, Y.-J., Ahn, C. H., Lee, M. B., and Choi, M.-S. (2011). Characteristics of electrospun $PVDF/SiO_2$ composite nanofiber membranes as polymer electrolyte. *Mater. Chem. Phys.*, *127*(1–2), 137–142. doi:10.1016/j.matchemphys.2011.01.046.

Kong, J., Liu, Z., Yang, Z., Tan, H. R., Xiong, S., Wong, S. Y. et al. (2012). Carbon/SnO_2/carbon core/shell/shell hybrid nanofibers: Tailored nanostructure for the anode of lithium ion batteries with high reversibility and rate capacity. *Nanoscale*, *4*(2), 525–530. doi:10.1039/c1nr10962f.

Kong, J., Yee, W. A., Wei, Y., Yang, L., Ang, J. M., Phua, S. L. et al. (2013). Silicon nanoparticles encapsulated in hollow graphitized carbon nanofibers for lithium ion battery anodes. *Nanoscale*, *5*(7), 2967–2973. doi:10.1039/c3nr34024d.

Landi, B. J., Ganter, M. J., Cress, C. D., DiLeo, R. A., and Raffaelle, R. P. (2009). Carbon nanotubes for lithium ion batteries. *Energy Environ. Sci.*, *2*(6), 638–654. doi:10.1039/b904116h.

Lang, L. and Xu, Z. (2013). In situ synthesis of porous Fe_3O_4/C microbelts and their enhanced electrochemical performance for lithium-ion batteries. *ACS Appl. Mater. Interfaces*, *5*(5), 1698–1703. doi:10.1021/am302753p.

Le Viet, A., Reddy, M. V., Jose, R., Chowdari, B. V. R., and Ramakrishna, S. (2011). Electrochemical properties of bare and Ta-substituted Nb_2O_5 nanostructures. *Electrochim. Acta*, *56*(3), 1518–1528. doi:10.1016/j.electacta.2010.10.047.

Lee, B.-S., Son, S.-B., Park, K.-M., Seo, J.-H., Lee, S.-H., Choi, I.-S. et al. (2012a). Fabrication of Si core/C shell nanofibers and their electrochemical performances as a lithium-ion battery anode. *J. Power Sources*, *206*, 267–273. doi:10.1016/j.jpowsour.2012.01.120.

Lee, B.-S., Son, S.-B., Park, K.-M., Yu, W.-R., Oh, K.-H., and Lee, S.-H. (2012b). Anodic properties of hollow carbon nanofibers for Li-ion battery. *J. Power Sources*, *199*, 53–60. doi:10.1016/j.jpowsour.2011.10.030.

Lee, H., Alcoutlabi, M., Watson, J. V., and Zhang, X. (2013a). Polyvinylidene fluoride-co-chlorotrifluoroethylene and polyvinylidene fluoride-co-hexafluoropropylene nanofiber-coated polypropylene microporous battery separator membranes. *J. Polym. Sci. B: Polym. Phys.*, *51*(5), 349–357. doi:10.1002/polb.23216.

Lee, J., Lee, C.-L., Park, K., and Kim, I.-D. (2014). Synthesis of an Al_2O_3-coated polyimide nanofiber mat and its electrochemical characteristics as a separator for lithium ion batteries. *J. Power Sources*, *248*, 1211–1217. doi:10.1016/j.jpowsour.2013.10.056.

Lee, S. W., Choi, S. W., Jo, S. M., Chin, B. D., Kim, D. Y., and Lee, K. Y. (2006). Electrochemical properties and cycle performance of electrospun poly(vinylidene fluoride)-based fibrous membrane electrolytes for Li-ion polymer battery. *J. Power Sources*, *163*(1), 41–46. doi:10.1016/j.jpowsour.2005.11.102.

Lee, Y.-I., Jang, D.-H., Kim, J.-W., Kim, W.-B., and Choa, Y.-H. (2013b). Electrospun $NiFe_2O_4$ nanofibers as high capacity anode materials for Li-ion batteries. *J. Nanosci. Nanotechnol.*, *13*(10), 7138–7141. doi:10.1166/jnn.2013.7698.

Lee, Y.-S., Jeong, Y. B., and Kim, D.-W. (2010). Cycling performance of lithium-ion batteries assembled with a hybrid composite membrane prepared by an electrospinning method. *J. Power Sources*, *195*(18), 6197–6201.

Li, H., Bai, Y., Wu, F., Li, Y., and Wu, C. (2015). Budding willow branches shaped $Na_3V_2(PO_4)_3$/C nanofibers synthesized via an electrospinning technique and used as cathode material for sodium ion batteries. *J. Power Sources*, *273*, 784–792. doi:10.1016/j.jpowsour.2014.09.153.

Li, L., Yin, X., Liu, S., Wang, Y., Chen, L., and Wang, T. (2010). Electrospun porous SnO_2 nanotubes as high capacity anode materials for lithium ion batteries. *Electrochem. Commun.*, *12*(10), 1383–1386. doi:10.1016/j.elecom.2010.07.026.

Li, X., Cheruvally, G., Kim, J.-K., Choi, J.-W., Ahn, J.-H., Kim, K.-W., and Ahn, H.-J. (2007). Polymer electrolytes based on an electrospun poly(vinylidene fluoride-co-hexafluoropropylene) membrane for lithium batteries. *J. Power Sources*, *167*(2), 491–498. doi:10.1016/j.jpowsour.2007.02.032.

Liang, Y., Ji, L., Guo, B., Lin, Z., Yao, Y., Li, Y. et al. (2011a). Preparation and electrochemical characterization of ionic-conducting lithium lanthanum titanate oxide/polyacrylonitrile submicron composite fiber-based lithium-ion battery separators. *J. Power Sources*, *196*(1), 436–441. doi:10.1016/j.jpowsour.2010.06.088.

Liang, Y., Lin, Z., Qiu, Y., and Zhang, X. (2011b). Fabrication and characterization of LATP/PAN composite fiber-based lithium-ion battery separators. *Electrochim. Acta*, *56*(18), 6474–6480. doi:10.1016/j.electacta.2011.05.007.

Lin, Z., Ji, L., Woodroof, M. D., and Zhang, X. (2010). Electrodeposited MnO_x/carbon nanofiber composites for use as anode materials in rechargeable lithium-ion batteries. *J. Power Sources*, *195*(15), 5025–5031. doi:10.1016/j.jpowsour.2010.02.004.

Liu, C., Li, F., Ma, L.-P., and Cheng, H.-M. (2010). Advanced materials for energy storage. *Adv. Mater.*, *22*(8), E28–E62.

Liu, J., Tang, K., Song, K., van Aken, P. A., Yu, Y., and Maier, J. (2013). Tiny $Li_4Ti_5O_{12}$ nanoparticles embedded in carbon nanofibers as high-capacity and long-life anode materials for both Li-ion and Na-ion batteries. *Phys. Chem. Chem. Phys.*, *15*(48), 20813–20818. doi:10.1039/c3cp53882f.

Liu, R., Duay, J., and Lee, S. B. (2011). Heterogeneous nanostructured electrode materials for electrochemical energy storage. *Chem. Commun.*, *47*(5), 1384–1404. doi:10.1039/c0cc03158e.

Liu, S., Long, Y. Z., Zhang, H. D., Sun, B., Tang, C. C., Li, H. L. et al. (2012). Preparation and electrochemical properties of $LiMn_2O_4$ nanofibers via electrospinning for lithium ion batteries. *Adv. Mater. Res.*, *562–564*, 799–802. doi:10.4028/www.scientific.net/AMR.562-564.799.

Lu, H.-W., Yu, L., Zeng, W., Li, Y.-S., and Fu, Z.-W. (2008). Fabrication and electrochemical properties of three-dimensional structure of $LiCoO_2$ fibers. *Electrochem. Solid State Lett.*, *11*(8), A140–A144. doi:10.1149/1.2932054.

Lu, H.-W., Zeng, W., Li, Y.-S., and Fu, Z.-W. (2007). Fabrication and electrochemical properties of three-dimensional net architectures of anatase TiO_2 and spinel $Li_4Ti_5O_{12}$ nanofibers. *J. Power Sources*, *164*(2), 874–879. doi:10.1016/j.jpowsour.2006.11.009.

Lu, Q., Hutchings, G. S., Zhou, Y., Xin, H. L., Zheng, H., and Jiao, F. (2014). Nanostructured flexible Mg-modified $LiMnPO_4$ matrix as high-rate cathode materials for Li-ion batteries. *J. Mater. Chem. A*, *2*(18), 6368–6373. doi:10.1039/c4ta00654b.

Luo, W., Hu, X., Sun, Y., and Huang, Y. (2012). Electrospun porous $ZnCo_2O_4$ nanotubes as a high-performance anode material for lithium-ion batteries. *J. Mater. Chem.*, *22*(18), 8916–8921. doi:10.1039/c2jm00094f.

Lv, H., Qiu, S., Lu, G., Fu, Y., Li, X., Hu, C., and Liu, J. (2015). Nanostructured antimony/carbon composite fibers as anode material for lithium-ion battery. *Electrochim. Acta*, *151*, 214–221. doi:10.1016/j.electacta.2014.11.013.

Mai, L., Xu, L., Han, C., Xu, X., Luo, Y., Zhao, S., and Zhao, Y. (2010). Electrospun ultralong hierarchical vanadium oxide nanowires with high performance for lithium ion batteries. *Nano Lett.*, *10*(11), 4750–4755. doi:10.1021/nl103343w.

Manthiram, A., Fu, Y., Chung, S.-H., Zu, C., and Su, Y.-S. (2014). Rechargeable lithium-sulfur batteries. *Chem. Rev.*, *114*(23), 11751–11787. doi:10.1021/cr500062v.

Manthiram, A., Murugan, A. V., Sarkar, A., and Muraliganth, T. (2008). Nanostructured electrode materials for electrochemical energy storage and conversion. *Energy Environ. Sci.*, *1*(6), 621–638. doi:10.1039/b811802g.

Melot, B. C. and Tarascon, J.-M. (2013). Design and preparation of materials for advanced electrochemical storage. *Acc. Chem. Res.*, *46*(5), 1226–1238. doi:10.1021/ar300088q.

Memarzadeh Lotfabad, E., Kalisvaart, P., Kohandehghan, A., Karpuzov, D., and Mitlin, D. (2014). Origin of non-SEI related coulombic efficiency loss in carbons tested against Na and Li. *J. Mater. Chem. A*, *2*(46), 19685–19695. doi:10.1039/C4TA04995K.

Miao, Y.-E., Zhu, G.-N., Hou, H., Xia, Y.-Y., and Liu, T. (2013). Electrospun polyimide nanofiber-based nonwoven separators for lithium-ion batteries. *J. Power Sources*, *226*, 82–86. doi:10.1016/j.jpowsour.2012.10.027.

Mizushima, K., Jones, P. C., Wiseman, P. J., and Goodenough, J. B. (1981). Li_xCoO_2 ($0 < x < 1$): A new cathode material for batteries of high energy density. *Solid State Ionics*, *3–4*, 171–174. doi:10.1016/0167-2738(81)90077-1.

Mukherjee, R., Krishnan, R., Lu, T.-M., and Koratkar, N. (2012). Nanostructured electrodes for high-power lithium ion batteries. *Nano Energy*, *1*(4), 518–533. doi:10.1016/j.nanoen.2012.04.001.

Nam, S. H., Shim, H.-S., Kim, Y.-S., Dar, M. A., Kim, J. G., and Kim, W. B. (2010). Ag or Au nanoparticle-embedded one-dimensional composite TiO_2 nanofibers prepared via electrospinning for use in lithium-ion batteries. *ACS Appl. Mater. Interfaces*, *2*(7), 2046–2052. doi:10.1021/am100319u.

Nan, D., Huang, Z.-H., Lv, R., Yang, L., Wang, J.-G., Shen, W. et al. (2014). Nitrogen-enriched electrospun porous carbon nanofiber networks as high-performance free-standing electrode materials. *J. Mater. Chem. A*, *2*(46), 19678–19684. doi:10.1039/C4TA03868A.

Ni, W., Cheng, J., Shi, L., Li, X., Wang, B., Guan, Q. et al. (2014). Integration of Sn/C yolk–shell nanostructures into free-standing conductive networks as hierarchical composite 3D electrodes and the Li-ion insertion/extraction properties in a gel-type lithium-ion battery thereof. *J. Mater. Chem. A*, *2*(45), 19122–19130. doi:10.1039/C4TA04554H.

Nishi, Y. (2001). Lithium ion secondary batteries; past 10 years and the future. *J. Power Sources*, *100*(1–2), 101–106.

Palacín, M. R. (2009). Recent advances in rechargeable battery materials: A chemist's perspective. *Chem. Soc. Rev.*, *38*(9), 2565–2575. doi:10.1039/b820555h.

Palomares, V., Serras, P., Villaluenga, I., Hueso, K. B., Carretero-González, J., and Rojo, T. (2012). Na-ion batteries, recent advances and present challenges to become low cost energy storage systems. *Energy Environ. Sci.*, *5*(3), 5884–5901. doi:10.1039/c2ee02781j.

Pitchai, R., Thavasi, V., Mhaisalkar, S. G., and Ramakrishna, S. (2011). Nanostructured cathode materials: A key for better performance in Li-ion batteries. *J. Mater. Chem.*, *21*(30), 11040–11051. doi:10.1039/C1JM10857C.

Qiu, Y., Geng, Y., Yu, J., and Zuo, X. (2013). High-capacity cathode for lithium-ion battery from LiFePO$_4$/(C + Fe$_2$P) composite nanofibers by electrospinning. *J. Mater. Sci.*, *49*(2), 504–509. doi:10.1007/s10853-013-7727-5.

Qu, H., Wei, S., and Guo, Z. (2013). Coaxial electrospun nanostructures and their applications. *J. Mater. Chem. A*, *1*(38), 11513–11528. doi:10.1039/c3ta12390a.

Raghavan, P., Choi, J.-W., Ahn, J.-H., Cheruvally, G., Chauhan, G. S., Ahn, H.-J., and Nah, C. (2008a). Novel electrospun poly(vinylidene fluoride-co-hexafluoropropylene)-in situ SiO$_2$ composite membrane-based polymer electrolyte for lithium batteries. *J. Power Sources*, *184*(2), 437–443. doi:10.1016/j.jpowsour.2008.03.027.

Raghavan, P., Manuel, J., Zhao, X., Kim, D.-S., Ahn, J.-H., and Nah, C. (2011). Preparation and electrochemical characterization of gel polymer electrolyte based on electrospun polyacrylonitrile nonwoven membranes for lithium batteries. *J. Power Sources*, *196*(16), 6742–6749. doi:10.1016/j.jpowsour.2010.10.089.

Raghavan, P., Zhao, X., Kim, J.-K., Manuel, J., Chauhan, G. S., Ahn, J.-H., and Nah, C. (2008b). Ionic conductivity and electrochemical properties of nanocomposite polymer electrolytes based on electrospun poly(vinylidene fluoride-co-hexafluoropropylene) with nano-sized ceramic fillers. *Electrochim. Acta*, *54*(2), 228–234. doi:10.1016/j.electacta.2008.08.007.

Raghavan, P., Zhao, X., Manuel, J., Shin, C., Heo, M.-Y., Ahn, J.-H. et al. (2010a). Electrochemical studies on polymer electrolytes based on poly(vinylidene fluoride-co-hexafluoropropylene) membranes prepared by electrospinning and phase inversion—A comparative study. *Mater. Res. Bull.*, *45*(3), 362–366. doi:10.1016/j.materresbull.2009.12.001.

Raghavan, P., Zhao, X., Shin, C., Baek, D.-H., Choi, J.-W., Manuel, J. et al. (2010b). Preparation and electrochemical characterization of polymer electrolytes based on electrospun poly(vinylidene fluoride-co-hexafluoropropylene)/polyacrylonitrile blend/composite membranes for lithium batteries. *J. Power Sources*, *195*(18), 6088–6094. doi:10.1016/j.jpowsour.2009.11.098.

Rao, M., Geng, X., Liao, Y., Hu, S., and Li, W. (2012). Preparation and performance of gel polymer electrolyte based on electrospun polymer membrane and ionic liquid for lithium ion battery. *J. Membr. Sci.*, *399–400*, 37–42. doi:10.1016/j.memsci.2012.01.021.

Reddy, M. V., Jose, R., Teng, T. H., Chowdari, B. V. R., and Ramakrishna, S. (2010). Preparation and electrochemical studies of electrospun TiO$_2$ nanofibers and molten salt method nanoparticles. *Electrochim. Acta*, *55*(9), 3109–3117. doi:10.1016/j.electacta.2009.12.095.

Reneker, D. H., Yarin, A. L., Zussman, E., and Xu, H. (2007). Electrospinning of nanofibers from polymer solutions and melts. *Adv. Appl. Mech.*, *41*, 44–197. doi:10.1016/S0065-2156(06)41002-4.

Ryu, M., Jung, K., Shin, K., Han, K., and Yoon, S. (2013). High performance N-doped mesoporous carbon decorated TiO$_2$ nanofibers as anode materials for lithium-ion batteries. *J. Phys. Chem. C*, *117*(16), 8092–8098. doi:10.1021/jp400757s.

Sahay, R., Suresh Kumar, P., Aravindan, V., Sundaramurthy, J., Chui Ling, W., Mhaisalkar, S. G. et al. (2012). High aspect ratio electrospun CuO nanofibers as anode material for lithium-ion batteries with superior cycleability. *J. Phys. Chem. C*, *116*(34), 18087–18092. doi:10.1021/jp3053949.

Sandhya, C. P., John, B., and Gouri, C. (2013). Synthesis and electrochemical characterisation of electrospun lithium titanate ultrafine fibres. *J. Mater. Sci.*, *48*(17), 5827–5832. doi:10.1007/s10853-013-7375-9.

Sankey, P., Clark, D. T., and Micheloto, S. (2010). The end of the oil age. 2011 and beyond: A reality check. Global Markets Research. Deutsche Bank Securities Inc.

Sawicka, K. M. and Gouma, P. (2006). Electrospun composite nanofibers for functional applications. *J. Nanopart. Res.*, *8*(6), 769–781. doi:10.1007/s11051-005-9026-9.

Scrosati, B. and Garche, J. (2010). Lithium batteries: Status, prospects and future. *J. Power Sources*, *195*(9), 430–2419.

Shubha, N., Prasanth, R., Hoon, H. H., and Srinivasan, M. (2013). Dual phase polymer gel electrolyte based on non-woven poly(vinylidenefluoride-co-hexafluoropropylene)-layered clay nanocomposite fibrous membranes for lithium ion batteries. *Mater. Res. Bull.*, *48*(2), 526–537. doi:10.1016/j.materresbull.2012.11.002.

Slater, M. D., Kim, D., Lee, E., and Johnson, C. S. (2013). Sodium-ion batteries. *Adv. Funct. Mater.*, *23*(8), 947–958. doi:10.1002/adfm.201200691.

Tarascon, J.-M. (2010). Is lithium the new gold? *Nat. Chem.*, *2*(6), 510. doi:10.1038/nchem.680.

Tarascon, J.-M. and Armand, M. (2001). Issues and challenges facing rechargeable lithium batteries. *Nature*, *414*(6861), 359–367. doi:10.1038/35104644.

Tarascon, J.-M., Gozdz, A. S., Schmutz, C. N., Shokoohi, F., and Warren, P. C. (1996). Performance of Bellcore's plastic rechargeable Li-ion batteries. *Solid State Ion.*, *86–88*(96), 49–54. doi:10.1016/0167-2738(96)00330-X.

Teh, P. F., Pramana, S. S., Sharma, Y., Ko, Y. W., and Madhavi, S. (2013). Electrospun $Zn_{1-x}Mn_xFe_2O_4$ nanofibers as anodes for lithium-ion batteries and the impact of mixed transition metallic oxides on battery performance. *ACS Appl. Mater. Interfaces*, *5*(12), 5461–5467. doi:10.1021/am400497v.

Thackeray, M. M., David, W. I. F., Bruce, P. G., and Goodenough, J. B. (1983). Lithium insertion into manganese spinels. *Mater. Res. Bull.*, *18*(4), 461–472. doi:http://dx.doi.org/10.1016/0025-5408(83)90138-1.

Thackeray, M. M., Wolverton, C., and Isaacs, E. D. (2012). Electrical energy storage for transportation—Approaching the limits of, and going beyond, lithium-ion batteries. *Energy Environ. Sci.*, *5*, 7854–7863. doi:10.1039/c2ee21892e.

Toprakci, O., Ji, L., Lin, Z., Toprakci, H. A. K., and Zhang, X. (2011). Fabrication and electrochemical characteristics of electrospun $LiFePO_4$/carbon composite fibers for lithium-ion batteries. *J. Power Sources*, *196*(18), 7692–7699. doi:10.1016/j.jpowsour.2011.04.031.

Toprakci, O., Toprakci, H. A. K., Ji, L., Xu, G., Lin, Z., and Zhang, X. (2012). Carbon nanotube-loaded electrospun $LiFePO_4$/carbon composite nanofibers as stable and binder-free cathodes for rechargeable lithium-ion batteries. *ACS Appl. Mater. Interfaces*, *4*(3), 1273–1280. doi:10.1021/am201527r.

Tran, T., McCormac, K., Li, J., Bi, Z., and Wu, J. (2014). Electrospun SnO_2 and TiO_2 composite nanofibers for lithium ion batteries. *Electrochim. Acta*, *117*, 68–75. doi:10.1016/j.electacta.2013.11.101.

Waluś, S., Barchasz, C., Colin, J.-F., Martin, J.-F., Elkaïm, E., Leprêtre, J.-C., and Alloin, F. (2013). New insight into the working mechanism of lithium-sulfur batteries: In situ and operando X-ray diffraction characterization. *Chem. Commun.*, *49*(72), 7899–7901. doi:10.1039/c3cc43766c.

Wang, B., Cheng, J., Wu, Y., Wang, D., and He, D. (2013). Electrochemical performance of carbon/Ni composite fibers from electrospinning as anode material for lithium ion batteries. *J. Mater. Chem. A*, *1*(4), 1368–1373. doi:10.1039/c2ta00487a.

Wang, H., Ma, D., Huang, X., Huang, Y., and Zhang, X. (2012). General and controllable synthesis strategy of metal oxide/TiO$_2$ hierarchical heterostructures with improved lithium-ion battery performance. *Sci. Rep.*, *2*, 1–8, 701. doi:10.1038/srep00701.

Wang, L., Ding, C. X., Zhang, L. C., Xu, H. W., Zhang, D. W., Cheng, T., and Chen, C. H. (2010a). A novel carbon–silicon composite nanofiber prepared via electrospinning as anode material for high energy-density lithium ion batteries. *J. Power Sources*, *195*(15), 5052–5056. doi:10.1016/j.jpowsour.2010.01.088.

Wang, L., Yu, Y., Chen, P. C., Zhang, D. W., and Chen, C. H. (2008). Electrospinning synthesis of C/Fe$_3$O$_4$ composite nanofibers and their application for high performance lithium-ion batteries. *J. Power Sources*, *183*(2), 717–723. doi:10.1016/j.jpowsour.2008.05.079.

Wang, Y. and Cao, G. (2008). Developments in nanostructured cathode materials for high-performance lithium-ion batteries. *Adv. Mater.*, *20*(12), 2251–2269. doi:10.1002/adma.200702242.

Wang, Y., Li, H., He, P., Hosono, E., and Zhou, H. (2010b). Nano active materials for lithium-ion batteries. *Nanoscale*, *2*(8), 1294–1305.

Wang, Y., Smarsly, B. M., and Djerdj, I. (2010c). Niobium doped TiO$_2$ with mesoporosity and its application for lithium insertion. *Chem. Mater.*, *22*(24), 6624–6631.

Whittingham, M. S. (1976). The role of ternary phases in cathode reactions. *J. Electrochem. Soc.*, *123*(3), 315–320. doi:10.1149/1.2132817.

Wright, P. V. (1975). Electrical conductivity in ionic complexes of poly(ethylene oxide). *Brit. Polym. J.*, *7*(5), 319–327. doi:10.1002/pi.4980070505.

Wu, H. and Cui, Y. (2012). Designing nanostructured Si anodes for high energy lithium ion batteries. *Nano Today*, *7*(5), 414–429. doi:10.1016/j.nantod.2012.08.004.

Wu, H., Zheng, G., Liu, N., Carney, T. J., Yang, Y., and Cui, Y. (2012). Engineering empty space between Si nanoparticles for lithium-ion battery anodes. *Nano Lett.*, *12*(2), 904–909. doi:10.1021/nl203967r.

Wu, L., Hu, X., Qian, J., Pei, F., Wu, F., Mao, R. et al. (2014a). Sb–C nanofibers with long cycle life as an anode material for high-performance sodium-ion batteries. *Energy Environ. Sci.*, *7*(1), 323–328. doi:10.1039/c3ee42944j.

Wu, Q., Tran, T., Lu, W., and Wu, J. (2014b). Electrospun silicon/carbon/titanium oxide composite nanofibers for lithium ion batteries. *J. Power Sources*, *258*, 39–45. doi:10.1016/j.jpowsour.2014.02.047.

Wu, Y., Reddy, M. V., Chowdari, B. V. R., and Ramakrishna, S. (2013). Long-term cycling studies on electrospun carbon nanofibers as anode material for lithium ion batteries. *ACS Appl. Mater. Interfaces*, *5*(22), 12175–12184. doi:10.1021/am404216j.

Wu, Y., Zhu, P., Reddy, M. V., Chowdari, B. V. R., and Ramakrishna, S. (2014c). Maghemite nanoparticles on electrospun CNFs template as prospective lithium-ion battery anode. *ACS Appl. Mater. Interfaces*, *6*(3), 1951–1958. doi:10.1021/am404939q.

Xu, G., Ding, B., Pan, J., Nie, P., Shen, L., and Zhang, X. (2014). High performance lithium–sulfur batteries: Advances and challenges. *J. Mater. Chem. A*, *2*(32), 12662–12676. doi:10.1039/C4TA02097A.

Xue, L., Xia, X., Tucker, T., Fu, K., Zhang, S., Li, S., and Zhang, X. (2013). A simple method to encapsulate SnSb nanoparticles into hollow carbon nanofibers with superior lithium-ion storage capability. *J. Mater. Chem. A*, *1*(44), 13807–13813. doi:10.1039/c3ta12921g.

Yabuuchi, N., Kubota, K., Dahbi, M., and Komaba, S. (2014). Research development on sodium-ion batteries. *Chem. Rev.*, *114*(23), 11636–11682. doi:10.1021/cr500192f.

Yang, C., Jia, Z., Guan, Z., and Wang, L. (2009a). Polyvinylidene fluoride membrane by novel electrospinning system for separator of Li-ion batteries. *J. Power Sources*, *189*(1), 716–720. doi:10.1016/j.jpowsour.2008.08.060.

Yang, Z., Cao, C., Liu, F., Chen, D., and Jiao, X. (2010a). Core–shell $Li(Ni_{1/3}Co_{1/3}Mn_{1/3})O_2$/ $Li(Ni_{1/2}Mn_{1/2})O_2$ fibers: Synthesis, characterization and electrochemical properties. *Solid State Ionics*, *181*(15–16), 678–683. doi:10.1016/j.ssi.2010.03.032.

Yang, Z., Choi, D., Kerisit, S., Rosso, K. M., Wang, D., Zhang, J. et al. (2009b). Nanostructures and lithium electrochemical reactivity of lithium titanites and titanium oxides: A review. *J. Power Sources*, *192*(2), 588–598.

Yang, Z., Du, G., Feng, C., Li, S., Chen, Z., Zhang, P. et al. (2010b). Synthesis of uniform polycrystalline tin dioxide nanofibers and electrochemical application in lithium-ion batteries. *Electrochim. Acta*, *55*(19), 5485–5491. doi:10.1016/j.electacta.2010.04.045.

Yanilmaz, M., Chen, C., and Zhang, X. (2013). Fabrication and characterization of SiO_2/ PVDF composite nanofiber-coated PP nonwoven separators for lithium-ion batteries. *J. Polym. Sci. B: Polym. Phys.*, *51*(23), 1719–1726. doi:10.1002/polb.23387.

Yanilmaz, M., Lu, Y., Dirican, M., Fu, K., and Zhang, X. (2014). Nanoparticle-on-nanofiber hybrid membrane separators for lithium-ion batteries via combining electrospraying and electrospinning techniques. *J. Membr. Sci.*, *456*, 57–65. doi:10.1016/j. memsci.2014.01.022.

Yao, H., Zheng, G., Hsu, P.-C., Kong, D., Cha, J. J., Li, W. et al. (2014). Improving lithium-sulphur batteries through spatial control of sulphur species deposition on a hybrid electrode surface. *Nat. Commun.*, *5*, 3943. doi:10.1038/ncomms4943.

Yin, Y.-X., Xin, S., Guo, Y.-G., and Wan, L.-J. (2013). Lithium-sulfur batteries: Electrochemistry, materials, and prospects. *Angew. Chem. Int. Ed.*, *52*(50), 13186–13200. doi:10.1002/ anie.201304762.

Yoo, J.-K., Kim, J., Jung, Y. S., and Kang, K. (2012). Scalable fabrication of silicon nanotubes and their application to energy storage. *Adv. Mater.*, *24*(40), 5452–5456. doi:10.1002/ adma.201201601.

Yoshino, A. (2012). The birth of the lithium-ion battery. *Angew. Chem. Int. Ed.*, *51*(24), 5798–5800. doi:10.1002/anie.201105006.

Yu, Y., Yang, Q., Teng, D., Yang, X., and Ryu, S. (2010). Reticular Sn nanoparticle-dispersed PAN-based carbon nanofibers for anode material in rechargeable lithium-ion batteries. *Electrochem. Commun.*, *12*(9), 1187–1190. doi:10.1016/j.elecom.2010.06.015.

Zhang, P., Guo, Z. P., Huang, Y., Jia, D., and Liu, H. K. (2011). Synthesis of Co_3O_4/carbon composite nanowires and their electrochemical properties. *J. Power Sources*, *196*(16), 6987–6991. doi:10.1016/j.jpowsour.2010.10.090.

Zhang, W. (2011). Structure and performance of $LiFePO_4$ cathode materials: A review. *J. Power Sources*, *196*(6), 2962–2970. doi:10.1016/j.jpowsour.2010.11.113.

Zhang, X., Liu, H., Petnikota, S., Ramakrishna, S., and Fan, H. J. (2014). Electrospun Fe_2O_3– carbon composite nanofibers as durable anode materials for lithium ion batteries. *J. Mater. Chem. A*, *2*(28), 10835–10841. doi:10.1039/c3ta15123a.

Zhang, X., Suresh Kumar, P., Aravindan, V., Liu, H. H., Sundaramurthy, J., Mhaisalkar, S. G. et al. (2012). Electrospun TiO_2–graphene composite nanofibers as a highly durable insertion anode for lithium ion batteries. *J. Phys. Chem. C*, *116*(28), 14780–14788. doi:10.1021/jp302574g.

Zhou, H., Ding, X., Yin, Z., Xu, G., Xue, Q., Li, J. et al. (2014). Fabrication and electrochemical characteristics of electrospun $LiMn_2O_4$ nanofiber cathode for Li-ion batteries. *Mater. Lett.*, *117*, 175–178. doi:10.1016/j.matlet.2013.11.086.

Zhu, C., Yu, Y., Gu, L., Weichert, K., and Maier, J. (2011). Electrospinning of highly electroactive carbon-coated single-crystalline $LiFePO_4$ nanowires. *Angew. Chem. Int. Ed.*, *50*(28), 6278–6282. doi:10.1002/anie.201005428.

Zhu, N., Liu, W., Xue, M., Xie, Z., Zhao, D., Zhang, M. et al. (2010a). Graphene as a conductive additive to enhance the high-rate capabilities of electrospun $Li_4Ti_5O_{12}$ for lithium-ion batteries. *Electrochim. Acta*, *55*(20), 5813–5818. doi:10.1016/j. electacta.2010.05.029.

Zhu, P., Wu, Y., Reddy, M. V., Sreekumaran Nair, A., Chowdari, B. V. R., and Ramakrishna, S. (2012). Long term cycling studies of electrospun TiO$_2$ nanostructures and their composites with MWCNTs for rechargeable Li-ion batteries. *RSC Adv.*, *2*(2), 531–537. doi:10.1039/c1ra00514f.

Zhu, Y., Han, X., Xu, Y., Liu, Y., Zheng, S., Xu, K. et al. (2013). Electrospun Sb/C fibers for a stable and fast sodium-ion battery anode. *ACS Nano*, *7*(7), 6378–6386. doi:10.1021/nn4025674.

Zhu, Z., Li, X., Zhao, Q., Qu, Z., Hou, Y., Zhao, L. et al. (2010b). FTIR study of the photocatalytic degradation of gaseous benzene over UV-irradiated TiO$_2$ nanoballs synthesized by hydrothermal treatment in alkaline solution. *Mater. Res. Bull.*, *45*(12), 1889–1893.

Zou, L., Gan, L., Kang, F., Wang, M., Shen, W., and Huang, Z. (2010). Sn/C non-woven film prepared by electrospinning as anode materials for lithium ion batteries. *J. Power Sources*, *195*(4), 1216–1220.

Zou, L., Gan, L., Lv, R., Wang, M., Huang, Z., Kang, F., and Shen, W. (2011). A film of porous carbon nanofibers that contain Sn/SnO$_x$ nanoparticles in the pores and its electrochemical performance as an anode material for lithium ion batteries. *Carbon*, *49*(1), 89–95. doi:10.1016/j.carbon.2010.08.046.

Zu, C. and Manthiram, A. (2013). Hydroxylated graphene-sulfur nanocomposites for high-rate lithium-sulfur batteries. *Adv. Energy Mater.*, *3*(8), 1008–1012. doi:10.1002/aenm.201201080.

7 Electrospinning Techniques and Electrospun Materials as Photoanodes for Dye-Sensitized Solar Cell Applications

Damien Joly, Ji-Won Jung, Il-Doo Kim, and Renaud Demadrille

CONTENTS

7.1 INTRODUCTION

In the last decade, nanomaterials have played a crucial role in the fabrication of micro- and nanoscale electronic devices and have attracted a tremendous attention from the academic and industrial research teams. Nanomaterials show unique physical and chemical properties, and they have found applications in various domains such as energy conversion and storage, information and communication technologies, sensors, biotechnologies, and health care [1]. In particular, in the recent years, nanostructured materials have contributed to produce lightweight batteries [2], highly responsive sensors [3], or efficient photovoltaic devices [4].

Among the different techniques and tools developed for the preparation of nano-materials, electrospinning appears nowadays as a powerful method allowing the fabrication of nanostructured and mesostructured inorganic, organic, and hybrid materials. This processing technique that was initially discovered in the early 1900s and patented by Coley [5] was further developed by Formhals in the 1930s [6]. Since the 1990s, electrospinning has gained a tremendous attention because of the unique capabilities of this technique in processing ceramic materials into nanostructures. Indeed, electrospinning is applicable to all condensable materials including polymer solutions, sol–gels, and molten materials. This remarkably simple method for generating nanofibers (NFs) of polymers provides a versatile technique for producing ceramic NFs with either a dense, porous, or hollow structure when it is combined with conventional sol–gel processing techniques.

In addition to the possibility to finely tune the chemical composition and the dimensions of the deposited nanomaterials, electrospinning allows to control the directional arrangement of the objects on various surfaces as well as the surface coverage. Besides, it is a cost-effective technique that remains an upscalable process compatible with industrial requirements. All these advantages render electrospinning techniques particularly useful and convenient for the fabrication of various electrode materials and notably for photovoltaic devices such as dye-sensitized solar cells (DSCs).

DSCs are photoelectrochemical devices designed to convert solar radiation directly into electricity. They consist of a light source, a photoanode (semiconductor and sensitizer), and a counter electrode (CE) separated by an electrolyte containing a redox couple (Figure 7.1). Their construction and operating principle are demonstrated in Figure 7.1a and b, respectively. The working principle can be classified as the following steps:

1. On illumination, the dye molecules absorb photons with energy (hv), which give rise to the electrons that become excited from the highest occupied molecular orbital to the lowest unoccupied molecular orbital.
2. The electrons from the excited state are injected into the conduction band of the semiconductor (generally, mesoporous TiO_2), leading to the oxidized state of sensitizer.
3. The electrolyte with redox mediator (I^-/I_3^-) supplies electrons to the dye for restoring it to the initial state.
4. The electrons in conduction band of the semiconductor flow through external load and arrive at the back side contact (CE).
5. The triiodide (I_3^-) in electrolyte is reduced by the electrons from the external circuit, which regenerate iodide (I^-).

The complete description of their basic working principles can be found elsewhere [7,8]. The breakthrough in the photon to electron conversion efficiency of DSCs was achieved in the early 1990s when Grätzel and O'Regan used a colloidal TiO_2 film to fabricate a mesoporous photoanode to replace bulky material [9]. Since this

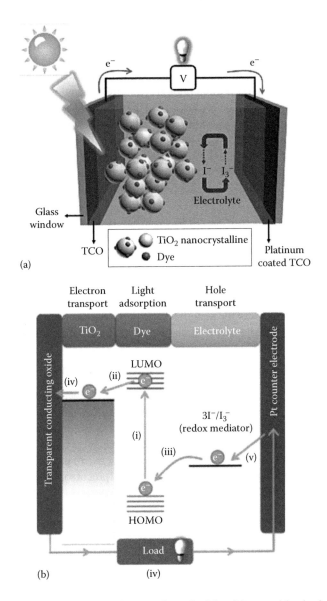

FIGURE 7.1 (a) Construction and (b) operating principle of dye-sensitized solar cells.

discovery, DSCs have attracted considerable interest due to their low-cost potential, short energy payback time, and their relatively high efficiency that reached over 12% in the last years [10]. In a DSC, the photoanode is a transparent conducting oxide (TCO) coated with a nanostructured film of a wide bandgap metal oxide semiconductor, and the CE is a TCO usually coated with a thin layer of platinum. In these devices, light is absorbed by a photosensitizer linked to the metal oxide surface through an anchoring function. Upon excitation, the electrons are injected from the excited state of the dye into the conduction band of the semiconductor and transported to the CE through an external circuit. In the meantime, the oxidized dye is regenerated thanks to the redox system present in the electrolyte, which is itself reduced at the CE, completing the circuit. Designing new highly absorbing sensitizers in order to harvest more photons from the sun and to increase the current densities delivered by the cells is one of the fruitful strategies to improve the performances of DSSCs [11]. With the same philosophy, another strategy is to develop new electrode materials and processing methods to create photoanode showing dramatically enhanced surface area and a high porosity in order to increase the number of sensitizing molecules attached to the surface. This second strategy sparks the strong interest of materials scientists for integrating electrospun materials into photovoltaic devices.

In this book chapter, we will briefly highlight some of the most interesting features associated with this technique for the preparation of various nanomaterials for application as photoanode or CE in DSCs. We will present the basic principles of the electrospinning process and discuss the various morphologies that can be developed in electrospun materials, which include NFs, nanobelts, nanorods, nanotubes, nanoflowers, and others. An overview of the various compositions of suitable binary or ternary wide band metal oxide semiconductors or even doped materials that can be obtained for integration in photovoltaic devices will also be presented.

7.2 ELECTROSPINNING BACKGROUND AND BASIC PRINCIPLES

Electrospinning is based on the use of a high voltage to induce the formation of liquid jet on the tip of a spinneret. A schematic illustration for electrospinning process is shown in Figure 7.2. The morphological features of electrospun fibers depend on various operating parameters (applied voltage, needle gauge, polymer molecular weight, the solution feeding rate, tip-to-collector distance), environmental parameters (such as temperature, humidity), and solution parameters (precursor type, polymer solubility/viscosity, dielectric constant of solvents). NF materials (e.g., inorganic/polymer composites, ceramics, metals, and their composites) have been used in various fields and recently employed as photoelectrodes for DSC, active electrode materials or membrane for electrochemical cells, catalysts. As illustrated in Figure 7.2c, metal

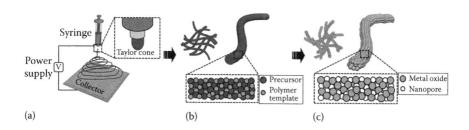

FIGURE 7.2   (a) Electrospinning equipment. (b) Inorganic precursor/polymer composite electrospun nanofiber ("as-spun nanofiber"). (c) Metal oxide nanofiber after high-temperature heat treatment in air atmosphere.

oxide NFs can be easily prepared by electrospinning inorganic precursor/polymer composite NFs followed by subsequent high-temperature heat treatment in air atmosphere. This leads to the decomposition of the matrix polymer while oxidation and crystallization of the inorganic precursors.

As an example, Figure 7.3 shows morphological evolution of Zn_2SnO_4 NFs with different packing density, that is, porous and dense NFs [12], which was synthesized based on phase separation between inorganic precursors and matrix polymer. First, the porous Zn_2SnO_4 NFs were synthesized using the electrospinning solution (containing zinc acetate ($Zn(OAc)_2$), tin (IV) acetate ($Sn(OAc)_4$), and polyvinyl acetate (PVAc) (which were chosen as a matrix template polymer and dissolved in dimethylformamide [DMF] solvent) as shown in Figure 7.3a through d). After electrospinning, the collected as-spun NFs including Zn and Sn precursors and PVAc were immediately bought into electrical furnace and were calcined at a high temperature in air atmosphere, and then polycrystalline Zn_2SnO_4 fibers were resulted from the decomposition of PVAc and crystallization of Zn and Sn precursors. The Zn_2SnO_4 fibers exhibited highly porous structure due to the miscibility between the Zn, Sn precursors, and PVAc, confirmed by the scanning transmission electron microscope analysis in Figure 7.3a. Normally, phase separation arises from inhomogeneous mixing properties between Sn-precursor and PVAc matrix, resulting in Sn-rich domain

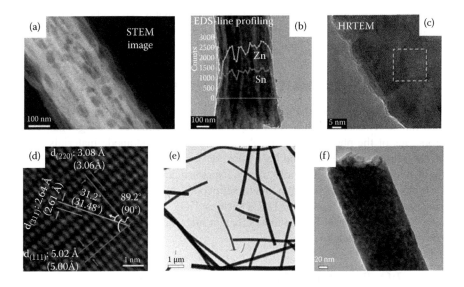

FIGURE 7.3 Morphologies and crystal structures of porous (a through d) and dense (e and f) Zn_2SnO_4 nanofibers prepared by using two different polymer matrixes, that is, PVAc and PVP. Porous Zn_2SnO_4 nanofibers: (a) scanning TEM image; (b) TEM image and EDS elemental mapping of Zn and Sn; (c) HR-TEM image; (d) lattice fringe of selected area in white-dotted frame of (c) (the values in parentheses correspond to the theoretical results). Dense Zn_2SnO_4 nanofibers: (e) TEM image and (f) magnified TEM image. (Reprinted from Choi, S.H., Hwang, I.S., Lee, J.H., Oh, S.G., and Kim, I.D., Microstructural control and selective C_2H_5OH sensing properties of Zn_2SnO_4 nanofibers prepared by electrospinning, *Chem. Commun.*, 47, 9315–9317, Copyright 2011. With permission of The Royal Society of Chemistry.)

and PVAc-rich domain. The oxidation of the Sn-precursor and the removal of the immiscible polymers take place during calcination step, and the precursor-rich domains were formed into Zn_2SnO_4 crystalline, and all of the existing PVAc-rich regions were converted into open pores, increasing the specific area of the Zn_2SnO_4 NFs. Figure 7.3b exhibits the compositional profile of Zn_2SnO_4 NFs, which confirmed Zn/Sn chemical composition ratio of 2:1. The high-resolution transmission electron microscopy image of Zn_2SnO_4 NF is shown in Figure 7.3c. The magnified transmission electron microscopy analysis of the dotted yellow frame showing lattice fringe in Figure 7.3d clearly indicates that the Zn_2SnO_4 NF has cubic inverse spinel structures. On the contrary, for Zn_2SnO_4 NFs synthesized from electrospinning solution containing a polyvinylpyrrolidone (PVP) and Zn/Sn precursors, the thermally treated NFs showed more dense inner structure composed of fine nanoparticles and smooth surface morphologies (Figure 7.3e and f). Compared with highly porous Zn_2SnO_4 NFs, as-spun fibers of Zn/Sn precursors and PVP polymer consist of more homogeneous mixture. During a high-temperature calcination step, residual organic composite burned out, and the Zn/Sn precursors were oxidized and crystallized, maintaining its unchanged dense inner structure even after thermal treatment. This indicates that miscibility control between the precursor and polymer plays a key role for tailoring the interior morphologies of electrospun metal oxide fibers. We understand that the miscibility control between inorganic precursor and matrix polymer can largely affect morphologies of electrospun metal oxide NFs, corroborating possibilities that various kinds of NFs with highly porous structure can be infinitely applied for advanced DSCs' materials.

7.3   ELECTROSPUN MATERIALS AS PHOTOANODES IN DSC

For the preparation of efficient DSC showing a high yield of conversion of the photons into electrons, one important requirement is to enhance the light-harvesting properties of the photoactive layer. As a consequence, the anode materials should reveal a large surface area and high porosity to facilitate the anchoring of a maximum amount of sensitizing molecules or nanoparticles [13].

Many binary wide bandgap oxide semiconductors have been investigated as potential electron acceptors for DSCs such as TiO_2, ZnO, or SnO_2. However, up to now, TiO_2 turned out to be the most efficient material leading to the highest power conversion efficiencies.

7.3.1   TiO₂ PHOTOANODES PREPARED BY ELECTROSPINNING

The mesoporous TiO_2 films that are used in DSCs as photo electrodes are fabricated by screen printing using colloidal nanoparticles prepared by the controlled sol–gel hydrolysis of Ti-alkoxides. After a high-temperature sintering step, the mesoporous layers consist of colloidal nanoparticles of 15–40 nm in diameter that are interconnected, and the thickness of the layers can be typically comprised between 3 and 15 μm. The effective surface area (for the grafting of the sensitizers), the porosity, and the pore volume (for the penetration of the redox electrolyte) of the electrodes are strongly dependent on the particles size.

In these layers, the photoinduced charge transfer that occurs between the dyes and the metal oxide and the transport of free electrons through the nanoparticles network are influenced by the surface states and the connectivity between the particles.

Recently one-dimensional (1D) nanostructures such as nanotubes [14] or nanorods [15] have been employed for the fabrication of DSC with liquid and quasi-solid-state electrolytes and have demonstrated an enhanced photocurrent generation compared to conventional nanoparticle-based systems.

One of the first examples of the use of electrospinning technique for the preparation of photoanode dates back to 2005 by Song et al. [15]. In this work, TiO_2 single-crystalline nanorods were prepared from electrospun NFs. A mechanical hot-pressing step was required to ensure adhesion of the as-spun Ti precursor/PVAc composite fibers to the fluorine doped tin oxide (FTO) and to produce each fibril giving rise to nanorods, which were converted to anatase single crystals after a calcination step. A DSC based on the TiO_2 nanorod electrode about 12 μm in thickness was sensitized with the ruthenium dye N3, and a poly(vinylidene fluoride-co-hexafluoropropylene) (PVDF-HFP) gel–based electrolyte was fabricated. The short-circuit current density (J_{sc}), the open-circuit photovoltage (V_{oc}), the fill factor (FF), and the overall conversion efficiency (η) of the TiO_2 nanorod electrode were found to be 14.77 mA cm^{-2}, 0.7 V, 60%, and 6.2%, respectively (Table 7.1). This first result paved the way for further improvements, and to date, the best efficiency reached 8.5% for TiO_2-based photoanode [16]. To improve the power conversion efficiency of their devices, the authors fabricated a multiscale porous NFs using a template-assisted techniques with SiO_2 nanoparticles. Among the TiO_2 NFs with different pore types, the surface area of the TiO_2 NFs fabricated with a polymer solution containing TiO_2 precursor and SiO_2 colloidal nanoparticles was the largest, reaching 150 m^2 g^{-1}. The surface area of this electrospun photoanode was nine times higher than the TiO_2 nanoparticle–based photoanode. This strategy led to a higher adsorption of the ruthenium dye (in this case, N719 was employed) and a better penetration of the electrolyte in the pores. Furthermore, the size of the NFs (i.e., 200–300 nm) induces a strong light scattering

TABLE 7.1
Photovoltaic Parameters of TiO_2-Based DSC

Photoanode	J_{sc} (mA cm^{-2})	V_{oc} (V)	FF (%)	η (%)	References
TiO_2 nanorods	14.77	0.70	60	6.20	[15]
TiO_2 nanofibers	10.70	0.71	62	4.80	[16]
Large pore TiO_2 nanofibers	12.60	0.72	65	6.00	
Small pore TiO_2 nanofibers	15.60	0.72	68	7.60	
Multiscale pore TiO_2 nanofibers	16.30	0.73	71	8.50	
TiO_2 nanoparticles	12.00	0.71	70	6.00	
TiO_2 nanofibers "rice grain"	8.20	0.96	62	5.10	[17]
TiO_2 nanotubes	6.01	0.80	69	3.33	[18]
TiO_2 Nanotubes/nanoparticle (8/2)	7.91	0.78	69	4.26	
TiO_2 nanoparticles	11.50	0.74	70	5.98	

that is beneficial. A maximum power conversion efficiency of 8.5% was obtained, which is significantly higher compared to 6.0% achieved using conventional TiO_2 nanoparticles (Table 7.1).

As mentioned earlier, the electrospinning technique affords the possibility to tune the shapes of the deposited materials by modifying the processing conditions. In 2010, Nair et al. used a photoanode, which consists of electrospun NFs made of TiO_2 with a "rice grain" shape [17]. Powder X-ray diffraction showed that these rice grain–shaped TiO_2 particles are mostly made of 90% anatase and 10% rutile phases. That later phase grows during the sintering process at 500°C even though rutile phase usually grows from anatase at temperature above 500°C. This phenomenon was attributed to the high density of the nearly spherical rice grain particles. The surface area of these electrodes was found to be slightly higher than the one of commercially available TiO_2 nanoparticles usually employed for DSC purpose (i.e., P25), 60 m^2 g^{-1} compared to 54 m^2 g^{-1}. These rice grain–based photoanodes were then sensitized with a ruthenium dye, that is, N3 dye, and used in combination with a liquid electrolyte. Overall efficiency was 5.1%, which is lower than the values obtained with standard devices. However, it should be underlined that this value was achieved without any $TiCl_4$ treatment of the electrode and without any scattering layer. It is also important to highlight the really high V_{oc} of these devices, 0.96 V, which is responsible for the relatively good efficiency despite of low J_{sc} and FF. However, the origin of this surprisingly high V_{oc} is not yet understood.

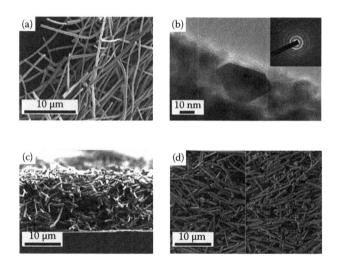

FIGURE 7.4 SEM image (a) and HR-TEM image (b) of electrospun TiO_2 nanotubes with the inset SAED pattern showing the anatase crystalline structure. (c) Cross-sectional view of TiO_2 nanotubes anode film on FTO glass substrate. (d) Plain view of DSC anode films of (left) pure TiO_2 nanotubes and (right) TiO_2 nanotubes blended with P25 TiO_2 nanoparticles. (Reprinted with permission from Wang, X., He, G., Fong, H., and Zhu, Z., Electron transport and recombination in photoanode of electrospun TiO_2 nanotubes for dye-sensitized solar cells, *J. Phys. Chem. C*, 117, 1641. Copyright 2013, American Chemical Society.)

In 2013, Wang et al. prepared TiO_2 nanotubes to be employed as photoanode [18]. The preparation of these TiO_2 nanotubes was carried out employing a coaxial electrospinning technique. The average inner diameter of these nanotubes was *circa* 275 nm and the wall thickness was around 115 nm (Figure 7.4). These nanostructures showed an anatase-type crystalline phase. For their sensitization, the well-known ruthenium complex N719 was used, but the devices performances were limited because of a lower dye loading compared to a classical nanoparticle-based photoanode. Nevertheless, this drawback was partially overcome by a faster electron transport in the nanotube-based electrodes, which improves charge collection and gives rise to a longer electron lifetime. As a consequence, high V_{oc} was achieved. Despite of the high V_{oc}, the maximum power conversion efficiency of the devices containing the nanotubes was found lower, compared to the one of devices containing nanoparticles, 3.33% versus 5.98% (Table 7.1). In order to improve efficiency and keep a high V_{oc}, a blend of nanoparticles and nanotubes was also studied leading to an increase of the performances up to 4.26%. Nevertheless, this illustrates that electrospinning can lead to structural diversity that can be promising for further developments.

7.3.2 DOPED TiO_2-BASED ELECTROSPUN PHOTOANODE

One practical advantage, of the electrospinning as mentioned before, is that multicompositional NFs can be prepared successfully by controlling the precursor component and the polymer matrix. Examples include heterostructures of TiO_2–Al_2O_3 NFs [19] or TiO_2 doped with various metals or carbon materials.

Naphade et al. took advantage of this possibility, which allows to start from both titanium and gold precursors, to grow in situ gold nanoparticles into TiO_2 NFs [20]. The gold content in the TiO_2 NFs was found to be close to 1 wt.%, and the size of the nanoparticles was in the range of 5–10 nm, while the diameter of NFs was around 150 nm (Figure 7.5). Those doped NFs were used efficiently as a scattering layer taking advantage of the plasmonic effect of the small nanoparticles and the superior scattering ability of NFs over nanoparticles. A remarkable enhancement in the efficiency by 25% was achieved with the Au-doped NF layer compared to the value achieved with the devices without any light-scattering layer (Table 7.2). The highest power conversion efficiency reached with these electrodes was 7.77%. Furthermore, it was found that the use of gold nanoparticles avoided charge recombination with the electrolyte. Indeed, the incorporation of Au nanoparticles in the TiO_2 enhanced charge trapping density of states in the conduction band and below the conduction band due to favorable band alignment of the Au/TiO_2 Schottky interfaces. As a result, J_{sc} and FF are improved compared to a cell employing no scattering layer or an NF-based scattering layer.

Archana et al. used the same processing technique to dope TiO_2 NFs with niobium [21]. They noticed that doping TiO_2 with niobium induces a decrease in grain size within the NFs despite being processed under the same experimental conditions as the nondoped ones. It was also noticed that doping with Nb reduced the active surface area. This phenomenon was attributed to larger pore size; as a consequence, the dye loading was reduced to 7.9×10^{-8} mol cm^{-2} for the Nb-doped electrodes (5% atoms of Nb) compared to 1.0×10^{-7} mol cm^{-2} for the nondoped NFs. Despite

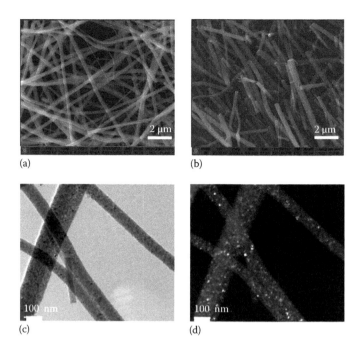

FIGURE 7.5   (a and b) The FE-SEM images of TiO$_2$ nanofibers and Au/TiO$_2$ nanofibers. (c) The dark field image of Au/TiO$_2$ nanofibers shown in (d). (Reprinted from Naphade, R.A., Tathavadekar, M., Jog, J.P., Agarkar, S., and Ogale, S., Plasmonic light harvesting of dye sensitized solar cells by Au-nanoparticle loaded TiO$_2$ nanofibers, *J. Mater. Chem. A.*, 2, 975–984, Copyright 2014. With permission of The Royal Society of Chemistry.)

TABLE 7.2

Photovoltaic Parameters of Doped TiO$_2$-Based DSC

Photoanode	J_{sc} (mA cm^{-2})	V_{oc} (V)	FF (%)	η (%)	References
TiO$_2$ nanofibers	13.90	0.76	64	6.76	[20]
Au/TiO$_2$ nanofibers	15.10	0.77	67	7.77	
TiO$_2$ nanoparticles	8.90	0.75	70	4.67	[21]
NbTiO$_2$ nanofibers (2%)	9.74	—	—	—	
Nb/TiO$_2$ nanofibers (5%)	10.00	—	—	—	

of lower dye content, J_{sc} of DSC with Nb-doped TiO$_2$ photoanode was found higher compared to the reference device (Table 7.2). The enhancement of the J_{sc} was attributed to a higher charge mobility (μ_n) and a higher charge diffusivity (D_n) induced by the doping. However, other parameters such as FF and V_{oc} were found lower because of a higher rate of recombination between the photoanode and the electrolyte leading to a lower power conversion efficiency (PCE). The reason given by authors is that the lattice strain induced by the doping favors the apparition of defects that could act as charge recombination centers.

7.3.3 TiO₂:Carbon Material–Based Electrospun Photoanode

With the emergence of carbon materials such as graphene and carbon nanotubes (CNTs) for optoelectronic applications, it was therefore appealing to create blends with TiO_2 in order to take advantage of the charges' transport capability of this class of materials, for the fabrication of highly conductive electrodes. In 2012, Madhavan et al. synthesized blends of graphene and TiO_2 NFs [22]. They found that a limited amount of graphene (between 0.2% and 1.0% in wt.) allows to keep consistency and stability for the NFs as their morphologies or diameters were unchanged (Figure 7.6). However, the length of NFs decreases as the amount of graphene increases. The shorter lengths may result from stress and breakage of the NFs due to the difference between the coefficients of thermal expansion of the two materials. Only TiO_2 NFs with content of 0.7% in wt. of graphene were tested. The V_{oc}, J_{sc}, FF, and efficiency for the graphene-loaded samples were found to be 0.71 V, 16.2 mA cm⁻², 66%, and 7.6% as compared to 0.71 V, 13.9 mA cm⁻², 63%, and 6.3% for pristine TiO_2 NFs, respectively (Table 7.3). A significant increase of 17% in the current density is observed, which is likely related to the creation of a junction between the wide bandgap semiconductor (TiO_2) and the narrowband gap semiconductor (graphene). Indeed, this junction induces deep depletion of the charge carriers at the interface with substantial band bending. This facilitates charge injection and prevents charge recombination even at low content of graphene.

(a)   (b)   (c)

(d)   (e)

FIGURE 7.6 SEM images of annealed electrospun: (a) TiO_2 fibers, (b) G-T-0.2, (c) G-T-0.5, (d) G-T-0.7, and (e) G-T-1.0. (Reprinted from Madhavan, A.A., Kalluri, S., Chacko, D.K., Arun, T.A., Nagarajan, S., Subramanian, K.R.V., Nair, A.S., Nair, S.V., and Balakrishnan, A., Electrical and optical properties of electrospun TiO_2-graphene composite nanofibers and its application as DSSC photo-anodes, *RSC Adv.*, 2, 13032–13037, Copyright 2012. With permission of The Royal Society of Chemistry.)

TABLE 7.3

Photovoltaic Parameters of TiO₂: Carbon Material–Based DSC

Photoanode	J_{sc} (mA cm⁻²)	V_{oc} (V)	FF (%)	η (%)	References
TiO₂ nanofibers	13.90	0.71	63	6.30	[22]
TiO₂/graphene 0.7%	16.20	0.71	66	7.60	
TiO₂ nanofibers (16.7 µm)	16.30	0.74	51	6.18	[23]
TiO₂/MWCNT (0.05%, 6.70 µm)	8.92	0.72	70	4.51	
TiO₂/MWCNT (0.05%, 14.47 µm)	14.20	0.76	72	7.80	
TiO₂/MWCNT (0.10%, 6.04 µm)	11.10	0.72	73	5.84	
TiO₂/MWCNT (0.10%, 14.64 µm)	18.53	0.75	74	10.24	
TiO₂/MWCNT (0.15%, 7.32 µm)	8.72	0.73	76	4.38	
TiO₂/MWCNT (0.15%, 13.97 µm)	12.11	0.78	76	7.13	

Yang and Leung used the electrospinning technique to prepare electrodes composed of nanorods of TiO₂ with carbon nanotubes [23]. The doping rate was controlled by adjusting the amount of nanotubes in the feeding solution, ranging from 0.05% to 0.15% in wt. These electrodes were sensitized with the N719 dye. As for graphene-loaded TiO₂, it was found that carbon nanotubes favor charge injection and reduce recombination processes. The optimum loading amount of carbon nanotubes was found to be 0.10% in wt., since higher concentrations induce a drop of the current densities delivered by the cells (by 22% for thin electrode and 35% for thicker electrode) (Figure 7.7). This phenomenon was explained by a lower absorption of the

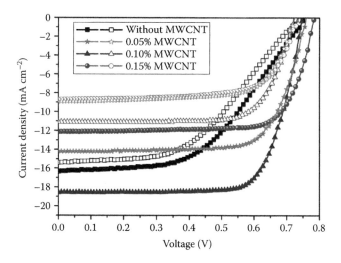

FIGURE 7.7 J–V characteristics of different photoanodes, hollow-symbol curve represents thinner photoanode with thickness about 6.6 ± 0.7 µm, and solid-symbol curve represents thicker photoanode with thickness about 14.3 ± 0.3 µm. (Yang, L. and Leung, W.W.-F.: Electrospun TiO₂ nanorods with carbon nanotubes for efficient electron collection in dye-sensitized solar cells. *Adv. Mater.* 2013. 25. 1792–1795. Copyright Wiley-VCH Verlag GmbH & Co. KGaA. Reprinted with permission.)

dye with increasing the amount of carbon nanotubes. For optimized devices, maximum efficiency reached 10.24%, which is the highest value ever reported for a DSC using an electrospun photoanode (Table 7.3).

7.3.4 ALTERNATIVE SEMICONDUCTING OXIDE FOR ELECTROSPUN PHOTOANODE

In recent years, researchers have undertaken the task of replacing the TiO_2 by other types of metal oxide semiconductor in DSC in order to improve the electron mobility within the mesostructured electrode [24].

Among the large variety of binary metal oxides, ZnO appeared as a promising alternative. Indeed, ZnO is a wide bandgap (*ca.* 3.3 eV) metal oxide that preferentially crystallizes in the hexagonal wurtzite phase. The conduction band edge of ZnO is close to that of TiO_2 (*ca.* −4.4 eV), and it shows a much higher electron mobility in the bulk (200–300 $cm^2 V^{-1} s^{-1}$) than TiO_2 (0.1 $cm^2 V^{-1} s^{-1}$) [25]. As a consequence, ZnO has been successfully investigated as a relevant alternative to TiO_2 for photoanodes in DSC applications [26]. Various shapes of ZnO nanostructures can be prepared such as spherical particles, rods, wires, or tetrapods [27]. Besides, this material can also be prepared by electrospinning using essentially the same workup as for TiO_2. In 2009, Zhang et al. synthesized electrospun ZnO photoanode of different thickness, ranging from 1.5 to 5.0 µm, by controlling the electrospinning duration (Figures 7.8(a) or 7.9(b)) [28]. After the calcination step, NFs showed rough surfaces and the overall film looked homogeneous without any cracks or defects, which can sometimes be observed due to mismatch between thermal expansion coefficient of ZnO and the glass substrate (Figure 7.8). In this work, the authors observed the spontaneous formation of a "self-relaxation layer" by disintegration of NFs that promote the adhesion to the FTO substrate. A buffer layer is formed that releases the tensile stress at the interface of the FTO substrate and ZnO NFs (Figure 7.8). This phenomenon prevents the ZnO NFs deposited subsequently from cracking and avoids the peeling off phenomenon from the substrate. Studies showed than the crystalline structure was pure wurtzite and the active area of the NFs was 30 $m^2 g^{-1}$. DSCs were fabricated with N719 as sensitizer, and the best efficiency reached 2.58% for a device with a 5.0 µm thick photoelectrode. Performances were further improved to 3.02% with a Zn(OAc) surface treatment (Table 7.4). This improvement was explained by two factors that both enhanced current density. First, the formation of a larger surface area highlighted by an increased dye loading, second, a more effective suppression of recombination processes between the ZnO and the electrolyte as confirmed by impedance spectroscopy measurements.

Apart from ZnO, SnO_2 is also a well-known and useful TCO for nanoelectronics that can be considered alternatively to TiO_2. This binary oxide also shows high electron mobility (10–125 $cm^2 V^{-1} s^{-1}$), and it possesses a wider bandgap (*ca.* 3.6 eV). However, its conduction band is located at a lower energy level than that of TiO_2. As a consequence, DSCs fabricated with a SnO_2 electrode usually lead to lower open-circuit voltages (V_{oc} below 600 mV). Nonetheless, Kumar et al. have investigated flower-shaped nanostructures of SnO_2, synthesized by an electrospinning by precisely controlling the precursor concentration in a polymeric solution [29]. These

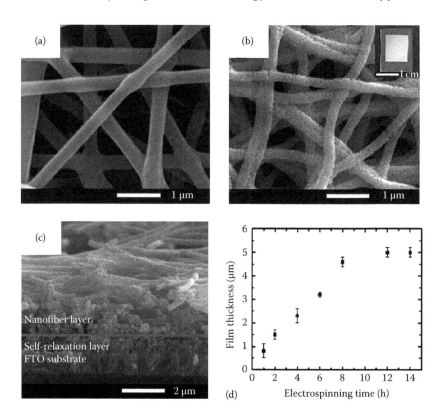

FIGURE 7.8 SEM images of the as-spun (a) and calcined (b) ZnO nanofibers on FTO substrates (inset photograph of calcined ZnO film on FTO substrate). (c) Typical cross-sectional SEM images of the calcined ZnO film on FTO substrate. (d) The variation in ZnO film thickness with different electrospinning time. (Reprinted with permission from Zhang, W., Zhu, R., Liu, X., Liu, B., and Ramakrishna, S., Facile construction of nanofibrous ZnO photoelectrode for dye-sensitized solar cell applications, *Appl. Phys. Lett.*, 95, 043304. Copyright 2009, American Institute of Physics.)

flower-shaped particles were made up of nanofibrils of diameters comprised between 70 and 100 nm, consisting of linear arrays of single-crystalline nanoparticles with a size of 20–30 nm (Figure 7.9).

Both flowers and NFs showed the same chemical composition and revealed a cassiterite phase even though the nanoflowers possess a superior crystallinity as demonstrated by the XPS peaks that were found sharper than the ones observed for the NFs. Mott–Schottky analysis showed that flowers have an order of magnitude higher electron density compared with the fibers as measured by electrochemical impedance spectroscopy. DSCs were fabricated using both fibers and flowers highlighting the better performances of the nanoflower-based device. J_{sc} was increased by *circa* 140%, V_{oc} by 16%, and FF by 55%, which is higher compared to the NF-based device (Table 7.4). This enhancement in J_{sc} was mainly attributed to a longer electron lifetime, as extracted from the Bode plot, which tends to show that the device using

TABLE 7.4
Photovoltaic Parameters of Electrospun Semiconducting Oxides
Alternative to TiO$_2$ for DSC

Photoanode	J$_{sc}$ (mA cm^{-2})	V$_{oc}$ (V)	FF (%)	η (%)	References
ZnO nanofibers (1.5 μm)	4.58	0.63	61	1.77	[28]
ZnO nanofibers (3.2 μm)	6.10	0.61	64	2.36	
ZnO nanofibers (5.0 μm)	6.62	0.59	66	2.58	
ZnO nanofibers (5.0 μm. Zn(Oac)$_2$ treatment)	9.14	0.57	58	3.02	
SnO$_2$ nanoflowers	7.30	0.70	60	3.00	[29]
SnO$_2$ nanofibers	3.00	0.60	38	0.71	
Zn$_2$SnO$_4$ nanofibers (1.5 μm)	6.39	0.73	59	2.77	[38]
Zn$_2$SnO$_4$ nanofibers (3 μm)	8.11	0.72	63	3.67	
Zn$_2$SnO$_4$ nanofibers (4 μm)	7.66	0.75	62	3.55	

nanoflowers inhibits back reaction between electrons from the conduction band and the iodide electrolyte. It is also worth noting that the dye loading for nanoflowers was slightly higher than for NFs and therefore this as well contributes to the enhanced J$_{sc}$. The improved V$_{oc}$ and FF were credited to high internal resistance for back electron transfer and higher electron density; as a consequence, nanoflower-based electrodes showed better charge collection and lower recombination, which allows to reach V$_{oc}$ as high as 700 mV.

For the replacement of TiO$_2$ in DSCs, most of the previous research has been focused on binary oxides such as ZnO, SnO$_2$, Nb$_2$O$_5$ [30,31], and In$_2$O$_3$ [32]. In contrast, the application of multication oxides has been rarely explored. In comparison with simple binary oxides, multication oxides have more freedom to tune the materials' chemical and physical properties by altering the compositions.

The only reported ternary oxides are SrTiO$_3$ [33] and Zn$_2$SnO$_4$ [34–37]. By varying the relative element ratio, the bandgap energy, the work function, and the electric resistivity of the ternary oxides can be readily tuned. Considering the availability of a wide range of multication oxides and their tunable properties, it is therefore interesting to investigate their applications in DSC and for their preparation electrospinning appears as a particularly well-adapted method.

In order to overcome the low charge mobility of the TiO$_2$ and the sensitivity to acidic conditions of ZnO, researchers have investigated the potential of Zn$_2$SnO$_4$. In 2013, Choi et al. used an electrospun Zn$_2$SnO$_4$ photoanode that was sensitized with N719 and also push–pull organic dyes [38]. The Zn$_2$SnO$_4$ photoanode was prepared by mixing Zn(OAc)$_2$ and Sn(OAc)$_4$ in the starting solution along with the PVAc. In order to ensure adhesion to the FTO surface, it was found to be mandatory to go through a hot-pressing step that partially melts the PVAc onto the FTO before calcination in order to avoid the peel-off of the NF mats from the substrate. Calcination at 450°C gives NFs that are porous and smooth, whereas calcination at 700°C tends to lead to rougher surface due to the formation of crystallite with an inverse spinel phase as confirmed

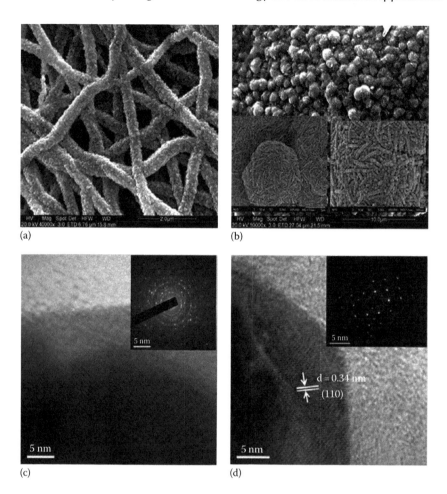

FIGURE 7.9 SEM images (a and b) and HR-TEM images (c and d) of fibers and flowers, respectively. Inserts: (b) magnified SEM images of the flower morphology; (c) selected area electron diffraction (SAED) pattern showing polycrystalline rings; and (d) SAED pattern of flowers showing single-crystalline spotty patterns. (Reprinted from Naveen Kumar, E., Jose, R., Archana, P.S., Vijila, C., Yusoff, M.M., and Ramakrishna, S., High performance dye-sensitized solar cells with record open circuit voltage using tin oxide nanoflowers developed by electrospinning, *Energy Environ. Sci.*, 5, 5401–5407, Copyright 2012. With permission of The Royal Society of Chemistry.)

by X-ray diffraction (Figure 7.10). Due to the limited heating temperature tolerance of glass substrate, calcination process was performed at 450°C for the fabrication of DSCs. This led to the formation of amorphous Zn_2SnO_4 fibers. The specific surface area for NF network calcinated at 450°C was found to be 124 m^2 g^{-1}, as measured by Brunauer–Emmett–Teller method, which is similar to conventional TiO_2 nanoparticle electrode. Comparison between organic and organometallic sensitizers clearly showed that Zn_2SnO_4 photoanodes give better performances when they are combined

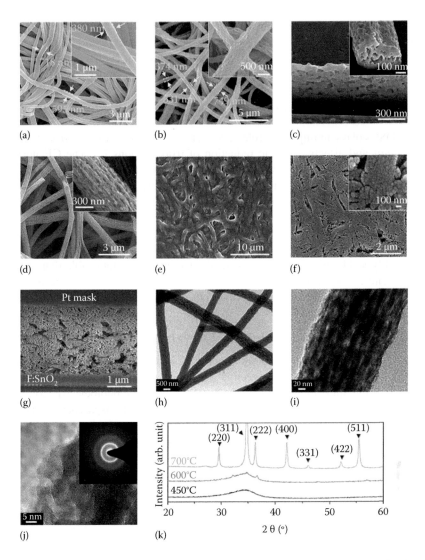

FIGURE 7.10 (a) SEM image of as-spun $Zn(OAc)_2$-$Sn(OAc)_4$/PVAc composite fibers; (b) SEM image of Zn_2SnO_4 fibers calcined at 500°C; (c) cross-sectional view of Zn_2SnO_4 fibers calcined at 500°C; (d) Zn_2SnO_4 fibers calcined at 700°C (reference); (e) hot-pressed $Zn(OAc)_2$-$Sn(OAc)_4$/PVAc composite fibers; (f) SEM image of Zn_2SnO_4 fibers calcined at 450°C after hot-pressing step; (g) cross-sectional view of Zn_2SnO_4 fibers calcined at 450°C after hot-pressing step; (h) TEM image of Zn_2SnO_4 fibers calcined at 450°C after hot-pressing step; (i) magnified TEM image in (h); (j) magnified TEM image of (i) and the inset shows the SAED pattern; (k) X-ray diffraction pattern of Zn_2SnO_4 fibers calcined at various temperatures of 450°C, 600°C, and 700°C. (Choi, S.-H., Hwang, D., Kim, D.-Y., Kervella, Y., Maldivi, P., Jang, S.-Y., Demadrille, R., and Kim, I.-D.: Amorphous Zinc Stannate (Zn_2SnO_4) nanofibers networks as photoelectrodes for organic dye-sensitized solar cells. *Adv. Funct. Mater.* 2013. 23. 3146–3155. Copyright Wiley-VCH Verlag GmbH & Co. KGaA. Reprinted with permission.)

to organic sensitizers. Performance with N719 was limited to 1.41% whereas organic sensitizer gave up to 3.67% for similar photoanode thickness (3 μm) (Table 7.4). This behavior was attributed to not only higher extinction molar coefficient of the organic sensitizers, but also to a better dye loading of organic dyes compared to N719.

7.4 ELECTROSPUN MATERIALS AS COUNTER ELECTRODE IN DSC

CE in DSCs plays an important role in collecting electrons stemming from the external circuit and affects catalytic reduction of triiodide ions I_3^-. The CE normally consists of thin platinum (Pt) layer coated on a TCO layer. This system is widely used in DSCs due to highly efficient electrocatalytic properties of Pt for reduction of (I_3^-) ions. Although Pt serves as a high-performing material in the CE, it is very expensive and prone to corrosion by I^-/I_3^- redox couple, which increases concern associated with long-term stability of DSCs due to reduced catalytic activity. Therefore, the major research in the CE for DSCs focuses on the development of low-cost, efficient, and highly stable materials. As this trend continued, it is noticeable that carbonaceous materials such as graphite, carbon black, and carbon NF can be promising candidate for CE among alternative materials.

7.4.1 Pt-Based Counter Electrode Prepared by Electrospinning

Pt NF, easily synthesized by electrospinning, can serve as a transparent CE without typical TCO layer. Kim et al. recently developed new design of DSC's electrode containing the multifunctional Pt NF, which exhibits high photoelectric conversion efficiency, outstanding catalytic effect, conductivity, and transparency as shown in Figure 7.11 [39].

This Pt NF-based CE is suitable for decreasing overall cost and amount of electrode material as well as improving long-term stability.

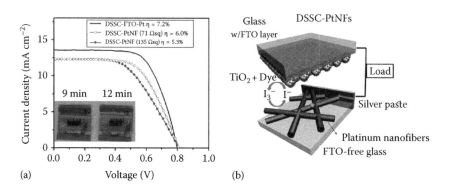

(a)   (b)

FIGURE 7.11 (a) Current–voltage curves of DSCs and (b) schematic illustration of DSCs prepared with Pt-spin-coated FTO counter electrode. (Reprinted with permission from Kim, J., Kang, J., Jeong, U., Kim, H., and Lee, H., Catalytic, conductive, and transparent platinum nanofiber webs for FTO-free dye-sensitized solar cells, *ACS Appl. Mater. Interfaces*, 5, 3176-3181. Copyright 2013, American Chemical Society.)

Figure 7.11 shows schematic architecture and the current–voltage plots for the Pt NF-based CE. Unlike typical DSC structure, the manipulated CE on the basis of Pt NFs was directly fabricated on a FTO-free glass substrate by electrospinning. The overall photoelectric conversion efficiency (η) for this material was 6% and 5.3% corresponding to the Pt NFs with sheet resistance (R_s) 71 Ω sq^{-1} (electrospinning time, t_{spin} = 12 min) or 135 Ω sq^{-1} (electrospinning time, t_{spin} = 9 min), respectively. The improved η was originated from the enhanced FF that was related to the internal series resistance of the cells, while V_{oc} and J_{sc} remain unchanged.

7.4.2 CARBON NANOFIBER-BASED COUNTER ELECTRODE PREPARED BY ELECTROSPINNING

Even though Pt-based CE reveals high performances, the use of Pt has been limited in terms of cost. To solve this problem, carbonaceous materials, as abundant and low-cost materials, have been suggested with goal of standing in for the expensive Pt. Kay and Grätzel first recommended carbon-based CE composed of Pt-free graphite and carbon black [40]. In particular, carbon nanofiber (CNF) has reached high efficiency of more than 7%, which was attributed to excellently porous structure and 1D conducting network with fast electron transfer to improve the photoelectric activity [41–43]. More recently, Sigdel et al. reported CNF/TiO$_2$ nanoparticle composite as CE materials for DSCs [44]. The electrospun carbon nanofiber (ECN) was coated with TiO$_2$ nanoparticle via spray coating technique, and the TiO$_2$ was applied to not only interconnect between CNFs but also bind the fiber to FTO glass substrate. Figure 7.12 exhibits the morphologies of pure CNF and TiO$_2$-coated CNFs and shows photovoltaic performances. The ECN mat was fabricated by electrospinning

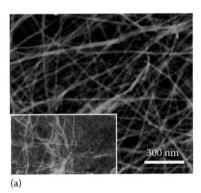
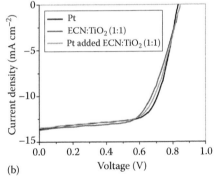

FIGURE 7.12 (a) Top view SEM image of pristine ECN mat (inset SEM image corresponds to crushed ECN/TiO$_2$ composite CE with a 1:1 wt.% mixing ratio). (b) Current–voltage curves for Pt, ECN/TiO$_2$ (1:1) and Pt-added ECN/TiO$_2$ (1:1)-based DSCs under AM 1.5 illumination at a light intensity of 100 mW cm^{-2}. (Reprinted from Sigdel, S., Dubey, A., Elbohy, H., Aboagye, A., Galipeau, D., Zhang, L., Fong, H., and Qiao, Q., Dye-sensitized solar cells based on spray-coated carbon nanofiber/TiO$_2$ nanoparticle composite counter electrodes, *J. Mater. Chem. A*, 2, 11448–11453, Copyright 2014. With permission of The Royal Society of Chemistry.)

method using polyacrylonitrile (PAN) dissolved in DMF solvent. After electrospinning, PAN NFs were overlaid on Al foil collector and stabilized by thermal treatment at 280°C for 6 h under air atmosphere. Then, the stabilized NFs were heated at 1200°C for 1 h for carbonization. In order to improve the performance of the pure ECN, the ECN mat was crushed into powder and mixed with TiO_2 nanoparticle. The composite was added in ethanol and subsequently coated on FTO glass substrate via spray coater, followed by heat treatment at 60°C for 1 h and 450°C for 30 min under air in both cases. As shown in Figure 7.11b, the open-circuit voltage (V_{oc}) of ECN/TiO_2 (1:1) (0.84 V) was higher than Pt electrode (0.83 V), and the short-circuit current densities (J_{sc}) were 13.62 mA cm^{-2}, 13.69 mA cm^{-2}, and 13.47 cm^{-2} corresponding to Pt, ECN/TiO_2, and Pt-added ECN/TiO_2-based cells, respectively. Above all, the photoelectric conversion efficiency (η%) for ECN/TiO_2 CE was 7.25%, which was comparable to the 7.57% of the conventional Pt and was attributed to synergetic performance of TiO_2/CNF composite. Besides this material, other CNF-based CEs, that is, CNFs combined with Pt nanoparticles and nitrogen-doped TiO_2 nanoparticle CNF composites [45,46], can be also promising candidate as CE materials for DSCs.

7.5 CONCLUSION AND PERSPECTIVES

The emergence of innovative technologies for energy conversion, capable to compete with the conventional silicon technology, requires the development of new functional materials with targeted nanostructures. One-dimensional nanostructured materials, such as NFs, nanowires, or nanorods, are expected to play a significant role in this field because of the correlation of their physical properties with the directionality. However, for practical applications, the perfect control of the chemical composition is also mandatory in these nanostructures.

Among the various fabrication methods for producing NFs or nanowires, electrospinning appears as a powerful tool since it allows a great control of the morphology and the orientation of the deposited nanostructures on substrates. In addition, the wonderful variety of electrospun materials in terms of chemical composition and functionalities accounts for the infatuation of the electrospinning technique for the fabrication of electrodes in energy-related applications and in particular in photovoltaic devices.

In the last decade, electrospinning has been employed to create mesoporous electrodes, with an excellent control of their nanoscale architectures and their chemical composition, applicable in DSC. This technique has been used to created photoanode of with binary and ternary metal oxides and power conversion efficiency up to 10% has been achieved. Electrospinning is now emerging as a quite attractive, versatile, and reliable technique, easy to transfer at the industrial scale for future developments of DSC technologies and more largely organic photovoltaic (OPV) and perovskites solar cells.

REFERENCES

1. Sahay, R., Suresh Kumar, P., Sridhar, R., Sundaramurthy, J., Venugopal, J., Mhaisalker, S. G., Ramakrishna, S. 2012. Electrospun composite nanofibers and their multifaceted applications. *J. Mater. Chem.* 22:12953–12971.

2. Cui, L. F., Yang, Y., Hsu, C. M., Cui, Y. 2009. Carbon–silicon core–shell nanowires as high capacity electrode for lithium ion batteries. *Nano Lett.* 9:3370–3374.

3. Dong, S. M., Chen, X., Gu, L., Zhang, L. X., Zhou, X. H., Liu, Z. H., Han, P. X. et al. 2011. A biocompatible titanium nitride nanorods derived nanostructured electrode for biosensing and bioelectrochemical energy conversion. *Biosens. Bioelectron.* 26:4088–4094.

4. (a) Ning, Z., Fu, Y., Tian, H. 2010. Improvement of dye-sensitized solar cells: What we know and what we need to know. *Energy Environ. Sci.* 3: 1170–1181; (b) Grätzel, M. 2009. Recent advances in sensitized mesoscopic solar cells. *Acc. Chem. Res.* 42:1788–1798.

5. Coley, J. F. 1900. Improved methods of and apparatus for electrically separating the relatively volatile liquid component from the component of relatively fixed substances of composite fluids. Patent GB 06385.

6. Formhals, A. 1934. Process and apparatus for preparing artificial threads. Patent U.S. 1,975,504.

7. Ardo, S., Meyer, G. J. 2009. Photodriven heterogenous charge transfer with transition metal compounds anchored to TiO_2 semiconductor surfaces. *Chem. Rev.* 38:115–164.

8. Hagfeldt, A., Boschloo, G., Sun, L. C., Kloo, L., Pettersson, H. 2010. Dye-sensitized solar cells. *Chem. Rev.* 110:6595–6663.

9. O'Regan, B., Grätzel, M. 1991. A low-cost, high-efficiency solar-cell based on dye sensitized colloidal TiO_2 films. *Nature* 353:737–740.

10. Hardin, B. E., Snaith, H. J., McGehee, M. D. 2012. The renaissance of dye-sensitized solar cells. *Nat. Photon.* 6:162–169.

11. Yum, J.-H., Baranoff, E., Wenger, S., Nazeeruddin, M. K., Grätzel, M. 2011. Panchromatic engineering for dye-sensitized solar cells. *Energy Environ. Sci.* 4:842–857.

12. Choi, S. H., Hwang, I. S., Lee, J. H., Oh, S. G., Kim, I. D. 2011. Microstructural control and selective C_2H_5OH sensing properties of Zn_2SnO_4 nanofibers prepared by electrospinning. *Chem. Commun.* 47:9315–9317.

13. Onozuka, K., Ding, B., Tsuge, Y., Naka, T., Yamazaki, M., Sugi, S., Ohno, S., Yoshikawa, M., Shiratori, S. 2006. Electrospinning processed nanofibrous TiO_2 membranes for photovoltaic applications. *Nanotechnology* 17:1026–1031.

14. Richter, C., Schmuttenmaer, C. A. 2010. Exciton-like trap states limit electron mobility in TiO_2 nanotubes. *Nat. Nanotechnol.* 5:769–772.

15. Song, M. Y., Ahn, Y. R., Jo, S. M., Kim, D. Y., Ahn, J.-P. 2005. TiO_2 single-crystalline nanorod electrode for quasi-solid-state dye-sensitized solar cells. *Appl. Phys. Lett.* 87:113113.

16. Hwang, S. H., Kim, C., Song, H., Son, S., Jang, J. 2012. Designed architecture of multiscale porous TiO_2 nanofibers for dye-sensitized solar cells photoanode. *ACS Appl. Mater. Interfaces* 4:5287–5292.

17. Nair, A. S., Shengyuan, Y., Peining, Z., Ramakrishna, S. 2010. Rice grain-shaped TiO_2 mesostructures by electrospinning for dye-sensitized solar cells. *Chem. Commun.* 46:7421–7423.

18. Wang, X., He, G., Fong, H., Zhu, Z. 2013. Electron transport and recombination in photoanode of electrospun TiO_2 nanotubes for dye-sensitized solar cells. *J. Phys. Chem. C* 117:1641–1646.

19. Tobin, J. S., Turinske, A. J., Stojilovic, N., Lotus, A. F., Chase, G. G. 2012. Temperature-induced changes in morphology and structure of TiO_2–Al_2O_3 fibers. *Curr. Appl. Phys.* 12:919–923.

20. Naphade, R. A., Tathavadekar, M., Jog, J. P., Agarkar, S., Ogale, S. 2014. Plasmonic light harvesting of dye sensitized solar cells by Au-nanoparticle loaded TiO_2 nanofibers. *J. Mater. Chem. A* 2:975–984.

21. Archana, P. S., Jose, R., Jin, T. M., Vijila, C., Yusoff, M. M., Ramakrishna, S. 2010. Structural and electrical properties of Nb-Doped anatase TiO_2 nanowires by electrospinning. *J. Am. Ceram. Soc.* 93: 4096–4102.
22. Madhavan, A. A., Kalluri, S., Chacko, D. K., Arun, T. A., Nagarajan, S., Subramanian, K. R. V., Nair, A. S., Nair, S. V., Balakrishnan, A. 2012. Electrical and optical properties of electrospun TiO_2-graphene composite nanofibers and its application as DSSC photo-anodes. *RSC Adv.* 2:13032–13037.
23. Yang, L., Leung, W. W.-F. 2013. Electrospun TiO_2 nanorods with carbon nanotubes for efficient electron collection in dye-sensitized solar cells. *Adv. Mater.* 25:1792–1795.
24. Jose, R., Thavasi, V., Ramakrishna, S. 2009. Metal oxides for dye-sensitized solar cells. *J. Am. Ceram. Soc.* 92:289–301.
25. Zhang, Q., Dandeneau, C. S., Zhou, X., Cao, C. 2009. ZnO nanostructures for dye-sensitized solar cells. *Adv. Mater.* 21:4087–4108.
26. Bouclé, J., Ackermann, J. 2012. Solid-state dye-sensitized and bulk heterojunction solar cells using TiO_2 and ZnO nanostructures: Recent progress and new concepts at the borderline. *Polym. Int.* 61:355–373.
27. Newton, M. C., Warburton, P. A. 2007. ZnO tetrapod nanocrystals. *Mater. Today* 10:50–54.
28. Zhang, W., Zhu, R., Liu, X., Liu, B., Ramakrishna, S. 2009. Facile construction of nanofibrous ZnO photoelectrode for dye-sensitized solar cell applications. *Appl. Phys. Lett.* 95: 043304.
29. Naveen Kumar, E., Jose, R., Archana, P. S., Vijila, C., Yusoff, M. M., Ramakrishna, S. 2012. High performance dye-sensitized solar cells with record open circuit voltage using tin oxide nanoflowers developed by electrospinning. *Energy Environ. Sci.* 5:5401–5407.
30. Guo, P., Aegerter, M. A. 1999. Ru(II) sensitized Nb_2O_5 solar cell made by the sol–gel process. *Thin Solid Films* 351:290–294.
31. Sayama, K., Sugihara, H., Arakawa, H. 1998. Photoelectrochemical properties of a porous Nb_2O_5 electrode sensitized by a ruthenium dye. *Chem. Mater.* 10:3825–3832.
32. Hara, K., Horiguchi, T., Kinoshita, T., Sayama, K., Sugihara, H., Arakawa, H. 2000. Highly efficient photon-to-electron conversion with mercurochrome-sensitized nanoporous oxide semiconductor solar cells. *Sol. Energy Mater. Sol. Cells* 64:115–134.
33. Burnside, S., Moser, J. E., Brooks, K., Gratzel, M., Cahen, D. 1999. Nanocrystalline mesoporous strontium titanate as photoelectrode material for photosensitized solar devices: Increasing photovoltage through flatband potential engineering. *J. Phys. Chem. B* 103:9328–9332.
34. Tan, B., Toman, E., Li, Y., Wu, Y. 2007. Zinc stannate (Zn_2SnO_4) dye-sensitized solar cells. *J. Am. Chem. Soc.* 129:4162–4163.
35. Li, Z., Zhou, Y., Bao, C., Xue, G., Zhang, J., Yu, T., Zou, Z. 2012. Vertically building Zn_2SnO_4 nanowire arrays on stainless steel mesh toward fabrication of large-area, flexible dye-sensitized solar cells. *Nanoscale* 4:3490–3494.
36. Huang, L., Jiang, L., Wei, M. 2010. Metal-free indoline dye sensitized solar cells based on nanocrystalline Zn_2SnO_4. *Electrochem. Commun.* 12:319–322.
37. Kim, D.-W., Shin, S.-S., Cho, I.-S., Lee, S., Kim, D.-H., Lee, C.-W., Jung, H.-S., Hong, K.-S. 2012. Synthesis and photovoltaic property of fine and uniform Zn_2SnO_4 nanoparticles. *Nanoscale* 4:557–562.
38. Choi, S.-H., Hwang, D., Kim, D.-Y., Kervella, Y., Maldivi, P., Jang, S.-Y., Demadrille, R., Kim, I.-D. 2013. Amorphous Zinc Stannate (Zn_2SnO_4) nanofibers networks as photoelectrodes for organic dye-sensitized solar cells. *Adv. Funct. Mater.* 23:3146–3155.
39. Kim, J., Kang, J., Jeong, U., Kim, H., Lee, H. 2013. Catalytic, conductive, and transparent platinum nanofiber webs for FTO-free dye-sensitized solar cells. *ACS Appl. Mater. Interfaces* 5:3176–3181.

40. Kay, A., Grätzel, M. 1996. Low cost photovoltaic modules based on dye sensitized nanocrystalline titanium dioxide and carbon powder. *Sol. Energy Mater. Sol. Cells* 44:99–117.

41. Veerappan, G., Kwon, W., Rhee, S.-W. 2011. Carbon-nanofiber counter electrodes for quasi-solid state dye-sensitized solar cells. *J. Power Sources* 196:10798–10805.

42. Joshi, P., Zhang, L., Chen, Q., Galipeau, D., Fong, H., Qiao, Q. 2010. Electrospun carbon nanofibers as low-cost counter electrode for dye-sensitized solar cells. *ACS Appl. Mater. Interfaces* 2:3572–3577.

43. Park, S.-H., Kim, B.-K., Lee, W.-J. 2013. Electrospun activated carbon nanofibers with hollow core/highly mesoporous shell structure as counter electrodes for dye-sensitized solar cells. *J. Power Sources* 239:122–127.

44. Sigdel, S., Dubey, A., Elbohy, H., Aboagye, A., Galipeau, D., Zhang, L., Fong, H., Qiao, Q. 2014. Dye-sensitized solar cells based on spray-coated carbon nanofiber/TiO_2 nanoparticle composite counter electrodes. *J. Mater. Chem. A* 2:11448–11453.

45. Noh, S. I., Seong, T.-Y., Ahn, H.-J. 2012. Carbon nanofibers combined with Pt nanoparticles for use as counter electrodes in dye-sensitized solar cells. *J. Ceramic. Process. Res.* 13:491–494.

46. An, H.-R., An, H. L., Kim, W.-B., Ahn, H.-J. 2014. Nitrogen-doped TiO_2 nanoparticle-carbon nanofiber composites as a counter electrode for Pt-free dye-sensitized solar cells. *ECS Solid State Lett.* 8:M33–M36.

8 Recent Developments of Electrospinning for Biosensors and Biocatalysis

*Ahsan Nazir, Nabyl Khenoussi,
Amir-Houshang Hekmati, Laurence Schacher,
and Dominique C. Adolphe*

CONTENTS

8.1 INTRODUCTION

Electrospinning has been successfully employed for the development of micro- and nanofibrous materials. The capability of this technique to produce versatile ultrafine fibers and nonwovens with large surface area is its major potential. Moreover, it offers an adaptable process, which can be used in many different applications, with possibility to produce the products at bulk scale. As mentioned in earlier sections, electrospun materials find applications in fields like filtration, tissue engineering, protective textiles, energy and biomedical products, pharmaceuticals, optics, electronics, and environmental engineering [1–4].

Among different domains of electrospun materials, their use in energy and environmental applications is increasingly gaining attention of researchers. These applications include different energy conversion, storage, and other electronic devices as well as a diverse set of filtration, sensing, and waste treatment products [5–7]. There has been a notable progress in the field with development of some commercially viable products.

The development of electrospun materials for biosensors and biocatalysis for different environmental applications is one of the aforementioned areas where important advancements have been made during the last decade. These developments have lead to a number of sensing and waste treatment systems from composite electrospun products.

Sensing devices are widely used in analysis of different agents in environment. Many sophisticated sensing systems have been employed since decades for analysis of different analytes in diverse media. These systems, however, have the disadvantages of high cost, limited sensing capability for determination of chemical composition, lower reliability for analysis of vital physiological compounds, and inability to analyze online. High sensitivity and precision of different biological receptors for specific substances have motivated many researches to develop biosensors. The initiative was taken by development of sensors based on simple protein systems that produced some outstanding results [8].

The sensing capability of biosensors is based on the conversion of analytes, by a biologically sensitive agent such as enzymes, to the compounds that can be sensed and a measureable signal can be generated. The conversion of analytes to newer products by biologically sensitive agent is basically carried out thanks to their ability to accelerate the chemical reactions. The capability of biosensors to accelerate only selective reactions allows them to work only on selective analytes. Thus, they act as selective biocatalysts, and this capability has been extensively explored to develop some novel products for analysis of a variety of analytes.

The biologically sensitive agents are normally able to generate a weak signal when they sense a specific analyte. Usually, a biological transducer is required to convert the results, produced by the biological sensing agent, into a more measureable signal that is presented by an electronic display system. The biotransducer is in contact with the biologically sensitive agent that undergoes a physicochemical change due to action of analyte on it. This change is measured by the biotransducer, which converts it into a signal. The intensity of signal depends on the magnitude of physicochemical change that, in turn, is dependent on the amount of analyte. So, the system is capable to selectively determine the concentration of analyte under study. A generalized working path of a biosensor with a specific example of enzyme-based biosensor is described in Figure 8.1. However, it is important to mention here that enzyme-based biosensors are of the classes of a wider classification system that include biosensors based on immunoglobulins, nucleic acids, aptamers, etc.

Biosensors may generally be named after the type of biotransducer they employ. For a biosensor system to work efficiently, the biological sensing agents must be

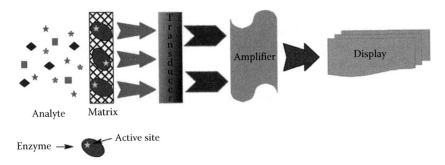

FIGURE 8.1   General working of a biosensor. (Reprinted from *Biosens. Bioelect.*, 23, Arya, S.K. et al., Recent advances in cholesterol biosensor, 1083–1100, Copyright 2008, with permission from Elsevier.)

stabilized/entrapped or coated in a specific pattern on a supporting media such as different porous substances, hydrogels, and nanofibers [10,11]. The properties of the stabilizing media have strong effects on performance of biosensor [12].

The following sections discuss various enzyme-carrying electrospun nanofibers, their development methods, enzyme binding systems, and possible applications.

8.2   ENZYME-CARRYING ELECTROSPUN FIBERS

As already stated, stabilization and immobilization of enzymes (or other bio active agents) is the prerequisite for development of biosensors. Enzymes can be immobilized and stabilized by attaching or entrapping them on several different types of media that may be micro- or nanosized. The nanosized media offer much more advantages as compared to microsized materials because of their high surface area. Different types of nanomaterial have been employed to immobilize enzymes; they include different type of nanoparticles, mesoporous nanomaterials such as mesoporous silica and nanofibers.

There have been extensive studies on immobilization of enzymes on nanofibers in the last decade. Nanofibers have been found to offer higher performance for carrying enzymes for biosensors and biocatalysis applications. Their surface area and porosity can be tailor made for specific enzymes. This allows them to offer low restriction to diffusion of enzymes and high efficiency in detection of analytes. Moreover, they can be modified, both physically and chemically, to achieve the required level of activity. Different chemical modifications may be made to fit the chemical nature of enzyme and the environment in which the biosensor has to perform. In addition, there is a large variety of materials that can be electrospun to nanofibers to be employed for biosensors development. With these advantages in mind, electrospun nanofibers can be used to produce highly efficient, tailor made nanosensors.

Enzymes have been the most frequently used bioactive agents for biosensors, and many of them have been studied for their bioactivity in biosensors developed from electrospun nanofibers. They can be immobilized on electrospun nanofibers either by attaching their molecules on fiber surface during/after electrospinning or by electrospinning the solution of polymer and enzyme along with some other additives [13]. Both these basic techniques have been studied and are concisely discussed in the following sections.

8.2.1 Electrospun Nanofibers with Covalently Immobilized Enzymes

Enzymes can be immobilized on nanofibers by bonding them covalently, using functional groups at their surface. Employing this methodology, a number of different enzymes have been immobilized on diverse polymeric materials. Some of these materials have been discussed in the following text.

8.2.1.1 Enzyme Immobilization on Polymers with Poor Biocompatibility

Enzymes, being "bio" species, have poor compatibility with no biocompatible species such as most of the synthetic polymers. However, due to excellent mechanical properties of synthetic polymers and presence of functional groups on them or their modified forms, they have widely been studied for their potential applications for supporting enzymes.

8.2.1.1.1 Polyacrylonitrile

Among a large number of members in polymer family, polyacrylonitrile (PAN) has been the focus of research in the field of enzyme immobilization. This is because of its chemical structure having functional groups that can react with a large number of species. As discussed in upcoming sections, a number of different enzymes have been immobilized using electrospun PAN fibers. A simple but effective technique was employed by Shen-Feng and coworkers for immobilization of lipase enzyme electrospun PAN. The covalent linkage between the fibers and lipase was developed by activating nitrile groups on PAN for amidation reaction. The enzyme was successfully immobilized on PAN nanofibers through amide linkages, as shown in Figure 8.2, and formed aggregates on PAN fiber surface. Good loading efficiency and high activity retention was observed to be obtained with this system [14].

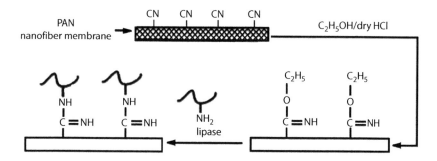

FIGURE 8.2 Schematic representation of covalent immobilized lipase on PAN by amidation reaction. (Reprinted from *J. Mol. Catal. B Enzym.*, 47, Li, S.-F., Chen, J.-P., and Wu, W.-T., Electrospun polyacrylonitrile nanofibrous membranes for lipase immobilization, 117–124, Copyright 2007, with permission from Elsevier.)

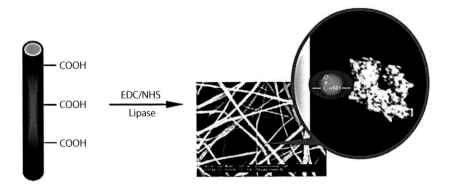

FIGURE 8.3 PAN/MA nanofibers immobilizing lipase through biomacromolecules. (Reprinted with permission from Ye, P. et al., Nanofibrous membranes containing reactive groups: Electrospinning from poly(acrylonitrile-co-maleic acid) for lipase immobilization. *Macromolecules*, 39, 1041. Copyright 2006 American Chemical Society.)

Similarly, poly(acrylonitrile-co-maleic acid) (PAN/MA) composite nanofibers have also been used for immobilization of lipase by formation of biocompatible layer of two macromolecules, that is, chitosan (CS) and gelatin [15]. The nanofibers were treated with 1-ethyl-3-(dimethylaminopropyl) carbodiimide hydrochloride/ *N*-hydroxyl succinimide in order to activate carboxyl groups on them for covalently linking the macromolecules. The enzyme was then attached to these modified fibers using glutaraldehyde, which has been used widely for addition of aldehyde groups to the polymer chain for covalent immobilization. The schematic diagram of this process has been shown in Figure 8.3. The nanocomposite was observed to have higher loading with improved activity retention as compared to that of unmodified PAN/MA nanocomposite. This allowed achieving the required biocompatibility along with mechanical strength that is somewhat difficult to achieve in biocompatible fibers such as CS and gelatin.

Another study by Wang et al. revealed that immobilized catalase enzyme shows superior performance in biosensors produced from poly(acrylonitrile-co-acrylic acid)/composite nanofiber with multiwall carbon nanotubes (MWCNT) as compared to one without MWCNT. The researchers covalently bonded the enzymes with polymer chains using dimethylaminopropyl-ethylcarbodiimide hydrochloride and hydroxysuccinimide, which activates the carboxyl groups for enzyme anchoring. The suggested reason was higher electron transfer efficiency by MWCNT. Moreover, the covalent binding of enzyme molecules directly with nanofibers reduced the distance between the MWCNT and enzyme thus allowing better sensing ability [16]. In a similar study, porphyrin (electron donor) was used along with carbon nanotubes (CNTs) (electron acceptor) in PAN nanofibers with catalase covalently bonded to them. This combination was showed to improve the activity and stability of enzyme [17].

In order to enhance the sensitivity of biosensor systems by improving the electron transfer from enzyme, Jose and coworkers developed electrospun gold nanofibers with CNT coating (for enzyme immobilization). The nanofibers were electrospun using a polyacrylonitrile–gold (PAN-Au) salt mixture. They were subsequently coated with carboxylated MWCNT that have the ability to covalently immobilize the enzyme glucose oxidase. So, a mediator-free biosensor, having conductive fibers very close to the enzyme, was developed. This allowed improving the electron transfer efficiency along with elimination of demerits related with redox active mediators. The sensor was found to have very good electron transfer rate and high linear sensitivity [18].

Prussian blue (PB) is an iron-hexacyanoferrate-based dye that has good reducing power. Fu and coworkers achieved the advantages of combining PB and MWCNTs within CS along with in situ covalent immobilization of glucose oxidase using 3-isocyanatopropyltriethoxysilane (ICPTES) as covalent coupling agent. Here, ICPTES also acted as sol–gel precursor. The PB used in this study was in close proximity of MWCNTs as they were deposited on them. The mixture of PB/MWCNT, ICPTES, CS, and glucose oxidase was electrospun on a glassy carbon electrode. The system was used for detection of glucose and was found to have good sensitivity and low detection limit [19].

In a study conducted by J. Shen and his coworkers, high electron transfer efficiency of gold was also explored to develop a biosensor for quantitative measurement of hydrogen peroxide (H_2O_2) with horseradish peroxidase incorporated on gold-coated electrospun silica nanofibers. The gold coating was used to enhance the electron conductivity through the system that otherwise is very poor and leads to poor performance of nanosensor. To develop the biosensor, gold–silica composite nanofibers dispersed in CS were spin coated on indium tin oxide, which was subsequently spin coated with enzyme solution in CS. The developed nanosensor was found to have good sensing capability for H_2O_2 with linear sensing range between 5×10^{-6} and 1×10^{-3} M and 2 µM detection limit [20].

A similar approach was used by H. Zhu et al. for the development of a biosensor for detection of glucose, H_2O_2, and glutathione with horseradish peroxidase in electrospun polyvinyl alcohol (PVA)/polyethyleneimine (PEI)/silver nanoparticles composite nanofibers. The nanofibers were produced from mixture of both the polymers. The composite nanofibers were functionalized and subsequently silver nanoparticles were produced on surface by reduction of silver nitrate. The enzyme was added afterward by treating the fibrous mat with enzyme solution. The composite system produced was found to be stable, recyclable with prompt response capability [21].

El-Aassar and his colleagues employed the idea of using spacer arm on electrospun poly(acrylonitrile-co-methyl methacrylate) or poly(AN-co-MMA) nanofibers. Inclusion of spacer between the solid support allows to reduce the steric hindrance by the immobilization media and thus provides an environment to bound enzyme molecules close to one enjoyed by free molecules. The researchers used carbonyl groups of poly(AN-co-MMA) to covalently attach a PEI arm to it. The PEI arm was then used to form covalent linkage with β-galactosidase using glutaraldehyde as a coupling agent. The technique allowed to improve the temperature stability of β-galactosidase [22]. Another such study was carried by Wang and Hsieh [23]. They immobilized lipase in electrospun cellulose fibers by adding polyethylene glycol (PEG) spacer arm.

8.2.1.1.2 Polystyrene

Polystyrene has been used for enzyme immobilization based on its mechanical properties and forming behavior. In an initial study in subjected area, modified polystyrene nanofibers were electrospun, and an enzyme alpha-chymotrypsin was attached chemically to their surface. The loading of enzyme was found to almost 10.4% weight percent with surface coverage of 27.4%. The developed system was evaluated for the hydrolytic activity of enzyme, and it was observed that the activity was reduced only by 35%, which means the enzyme can perform well even after getting immobilized [24]. In a similar study, Kim et al. improved the stability and activity of lysozyme by attaching it covalently on the surface of electrospun polystyrene/poly(styrene-co-maleic anhydride) composite nanofibers. The enzyme was cross-linked with glutaraldehyde treatment that improved not only enzyme loading by cross-linking additional enzyme molecules but also their stability. The enzyme activity was found to increase nine times as compared to that without any cross-linking treatment. Moreover, it had high stability at high temperature and high pH [25].

Lower loading of enzymes on polystyrene nanofibers, because of their hydrophobicity, limits their application in the field. However, some modifications may make it possible to overcome this problem. Nair and coworkers resolved the problem just by treating maleic anhydride–modified polystyrene nanofibers with alcohol [26]. This allowed the fibers to remain dispersed during loading treatment and molecules of lipase easily accessed the covalent sites to form covalent bond with the fibers. This technique improved the loading capacity by eight times as compared to enzyme-loaded polystyrene fibers with aqueous alcoholic treatments.

8.2.1.1.3 Polyamide

There are a few studies on immobilization of enzymes on polyamide fibers. Uzun and coworkers covalently immobilized glucose oxidase in nylon 6,6/MWCNT/(poly-4-(4,7-di(thiophen-2-yl)-1H-benzo[d]imidazol-2-yl)benzaldehyde) (PBIBA) composite nanofibers. PBIBA, a conductive polymer, was used as covalent immobilization media with good electroactivity. Thus, fiber surface with high electroactivity was obtained on otherwise poor electroactive nylon 6,6 nanofibers. The resultant composite system was able to provide good sensitivity with high electron transfer [27].

8.2.1.1.4 Polymethylmethacrylate

Kumar and his colleagues functionalized the surface of electrospun polymethyl methacrylate nanofibers with phenylenediamine and glutaraldehyde to make it able to immobilize the enzyme covalently and increase its activity and stability. They attached xylanase to fiber surface to observe its efficiency and stability. The attached enzyme retained 80% of its activity even after 11 reaction cycles. Moreover, the enzyme was also found to have good thermal stability, so the nanocomposite may find application in industrial processes [28].

8.2.1.2 Enzyme Immobilization on Biocompatible/ Biodegradable Nanofibers

Immobilization of enzymes on biodegradable polymer is an area with extensive potential applications. Different enzymes have been immobilized on biodegradable fibers keeping in view the potential application of the nanocomposite formed thereof. Kim and Park immobilized lysozyme on poly(ε-caprolactone)/poly(lactic-co-glycolic acid) (PLGA)/poly(ethylene glycol)-NH$_2$ composite nanofibers. The amine groups on composite were used to immobilize the model enzyme using a homobifunctional coupler. High enzyme loading and higher activity retention of resulting nanocomposite were outcomes that made the product superior to that containing lysozyme in cast films [29]. Similarly, Wang and Hsieh added a PEG arm to electrospun nanofibers in order to immobilize lipase [23]. Addition of PEG simultaneously added amphiphilic spacers and reactive groups for coupling reaction with lipase. The composite structure allowed lipase to show much higher activity in different conditions including acidic/basic environments and high temperature.

Electrospun poly(vinyl alcohol)/poly(acrylic acid) or PVA/PAA nanofibers were used to immobilize α-amylase by Basturk et al. The enzyme was immobilized by using 1,1′-carbonyldiimidazole to activate the amine groups on PVA/PAA nanocomposite. A significant improvement in temperature and pH stability of α-amylase was achieved through this development. Also, the reusability of enzyme also improved using this technique [30].

Because of structure similarities between hemoglobin (Hb) and peroxidase, Hb could be used for reduction of H$_2$O$_2$, and hence its detection in certain media is encountered in environmental controls. Li and coworkers used Hb instead of peroxidase in collagen nanofibers containing CNTs. The biosensor offered the advantages of good biocompatibility and stability along with high sensitivity for H$_2$O$_2$ detection [31]. In another study, PB-coated CS/PVA biosensor for detection of glucose has been developed by Wu and Yin. The sensor, containing immobilized glucose oxidase, performed well for detection of glucose through quantitative measurement of H$_2$O$_2$, which is produced intermediately during oxidation of glucose by glucose oxidase. The sensor was developed by electrospinning CS/PVA alcohol composite nanofibers on a glass plate coated with indium tin oxide, which has been electrodeposited, beforehand, with PB film. The nanofiber indium tin oxide composite was again electrodeposited with PB. Glucose oxidase in phosphate buffer solution was added to nanofibrous mat and stored overnight to stabilize the electrostatic attraction between the two. PB was employed here because of its outstanding electron transfer capability and its potential to reduce H$_2$O$_2$. However, its instability in alkaline pH has been a problem that has been well addressed by current development. The developed composite biosensor was found to perform well with much narrow measurement limits [32].

8.2.1.3 Electrospun Nanofibers with Biomimetic Composite Nanofibers

Lower mechanical strength of natural polymers such as CS limits their applications in the form of nanofiber for biosensors and biocatalysis. However, the approaches as that applied by Ye et al. (discussed in earlier sections) [15] need to be used for getting composite nanofibers with required level of biocompatibility and strength. Another similar study by Wang and coworkers has been discussed in the following sections [33].

8.2.2 ELECTROSPUN NANOFIBERS WITH ENZYMES ATTACHED TO THEM BY ADSORPTION

Adsorption is another possible route for immobilization of enzymes on nanofibrous webs. However, there are limited numbers of studies on immobilization of enzymes through adsorption. They are discussed in the following paragraphs.

Combining good mechanical strength with functional properties is normally a complex task for electrospun nanofibers. Wang and coworkers achieved this by immobilizing lipase on poly(*bis*(methylphenoxy))phosphazene (PMPPH)/PAN nanofibers having a core–sheath arrangement. Lipase was adsorbed on the PMPPH outer shell, as shown in Figure 8.4. Higher adsorption capacity and activity retention was observed when compared with PAN fibers alone. So, the composite obtained in study allowed to achieve the required properties as well as easy processability that other would have not been possible with a single polymer [34].

Hydrophobic immobilizers are of particular interest for lipases. This is because of improved activity of lipase hydrophobic environment. To take advantage of this property of lipase, it was immobilized on polysulfone nanofibers by blending it with polyvinylpyrrolidone and PEG. These additives were added to increase the compatibility of nanofibers with enzymes and to reduce their hydrophobicity. The blended nanofibers were able to adsorb enzyme molecules from its solution. It was observed that the blending polysulfone with biocompatible components improved the activity of enzyme [33].

Another study exploring the immobilization of enzymes through adsorption was accomplished by Chen and Hsieh [35]. They developed poly(acrylic acid) grafts on electrospun cellulose fibers and adsorbed lipase enzyme onto them. The adsorbed

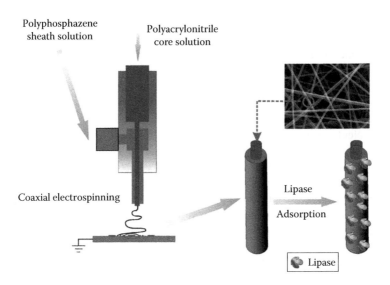

FIGURE 8.4 Coaxial electrospinning arrangement for immobilization of lipase in PMPPH/PAN composite nanofibers. (From Wang, S.G. et al., *Int. J. Mol. Sci.*, 13, 14136, 2012.)

enzymes were found to offer better activity and organic solvent stability as compared to free enzymes. Furthermore, a similar study was carried out by Sakai and coworkers who immobilized lipase in electrospun PAN nanofibers through adsorption [36]. The immobilized enzymes were found to show improved activity for catalyzing transesterification reactions as one used for conversion of glycidol to glycidyl *n*-butyrate with vinyl *n*-butyrate.

8.2.3 ELECTROSPUN NANOFIBERS ENCAPSULATING ENZYMES

There have been only a few studies on development of enzyme encapsulated in nanofibers. This is mainly because of lower compatibility of enzymes with the electrospinning solution. As most of the enzymes are soluble in water, a limited number of polymers can be co-electrospun with them. However, some water-soluble polymers have been studied for their enzyme immobilization capability, but they also lose the enzyme molecules by swelling when they come in contact with water. A possible way out to this is the use of cross-linkers, but they also have disadvantages associated with them, such as reduction in porosity and enzyme activity [37,38].

PVA has been extensively used for encapsulation of enzymes, particularly, because of its water solubility. Glucose oxidase has been immobilized in PVA nanofibers for development of amperometric biosensors [39]. The enzyme was blended in electrospinning solution, which allowed to successfully encapsulating it in PVA nanofibers after electrospinning, as shown in Figure 8.5. The nanocomposite was found to show low response time and low detection limit for glucose sensing and provided a simple approach for development of glucose biosensors. In another study, a similar approach was employed for incorporation of cellulase in PVA nanofibers [37]. The fibers were subsequently cross-linked with glutaraldehyde vapors. The nanofibrous web showed superior efficiency for biotransformations as compared to cast membranes.

Lipase was also loaded in electrospun PVA fibers by electrospinning an aqueous solution containing the both [38]. The enzyme in nanocomposite showed excellent activity, especially at higher temperatures, as compared to free enzyme. To enhance the stability of enzyme, the nanocomposite was cross-linked with glutaraldehyde; this, however, resulted in a reduced activity.

Based on shortcomings of physical entrapment of enzymes in nanofibers through electrospinning of aqueous solutions, different approaches, for example, addition of additives like biocompatible polymers and surfactants, have been employed to disperse enzymes in a solvent and electrospin them [40]. Patel et al. encapsulated horseradish peroxidase in electrospun mesoporous silica fibers. The nanofibers were electrospun from a mixture of an orthosilicate, enzyme, glucose, and PVA. The composite nanofibers were found to have good porosity, with pore size ranging between 2 and 4 nm. Moreover, the fibers were found to be mechanically flexible and rechargeable thus widening their application areas in different fields [41]. The loss of enzymes during contact with fluids, however, remains a potential disadvantage.

Coaxial electrospinning is another approach that allows encapsulating and stabilizing enzymes in nanofibers by keeping the enzyme in core with a polymer sheathed by another polymer. However, more studies are required to get a detailed insight

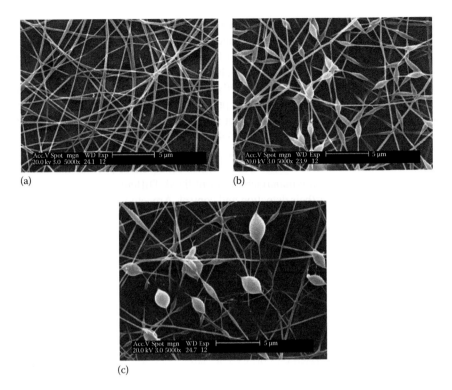

FIGURE 8.5   (a) Electrospun PVA nanofibers, (b and c) electrospun PVA/glucose oxidase nanofibers. (Reprinted from *React. Funct. Polym.*, 66, Ren, G. et al., Electrospun poly(vinyl alcohol)/glucose oxidase biocomposite membranes for biosensor applications, 1559–1564, Copyright 2006, with permission from Elsevier.)

into the process of enzyme immobilization and its activity retention from coaxial electrospun nanofibers [42].

8.3   ELECTROSPUN BIOSENSORS AND BIOCATALYSTS: ENVIRONMENTAL APPLICATIONS

Electrospun biosensors and biocatalysts have also been explored for their potential environmental applications. An overview on the environmental applications of electrospun biosensors and biocatalysts is presented in the following paragraphs.

Hydrogen peroxide is a major constituent of effluents from many industries. Its drainage in wastewater must be kept below certain limits and is regularized by different environmental regulations. Different types of sensors have conventionally been utilized for detection of H_2O_2 in wastewater. Based on advantages offered by electrospun nanofibers, different researchers have attempted to develop sensors for detection of H_2O_2. As discussed in the preceding sections, quantitative measurement of H_2O_2 with different types of electrospun biosensors has been studied, and most of them were found to offer interesting results that can be used for the development of commercially viable solutions to control the level of H_2O_2 in industrial wastes. For example, horseradish peroxidase incorporated on gold-coated silica nanofibers has been utilized for detection of H_2O_2. Similarly, electrospun PVA/PEI/silver nanoparticle composite nanofibers have also been employed for its detection [20,21]. In another study, hemoglobin was directly electrospun on surface of glassy carbon electrodes through electrospinning. The produced biosensor showed high sensitivity not only for H_2O_2 but also for nitrites that are considered to be hazardous for marine life and also affect the nitrogen cycle [43,44].

Phenols, also called carbolic acids, are corrosive to human skin, eyes, and respiratory tract and can affect the nervous system. Safe levels of phenols must be kept in facilities where they are manufactured or utilized. Different sensing systems are conventionally used for their detection and control. Arecchi and coworkers, inspired from the properties of electrospun biosensors, developed a novel biosensor for detection of phenols by drop-coating tyrosinase on polyamide-6 (nylon-6) nanofibers deposited on glassy carbon electrode. The biosensor showed a high sensitivity and low response time for detection of phenols [45,46]. In a similar work, PVA nanofibers encapsulating laccase have been found to produce highly sensitive biosensors for detection and monitoring of chlorophenols, which are potential carcinogens and are likely to be released into the environment through agricultural products, dyes, and pharmaceuticals [47,48].

Chloramphenicol, an antibiotic, is known for its bone marrow toxicity. For its detection, a biosensor based on polystyrene nanofibers prepared by electrospinning was developed. The electrospun nanocomposites, containing antibody for chloramphenicol, were deposited on a quartz crystal electrode. The antibody was, subsequently, functionalized with mercaptopropionic acid. The biosensor was found to show high sensitivity, good response time, and high capturing selectivity for chloramphenicol [49,50].

Control of different pathogenic pollutants is also an important environmental application of electrospun biosensors. In an effort to do that, *Escherichia coli* antibody was immobilized on electrospun cellulose nitrate nanofibers loaded with gold nanoparticles. The biosensor was used for detection of *E. coli* O157:H7 bacteria and was found to have excellent sensitivity for it. Moreover, its low detection time enables it to be used in on-field detection for many environmental, safety, and health applications [51].

Electrospun biocatalysts also offer many advantages over the conventional systems for the treatment of different pollutants, and they have been studied in recent years for their efficiency and comparative advantages. As, for example, Niu and colleagues explored the idea of encapsulating horseradish peroxidase in electrospun poly(lactide-co-glycolide) nanofibers for degradation of pentachlorophenol, a highly toxic agent used for wood preservation. Encapsulation of enzyme in nanofibers greatly improved the adsorption and hence the degradation of pentachlorophenol by enzyme. The resulting system was found to have good pentachlorophenol degradation capability along with stability in operational media [52].

Laccases are enzymes that catalytically oxidize many substrates. To explore this potential of laccase, different researchers encapsulated them in electrospun micro- and nanofibers to increase their efficacy. Electrospun laccase–loaded PLGA nanofibers have been studied by Dai and colleagues for their potential application in degradation of polycyclic aromatic hydrocarbons in soils. They developed a core–shell arrangement with laccase in the core encapsulated by porous polymer shell. The composite was found to efficiently adsorb and degrade different polycyclic aromatic hydrocarbons in soils with much lower operation time as compared to conventional systems [53]. In a similar study, laccase loaded PLGA nanofibers were used for removal of diclofenac from water. Diclofenac, an anti-inflammatory drug, has been found to biomagnify the food chain, and thus its removal from effluents of pharmaceutical wastes is a crucial environmental issue. Laccase was loaded on PLGA nanofibers after they were electrospun using glutaraldehyde as a cross-linker. The resulting system was found to be an efficient, stable, reusable, and low-cost solution for removal of diclofenac from wastewater [54,55].

8.4   TECHNICAL CHALLENGES AND PERSPECTIVES

As discussed in previous sections, nanofibers have been employed successfully for immobilization of enzymes for biosensor applications. Many studies have developed solutions that may be viable for commercial nanosensor products. There are, however, some areas that need more investigations in future to get the maximum output of the electrospinning technology for biosensor and biocatalysis applications.

Electron mediators play an important role in sensitivity of a biosensor and they must be located on/near the surface of the devices. So, there is a need to investigate novel techniques to locate electron mediators such as MWCNT and metal nanoparticles closer to fiber surface.

Avoiding electron mediator, without affecting the sensor activity, is another prospective field that needs to be addressed in future. This is because mediators are expected to leach out of the surface of the fiber during biosensing and may produce some toxic results, for example, for applications in biomedical field. If such an issue could be overcome, the development of blood-powered implantable devices could be accomplished easily [18].

Another field that needs extensive research is the reduced efficiency of immobilized enzymes. Immobilization of enzymes on nanofibers results in decrease of their activity because of their interaction and binding with the fiber-forming polymers. Normally higher loading compensates the decrease in activity, but still the activity remains lower than that expected. Therefore, more studies are needed, both in fields of enzyme synthesis/modification and fibers synthesis/modifications to achieve strong binding and high loading along with excellent activity [13].

Lower enzyme loading has also been a problem for application of enzyme mobilized nanofibers in biosensors and biocatalysis [13]. Particularly, the enzyme loading on hydrophobic fibers is quite a hectic task. Some studies have focused on techniques to increase the enzyme loading such as that by Kim et al. [25]. Similarly, enzyme loading was increased on electrospun maleic anhydride modified polystyrene nanofibers by dispersing them in water through alcohol treatment [26]. However, simpler techniques applicable to multiple substrates are still the need of day and are expected to result in industrially viable applications in the field.

Based on advantages offered by biomimetic electrospun nanofibers, there is a need of more studies in the area in order to have a better understanding of the interaction between enzymes and biomacromolecules, when they are present on nanofibers [13].

And finally, there have been a considerable progress in development of electrospun biosensors and biocatalysts for environmental applications, but extensive studies are still needed to explore the full potential of these materials. With increasing number of harmful and uncontrollable pollutants, more and more modifications and extensions of electrospun biosensors and biocatalysts need to be explored and commercialized.

8.5 SUMMARY

Electrospun nanofibers have been the focus of research for last almost one decade because of their extraordinary properties. One of the potential application areas of electrospun nanofiber is immobilization of enzymes for biosensors and biocatalysis. Enzymes have been immobilized in electrospun nanofibers by many different routes such as covalent immobilization, immobilization by adsorption and encapsulation. Each route has its own benefits, but the route most widely adopted is covalent immobilization. It allows bringing many polymers into field that cannot, otherwise, be used for enzyme mobilization. However, each of the techniques has a set of advantages and disadvantages related to it, and continued research can help reduce the demerits related with each of them.

REFERENCES

1. Yun, K. M. et al. Morphology optimization of polymer nanofiber for applications in aerosol particle filtration. *Separation and Purification Technology* 75, 340–345 (2010).
2. Lee, K. Y., Jeong, L., Kang, Y. O., Lee, S. J., and Park, W. H. Electrospinning of polysaccharides for regenerative medicine. *Advanced Drug Delivery Reviews* 61, 1020–1032 (2009).
3. Almuhamed, S. et al. Measuring of electrical properties of MWNT-reinforced PAN nanocomposites. *Journal of Nanomaterials* 2012, 1–7 (2012).
4. Almuhamed, S. et al. Electrospinning of PAN nanofibers incorporating SBA-15-type ordered mesoporous silica particles. *European Polymer Journal* 54, 71–78 (2014).
5. Hekmati, A. H. et al. Effect of nanofiber diameter on water absorption properties and pore size of polyamide electrospun-6 nanoweb. *Textile Research Journal* 84, 2045–2055 (2014).
6. Hekmati, A. H. et al. Effect of needle length, electrospinning distance, and solution concentration on morphological properties of polyamide-6 electrospun nanowebs. *Textile Research Journal* 83, 1452–1466 (2013).
7. Thavasi, V., Singh, G., and Ramakrishna, S. Electrospun nanofibers in energy and environmental applications. *Energy and Environmental Science* 1, 205–221 (2008).
8. Scheller, F. and Schubert, F. *Biosensors: Techniques and Instrumentation in Analytical Chemistry.* Elsevier Science Publishing, the Netherlands (1992).
9. Arya, S. K., Datta, M., and Malhotra, B. D. Recent advances in cholesterol biosensor. *Biosensors and Bioelectronics* 23, 1083–1100 (2008).
10. Banica, F.G. *Chemical Sensors and Biosensors: Fundamentals and Applications.* John Wiley & Sons, U.K. (2012).

11. Vo-Dinh, T. and Cullum, B. Biosensors and biochips: Advances in biological and medical diagnostics. *Fresenius' Journal of Analytical Chemistry* 366, 540–551 (2000).
12. Zhao, M. et al. The application of porous ZnO 3D framework to assemble enzyme for rapid and ultrahigh sensitive biosensors. *Ceramics International* 39, 9319–9323 (2013).
13. Wang, Z.-G., Wan, L.-S., Liu, Z.-M., Huang, X.-J., and Xu, Z.-K. Enzyme immobilization on electrospun polymer nanofibers: An overview. *Journal of Molecular Catalysis B: Enzymatic* 56, 189–195 (2009).
14. Li, S.-F., Chen, J.-P., and Wu, W.-T. Electrospun polyacrylonitrile nanofibrous membranes for lipase immobilization. *Journal of Molecular Catalysis B: Enzymatic* 47, 117–124 (2007).
15. Ye, P., Xu, Z.-K., Wu, J., Innocent, C., and Seta, P. Nanofibrous membranes containing reactive groups: Electrospinning from poly(acrylonitrile-co-maleic acid) for lipase immobilization. *Macromolecules* 39, 1041–1045 (2006).
16. Wang, Z.-G., Ke, B.-B., and Xu, Z.-K. Covalent immobilization of redox enzyme on electrospun nonwoven poly(acrylonitrile-co-acrylic acid) nanofiber mesh filled with carbon nanotubes: A comprehensive study. *Biotechnology and Bioengineering* 97, 708–720 (2007).
17. Wan, L.-S., Ke, B.-B., Wu, J., and Xu, Z.-K. Catalase immobilization on electrospun nanofibers: Effects of porphyrin pendants and carbon nanotubes. *The Journal of Physical Chemistry C* 111, 14091–14097 (2007).
18. Jose, M. V., Marx, S., Murata, H., Koepsel, R. R., and Russell, A. J. Direct electron transfer in a mediator-free glucose oxidase-based carbon nanotube-coated biosensor. *Carbon* 50, 4010–4020 (2012).
19. Fu, G., Yue, X., and Dai, Z. Glucose biosensor based on covalent immobilization of enzyme in sol–gel composite film combined with Prussian blue/carbon nanotubes hybrid. *Biosensors and Bioelectronics* 26, 3973–3976 (2011).
20. Shen, J. et al. Gold-coated silica-fiber hybrid materials for application in a novel hydrogen peroxide biosensor. *Biosensors and Bioelectronics* 34, 132–136 (2012).
21. Zhu, H. et al. Facile fabrication of AgNPs/(PVA/PEI) nanofibers: High electrochemical efficiency and durability for biosensors. *Biosensors and Bioelectronics* 49, 210–215 (2013).
22. El-Aassar, M. R., Al-Deyab, S. S., and Kenawy, E.-R. Covalent immobilization of β-galactosidase onto electrospun nanofibers of poly(AN-co-MMA) copolymer. *Journal of Applied Polymer Science* 127, 1873–1884 (2013).
23. Wang, Y. and Hsieh, Y. L. Enzyme immobilization to ultra fine cellulose fibers via amphiphilic polyethylene glycol spacers. *Journal of Polymer Science Part A: Polymer Chemistry* 42, 4289–4299 (2004).
24. Jia, H. et al. Enzyme-carrying polymeric nanofibers prepared via electrospinning for use as unique biocatalysts. *Biotechnology Progress* 18, 1027–1032 (2002).
25. Kim, B. C. et al. Preparation of biocatalytic nanofibres with high activity and stability via enzyme aggregate coating on polymer nanofibres. *Nanotechnology* 16, S382 (2005).
26. Nair, S., Kim, J., Crawford, B., and Kim, S. H. Improving biocatalytic activity of enzyme-loaded nanofibers by dispersing entangled nanofiber structure. *Biomacromolecules* 8, 1266–1270 (2007).
27. Uzun, S. D., Kayaci, F., Uyar, T., Timur, S., and Toppare, L. Bioactive surface design based on functional composite electrospun nanofibers for biomolecule immobilization and biosensor applications. *ACS Applied Materials and Interfaces* 6, 5235–5243 (2014).
28. Kumar, P. et al. Covalent immobilization of xylanase produced from *Bacillus pumilus* SV-85S on electrospun polymethyl methacrylate nanofiber membrane. *Biotechnology and Applied Biochemistry* 60, 162–169 (2013).

29. Kim, T. G. and Park, T. G. Surface functionalized electrospun biodegradable nanofibers for immobilization of bioactive molecules. *Biotechnology Progress* 22, 1108–1113 (2006).

30. Baştürk, E., Demir, S., Danış, Ö., and Kahraman, M. V. Covalent immobilization of α-amylase onto thermally crosslinked electrospun PVA/PAA nanofibrous hybrid membranes. *Journal of Applied Polymer Science* 127, 349–355 (2013).

31. Li, J. et al. A novel hydrogen peroxide biosensor based on hemoglobin-collagen-CNTs composite nanofibers. *Colloids and Surfaces B: Biointerfaces* 118, 77–82 (2014).

32. Wu, J. and Yin, F. Sensitive enzymatic glucose biosensor fabricated by electrospinning composite nanofibers and electrodepositing Prussian blue film. *Journal of Electroanalytical Chemistry* 694, 1–5 (2013).

33. Wang, Z.-G., Wang, J.-Q., and Xu, Z.-K. Immobilization of lipase from *Candida rugosa* on electrospun polysulfone nanofibrous membranes by adsorption. *Journal of Molecular Catalysis B: Enzymatic* 42, 45–51 (2006).

34. Wang, S.-G., Jiang, X., Chen, P.-C., Yu, A.-G., and Huang, X.-J. Preparation of coaxial-electrospun poly[*bis*(p-methylphenoxy)]phosphazene nanofiber membrane for enzyme immobilization. *International Journal of Molecular Sciences* 13, 14136–14148 (2012).

35. Chen, H. and Hsieh, Y. L. Enzyme immobilization on ultrafine cellulose fibers via poly(acrylic acid) electrolyte grafts. *Biotechnology and Bioengineering* 90, 405–413 (2005).

36. Sakai, S. et al. Immobilization of *Pseudomonas cepacia* lipase onto electrospun polyacrylonitrile fibers through physical adsorption and application to transesterification in nonaqueous solvent. *Biotechnology Letters* 32, 1059–1062 (2010).

37. Wu, L., Yuan, X., and Sheng, J. Immobilization of cellulase in nanofibrous PVA membranes by electrospinning. *Journal of Membrane Science* 250, 167–173 (2005).

38. Wang, Y. and Hsieh, Y.-L. Immobilization of lipase enzyme in polyvinyl alcohol (PVA) nanofibrous membranes. *Journal of Membrane Science* 309, 73–81 (2008).

39. Ren, G. et al. Electrospun poly(vinyl alcohol)/glucose oxidase biocomposite membranes for biosensor applications. *Reactive and Functional Polymers* 66, 1559–1564 (2006).

40. Herricks, T. E. et al. Direct fabrication of enzyme-carrying polymer nanofibers by electrospinning. *Journal of Materials Chemistry* 15, 3241–3245 (2005).

41. Patel, A. C., Li, S., Yuan, J.-M., and Wei, Y. In situ encapsulation of horseradish peroxidase in electrospun porous silica fibers for potential biosensor applications. *Nano Letters* 6, 1042–1046 (2006).

42. Jiang, H. et al. A facile technique to prepare biodegradable coaxial electrospun nanofibers for controlled release of bioactive agents. *Journal of Controlled Release* 108, 237–243 (2005).

43. Ding, Y., Wang, Y., Li, B., and Lei, Y. Electrospun hemoglobin microbelts based biosensor for sensitive detection of hydrogen peroxide and nitrite. *Biosensors and Bioelectronics* 25, 2009–2015 (2010).

44. Moorcroft, M. J., Davis, J., and Compton, R. G. Detection and determination of nitrate and nitrite: A review. *Talanta* 54, 785–803 (2001).

45. Arecchi, A., Scampicchio, M., Drusch, S., and Mannino, S. Nanofibrous membrane based tyrosinase-biosensor for the detection of phenolic compounds. *Analytica Chimica Acta* 659, 133–136 (2010).

46. Lin, S.-H. and Juang, R.-S. Adsorption of phenol and its derivatives from water using synthetic resins and low-cost natural adsorbents: A review. *Journal of Environmental Management* 90, 1336–1349 (2009).

47. Igbinosa, E. O. et al. Toxicological profile of chlorophenols and their derivatives in the environment: The public health perspective. *The Scientific World Journal* 2013, 1–11 (2013).

48. Liu, J., Niu, J., Yin, L., and Jiang, F. In situ encapsulation of laccase in nanofibers by electrospinning for development of enzyme biosensors for chlorophenol monitoring. *Analyst* 136, 4802–4808 (2011).

49. Sun, M., Ding, B., Lin, J., Yu, J., and Sun, G. Three-dimensional sensing membrane functionalized quartz crystal microbalance biosensor for chloramphenicol detection in real time. *Sensors and Actuators B: Chemical* 160, 428–434 (2011).

50. Yunis, A. Chloramphenicol toxicity: 25 years of research. *The American Journal of Medicine* 87, 44N–48N (1989).

51. Luo, Y. et al. Novel biosensor based on electrospun nanofiber and magnetic nanoparticles for the detection of *E. coli* O157: H7. *IEEE Transactions on Nanotechnology* 11, 676–681 (2012).

52. Niu, J. et al. Immobilization of horseradish peroxidase by electrospun fibrous membranes for adsorption and degradation of pentachlorophenol in water. *Journal of Hazardous Materials* 246, 119–125 (2013).

53. Dai, Y., Yin, L., and Niu, J. Laccase-carrying electrospun fibrous membranes for adsorption and degradation of PAHs in shoal soils. *Environmental Science and Technology* 45, 10611–10618 (2011).

54. Sathishkumar, P. et al. Laccase-poly(lactic-co-glycolic acid)(PLGA) nanofiber: Highly stable, reusable, and efficacious for the transformation of diclofenac. *Enzyme and Microbial Technology* 51, 113–118 (2012).

55. Ilic, S. et al. Pentadecapeptide BPC 157 and its effects on a NSAID toxicity model: Diclofenac-induced gastrointestinal, liver, and encephalopathy lesions. *Life Sciences* 88, 535–542 (2011).

9 Eco-Friendly Electrospun Membranes Made of Biodegradable Polymers for Wastewater Treatment

Annalisa Aluigi, Giovanna Sotgiu,
Paolo Dambruoso, Andrea Guerrini,
Marco Ballestri, Claudia Ferroni, and Greta Varchi

CONTENTS

9.1 INTRODUCTION

The rapid population growth, associated with a rapid worldwide industrialization, is leading to an uncontrollable increase in environmental pollution of air and water. The World Water Council estimates that by 2030, 3.9 billion people will live in regions characterized as "water scarce,"[1] and water scarcity is gradually becoming a big threat to food security, human health, and natural ecosystems.[2]

Water is essential in feeding nutrients to living systems, and many people are currently suffering from water shortage due to the high pollution. Moreover, research has demonstrated that there is a clear correlation between diseases and water quality.[3] The 88% of four billion cases of acute gastrointestinal disease annually worldwide occurring is caused by unsafe water; it has been estimated that 94% of diarrheal cases can be prevented by increasing the availability of clean water[4]; intestinal parasitic infections and diarrheal diseases caused by waterborne bacteria and enteric viruses have become a leading cause of malnutrition owing to poor digestion of the food eaten by people sickened by water.[5]

Even if the scientific and technological research is carrying out many activities related to the water treatment, more effective and robust methods are needed. In fact, conventional methods of wastewater treatment include physical separation techniques for particle removal, biological and chemical treatments to remove suspended solids, organic matter and dissolved pollutants or toxins, evaporative techniques, and other physical and mechanical methods. These methods are chemically, energetically, and operationally intensive, and they often require considerable infusion of capital, engineering expertise, and dedicated/costly infrastructures. For the aforementioned reasons, the scientific and technological research in wastewater treatment is moving toward more handy methodologies capable to mitigate the environmental impact.[6]

Compared to existing ones, the new approaches should involve low cost, high durability, and high efficiency without further stressing the environment or endangering human health by the treatment itself. In wastewater treatment, membrane technology is becoming increasingly important. Membrane-mediated filtration processes were introduced in the 1950s mainly for niche applications such as drinking water treatment such as desalination of seawater. However, advances in materials, with particular regard to the rapid development of nanomaterials and nanotechnologies, have opened the door to affordable implementation of engineered membranes that can find applications on the overall water cycle, for example, production of drinking and process water, industrial process and wastewater treatment, municipal sewage treatment, product recovery from aqueous streams, water loop closure, and treatment of groundwater, agricultural waste streams, and percolation waters.

In this context, several nanotechnology-based membranes were developed in the last years. Among them, zeolitic nanoparticle-coated ceramic membranes may be applicable for reverse osmosis (RO) offering clear advantages in mechanical stability under high pressures, chemical stability toward disinfectants, and fouling resistance[7]; on the other hand, photocatalytic surfaces, for example, titania, zinc oxide, or ferric oxide nanoparticle-coated membranes, can be activated by UV or sunlight to engage redox processes for the degradation of organic compounds.[8] Generally, catalysts are immobilized on polymeric membranes in order to create reactive surfaces eliminating the problem of catalyst recovery. The major limitation is that, when immobilized on polymeric membranes, the active area of catalysts is reduced as well as their photoactivity.[9]

Hybrid inorganic–organic nanocomposite membranes, such as polymer matrices comprising inorganic fillers or surface functionalized via nanoparticles' self-assembly, were developed for a variety of goals as targeted degradation, enhanced separation performances, increased thermal and mechanical stability, enhanced antimicrobial activity, and reduced fouling.[10]

Other nanostructured membranes that have attracted attention for their unique properties are aligned nanotube membranes and isoporous block membranes. Aligned nanotube membranes exhibit a fast mass transfer allowing, in this way, reduced hydraulic driving pressure and lower energy cost[11]; isoporous copolymers block membranes made of multiple block polymeric species that has the ability to self-assemble into highly ordered structures providing the opportunity for tunable narrow pore size distribution, high porosity, and, sharp molecular weight cut-off.[2]

Through many nanostructured membranes under investigation, also nanofibrous mats obtained by electrospinning (ES) have shown promising results in the field of water depuration. Moreover, nowadays, innovative technologies are focused around bio-based materials in order to decrease dependencies on limited fossil fuel.[13] In this regard, although several nanofibrous mats made of synthetic polymers have been developed as efficient membranes for the treatment of wastewater, in recent years, there has been a growing interest in developing bio-based materials to produce economic and environmentally sustainable filters for large-scale uses. Biopolymers from sustainable resources are being considered as alternatives to commodity synthetic polymers because of their biodegradability and eco-sustainability.

In this chapter, an overview about the applications of electrospun nanofibrous membranes (ENMs) on the water remediation processes will be presented focusing the attention on the biopolymer-based electrospun membranes.

9.2 MEMBRANE TECHNOLOGY FOR WASTEWATER TREATMENT

The treatment of water streams strongly depends on their composition and their purity grade that is essential for reuse. In general, wastewater treatment consists a cascade of different processes directed to contaminant removal or reduction, in order to ensure a water quality acceptable for an end use, which can be drank or returned safely to the environment (sewage treatment). Substances that must be removed during water purification include coarser particles, suspended solids, bacteria, algae, viruses, fungi, manganese and sulfur, heavy metals, fertilizers, and other pollutant organic compounds. Traditional processes used for water depuration can be classified into physical separation techniques for particle removal (as filtration and sedimentation), chemical treatment to remove suspended solids (coagulation, flocculation, ion-exchange carbon adsorption, etc.), organic matter and dissolved pollutants or toxins, and biological treatment such as aerobic- and anaerobic-activated sludge processes. Membranes can replace or supplement the aforementioned techniques. A membrane is a semipermeable thin layer that acts as a barrier able to separate contaminants present in one phase (feed) and to give back the cleaned phase (permeate) under a driving force such as pressure or concentration gradient. The nature of the material passing through the membrane is determined by the size and the chemical characteristics of the membrane.

The membrane performances are defined by two key parameters: flux and selectivity. Flux indicates the rate of water passing through the membrane per unit of surface area, while selectivity (or retention) is the concentration ratio of a component between the filtered permeate and the input feed water. These two factors are influenced by several structural and morphological membrane properties such as porosity, pore size and distribution, wettability, thickness, and pressure drop across the membrane.

Membranes used for water filtration are classified into sieving membranes, characterized in terms of pore sizes and functionalized or affinity membranes.

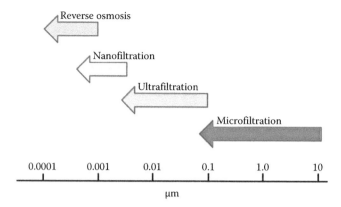

FIGURE 9.1   Membrane separation processes.

Based on the pores size and on particles' dimensions that they can retain, sieving membranes can be used for microfiltration (MF), ultrafiltration (UF), nanofiltration (NF), and RO processes (Figure 9.1).[14,15]

In general, *MF* is used as prefiltration for other water treatment and operates with pressures ranging from 0.1 to 2 bar. The size of the particles retained by MF ranges between 0.1 and 10 μm; therefore, the typical contaminants removed in the MF are suspended solids, bacteria, and protozoa (Figure 9.2a).

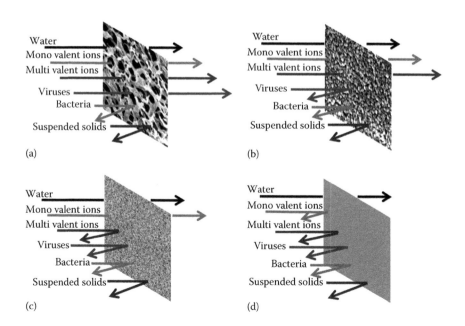

FIGURE 9.2   Filtration processes. (a) Microfiltration (MF), (b) ultrafiltration (UF), (c) nanofiltration (NF), and (d) Reverse osmosis (RO).

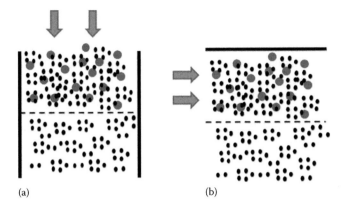

(a)   (b)

FIGURE 9.3   (a) Dead-end and (b) cross-flow filtration.

UF is used to remove colloidal dispersion, bacteria, and viruses from raw water to produce potable water (Figure 9.2b). It is applied in cross-flow (also known as tangential flow filtration) with pressures in the range of 1–5 bar or dead-end mode with pressures in the range of 0.2–0.3 bar (Figure 9.3).

NF is a relatively recent process, which uses membranes having pore sizes from 1 to 10 Å. Within wastewater treatment, NF is used for water softening because the membranes are able to retain hydrated divalent ions such as calcium (Ca^{2+}) and magnesium (Mg^{2+}) pushing through the hydrated monovalent ions. The operating pressure ranges between 5 and 20 bar (Figure 9.2c).

Finally, the *RO* is a process in which an applied pressure is used to overcome osmotic pressure allowing the retention of the solute on the pressured size of the membrane and the passage of the solvent to the other side of the membrane. RO is used for desalination and ultrapure water production. This process requires a working pressure between 10 and 100 bar (Figure 9.2d).

Instead, affinity membranes, principally developed to overcome the limited specificity of membranes for UF and MF processes, are able to selectively bind molecules, ions, or bacteria by means of particular functional groups grafted on their surface.[16]

Compared to the aforementioned technologies for wastewater treatment, pressure-driven membrane processes show several advantages, in that their industrial applications have considerably expanded in the last 50 years. In fact, membrane processes do not require chemical additives or thermal inputs, and they work with relatively high-efficiency and low-energy consumption without generating a large amount of sludge, thus avoiding the formation of secondary pollution and membranes' regeneration problems.[17,18] Moreover, membrane-based processes guarantee constant water quality regardless of feed quality; they require low cost and easy plant design, since membranes can be installed in compact and automated modular units while offering solutions for large-scale water treatments.

An ideal membrane for wastewater treatment should show mechanical, chemical, and thermal stability, high water flux, high solute rejection, long-term stability of water flux, and low-fouling performances. Moreover, it should be processed into large-scale membranes and modules while being inexpensive. Even if membrane's technology is a well-established and effective technology for environmental problems' remediation, especially water, commercially available membranes show several drawbacks.

The most common filtration media are nonwovens of microfibers. The main advantages of these filters are the high internal surface area and the high dirt loading capacity due to the presence of a fibrous networks and a good mechanical strength. However, nonwoven filter media showed also several drawbacks: (1) They can remove particles between 10 and 200 μm mean diameter due to the considerably larger average pore size generated by the micron-size fibers. (2) Particles are easily trapped and lodged within the tortuous path of the nonwoven media; therefore, they are not easily cleanable and reusable. (3) A thicker fibrous layer is required to reduce the average pore size of the media in order to separate particles less than 10 μm resulting in a flux drop. For the aforementioned reasons, there is an urgent need to develop new generation of membranes with improved productivity, selectivity, fouling resistance, and stability, available at lower cost and with fewer manufacturing defects.

9.3   APPLICATIONS OF ENMs IN WATER DEPURATION

Over the past decade, the progresses and advances of nanotechnologies have offered important opportunities to develop next-generation water treatment systems enabling new functionalities such as high permeability, catalytic reactivity, and fouling resistance. In the field of membrane filtration, among the variety of nanostructured membranes,[19–21] ENMs are one of the most important materials studied for wastewater treatment.

Even if several methods have been developed for nanofiber production, such as vapor-phase approach,[22] solution–liquid–solid technique,[23] solvothermal synthesis,[24] or self-assembly,[25] ES has been demonstrated to be the most efficient technique due to its easiness, high versatility, high production rate, low cost,[26] and easy scale-up for further process industrialization.

Briefly, a basic ES apparatus is made of a high voltage supplier, a syringe filled with the polymer fluid and connected to a spinneret, a syringe pump for flow rate controller, and a metal collector. In the ES process, a high voltage is applied to the polymer droplet at the capillary tip, and when the electrostatic forces overcome the solution surface tension, a charged jet is ejected toward the collector. Before reaching the collector, the solvent evaporates, and the polymer is collected as an interconnected web of nanofibers. The formation and the morphologies of nanofibers are the function of several parameters classified into solution parameters such as viscosity, surface tension, and conductivity, process parameters such as electric field and flow rate, and environmental conditions such as temperature and relative humidity.

A great advantage of ES is its capability to control the fiber diameter from tens of nanometers to a few micrometers, by simply varying solution properties and process parameters.[27,28] The change in fiber diameter provides the opportunity to tune membrane's thickness and porosity.

Nanofibrous membranes, due to their several attractive attributes as high specific surface, high porosity, fine-tunable pore size, interconnected open-pore structure, and high permeability to gas and fluid, are widely considered as filtering media for air and water.

Although several companies such as Donaldson Company Inc.,[29] Elmarco,[30] and Ahlstrom Corporation[31] employ ENMs for air filtration, their potential in water treatment is still largely unexploited.

In particular, the key advantages of ENMs for water filtration are the high loading capacity due to the large internal surface area and their capability to support high fluxes due to their high porosity. Moreover, ENMs can be easily functionalized through surface modification or encapsulation of functional molecules in order to obtain sieving membranes that can be used also as adsorbents able to bind toxic substances (affinity membranes). Beyond these peculiar characteristics, electrospun membranes show some drawbacks that need to be addressed: they are difficult to handle, and they show poor mechanical properties if compared to cast membranes or microfiber-based nonwovens made of the same polymer.

The most common strategy proposed to overcome the poor mechanical properties of nanofibrous membranes is the deposition of nanofibrous into conventional nonwoven filters (e.g., nonwovens of poly(ethylene terephthalate) [PET] microfibers).[32] However, several authors showed how the mechanical properties of electrospun membranes can be improved also by inducing a fiber alignment during the ES process.[33,34] On the other hand, Lijo et al. demonstrated that the addition of a smaller amount of TiO_2 nanoparticles increased the tensile strength of polyimide nanofibrous membranes from 0.36 to 0.65 MPa.[35]

Therefore, the improvement in the ES technology and nanofiber production has paved the way to use ENMs in several steps of wastewater treatments. Several research groups have proposed the use of ENMs for pressure-driven membrane processes such as MF, UF, NF, and desalination (Table 9.1).

Gopal et al. demonstrated that polysulfone (PSU) electrospun membranes (nanofiber mean diameter 470 nm) hold a potential as prefilters for particulate removal from water. The membranes were tested as prefilter for the removal of particles of several dimensions using a dead-end filtration system and evaluating the flux recovery, that is, the ratio of water flux after and before the separation of a microparticles and the separation factor (Equation 9.1):

$$\text{Separation factor} (\%) = \left(1 - \frac{C_{permeate}}{C_{feed}}\right) \times 100 \qquad (9.1)$$

where $C_{permeate}$ and C_{feed} are the particle concentration of collected permeate and original feed, respectively.

The separation factor for 10, 8, and 7 μm particle solutions was above 99%, while the flux recovery was of 100%, while a drop in both separation factor and flux recovery was observed when separating 2 and 1 μm particles, indicating that the membrane is severely and irreversibly fouled.[36] In another work of Aussawasathien et al., electrospun membranes made of nylon-6 nanofibers having diameters between

TABLE 9.1

Applications of Electrospun Nanofibrous Membranes in Water Depuration

ENMs	Applications	References
Polysulfone	Prefilters for particulate removal from water	[36]
Nylon-6	Prefilters for particulate removal from water	[37]
Polyacrylonitrile	High flux ultrafiltration/ nanofiltration membranes Adsorbent of soybean oil	[38]
Polyvinylidene fluoride	Membrane distillation	[40]
Poly methyl methacrylate–grafted polysulfone nanofibers	Adsorbent of toluidine blue O dye	[41]
Poly methyl methacrylate nanofibers functionalized with phenylcarbomylated and azido-phenylcarbomylated β-cyclodextrins	Adsorbent of phenolphthalein	[42]
Oxolane-2,5-dione-modified electrospun cellulose nanofibers	Adsorbent for Pb(II) and Cd(II)	[51]
PMMA-functionalized cellulose acetate	Adsorbent for Cu(II), Hg(II) and Cd(II)	[54]
Cellulose acetate nanofibers surface functionalized with fluorinated polybenzoxazine layer incorporated with silica nanoparticles	Superhydrophobic/ superoleophilic membranes for separation of oil–water mixtures	[55]
Chitosan	Removal of fine particles, Cu(II), Pb(II), Cr(VI), Fe(III), Cd(II), Ag(I), and bacteria	[56–59]
Silk fibroin/cellulose acetate blend electrospun membranes	Adsorbent for Cu(II)	[71]
Silk fibroin nanofibers	Adsorbent for Cu(II)	[72]
Silk fibroin/keratose blend nanofibers	Adsorbent for Cu(II)	[72]
Cysteine-S-sulfonate keratin nanofibrous membranes	Adsorbent for Cu(II), Ni(II), and Co(II), methylene blue	[76,77]
Polyvinyl alcohol/chitosan blend nanofibers	Adsorbent for Cr(III) and Cr(VI)	[79]
Polyvinyl alcohol/cyanobacterial extracellular polymeric substances blend nanofibers	Adsorbent for Cr(III) and Cr(VI)	[79]
Polyvinyl alcohol/tetraethyl orthosilicate/ aminopropyltriethoxysilane electrospun membranes	Adsorbent for uranium	[80]
PVA/SiO$_2$ composite nanofibers functionalized with mercapto groups	Adsorbent for Cu(II)	[81]
Poly-(lactic acid)	Prefilters for particulate removal from water	[86]
Polycaprolactone/chitosan blend nanofibers	Removal of fine particles and bacteria	[89]
Poly-3-hydroxybutyrate/TiO$_2$ nanocomposite nanofibers	Adsorption and photocatalytic degradation of malachite green dye	[94]

30 and 110 nm showed a separation factor higher than 95% also for particles size higher than 1 μm, while the separation of 0.5 μm particles was not as good exhibiting around 84.5% separation factor.[37]

ENMs have been also proposed for the design of a new type of high-flux UF/NF membranes: these membranes are made of polyacrylonitrile (PAN) nanofibers with an average diameter from 124 to 720 nm and a porosity of about 70%, together with a chitosan (CS) top layer of 1 μm thickness displaying a much higher flux rate than conventional membranes for NF, while maintaining the same rejection efficiency (>99.9%) for oily wastewater filtration.[38]

In the past few years, electrospun nanofibers have gained much interest also for application in membrane distillation (MD), an emerging technology for water desalination and purification that can compete with the common desalination technology such as RO. The driving force in the MD process is the partial vapor pressure commonly triggered by a temperature difference, taking place between the two sides of a porous membrane.[39] C. Feng et al. report for the first time the use of electrospun polyvinylidene fluoride nanofibrous membrane (500 nm mean diameter) for air-gap MD to produce potable water from a saline water of sodium chloride (NaCl). The NaCl separation was in a range of 98.7%–99.9%, and the permeate NaCl concentration was found to be in the range of 110–280 ppm, below the salt concentration limit of drinking water.[40]

Moreover, through the ES of reactive polymers or a post-surface functionalization of nanofibers with chemical groups able to bind specific substances, it is possible to prepare membranes that can offer both physical filtration and chemical adsorption of toxic agents such as heavy metals, dyes, or organic compounds. In general, the adsorption performances of a membrane (adsorbent) toward a toxic substance (adsorbate) are determined by measuring the adsorption capacity q (mg/g) described as follows:

$$q = \frac{(C_0 - C_f) * V}{m} \tag{9.2}$$

where
   C_0 (mg/L) and C_f (mg/L) are the initial and final concentration of the adsorbate, respectively
   V (L) is the volume of the water solution to be purified
   m (g) is the membrane mass

Several nanofibrous membranes were developed and tested as adsorbent of heavy metals, dyes, and organic compounds. For example, Ma et al., developed poly methyl methacrylate (PMMA)–grafted PSU nanofibers having an adsorption capacity toward toluidine blue O dye of about 380 mmol/g.[41] Functionalized nanofibers have been also proposed for the removal of organic molecules from water. As an example, PMMA nanofibers functionalized with phenyl-carbomylated and azido-phenyl carbomylated β-cyclodextrins have been demonstrated to be able to capture phenolphthalein molecules from water.[42] Oil coming from industrial waste is also an organic material occasionally found in water sources. It has been demonstrated that a membrane system made of a three-tier membrane stack, with the middle tier composed of PAN nanofibers, is able to remove a high percentage of soybean oil at a higher permeation rate compared with conventional UF membranes.[38]

9.4 BIODEGRADABLE POLYMER-BASED ENMs FOR WATER DEPURATION

Eco-efficient polymers display several advantages in terms of greenhouse gas emission, fossil energy usage, and management of waste treatment[43,44]; nevertheless, these materials are often more expensive and less familiar to consumers as compared with common plastics. Therefore, the successful applications of biodegradable membranes, made of biodegradable polymers obtained from low-cost renewable source, could have a significant positive environmental impact. In this review, biodegradable polymers are intended as those that are degraded in biological environments (such as soils, rivers, or human body) through enzymatic or nonenzymatic hydrolysis and not through thermal oxidation, photolysis, or radiolysis.[45,46] The major applications of these biodegradable polymers are in the field of biomedical and ecological polymers that keep the environment clean. The classification of biodegradable polymers is based on their origin (Figure 9.4) that can be natural or synthetic. Several nanofibrous membranes based on biodegradable materials have been proposed as next-generation environmental-friendly filter media. This paragraph details the latest developments about different classes of biopolymer nanofibrous membranes for water treatment (Table 9.1).

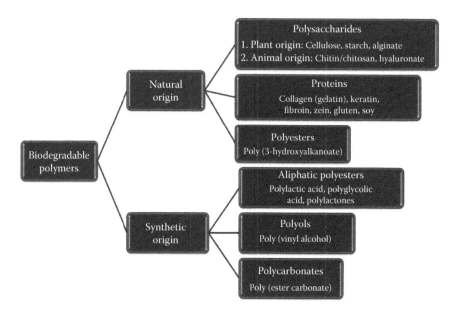

FIGURE 9.4   Biodegradable polymer classification.

9.4.1 POLYSACCHARIDES

The most abundant natural, renewable, and biodegradable polysaccharide is cellulose, a polydispersed linear biopolymer of poly-β-(1,4)-D-glucose units with asyndiotactic configuration.

Cellulose membranes are widely used in membrane preparation due to their hydrophilicity; however, the main difficulty for cellulose ES is related to its poor solubility that is limited to few solvents basically not suitable for this technique such as dimethyl sulfoxide/paraformaldehyde[47] or sulfur dioxide.[48]

However, electrospun membranes entirely composed of cellulose were obtained from 3% cellulose solutions in lithium chloride/N,N-dimethylacetamide solvent system by using a water coagulation bath (for removal of LiCl) and heated collectors (for the organic solvent removal).[49]

Nevertheless, the common strategy used to prepare cellulose nanofibers is the ES of cellulose derivatives like cellulose acetate (CA), hydroxyl-propyl cellulose, hydroxyl-propyl methylcellulose, and ethyl-cyanoethyl cellulose, which are readily soluble in volatile solvents, followed by conversion into cellulose through hydrolysis.[50] In a work of Stephen et al., cellulose nanofibers were regenerated from electrospun CA through deacetylation and then modified with oxolane-2,5-dione. The functionalized cellulose electrospun membranes were made of nanofibers having a mean diameter of about 385 nm, and they showed a surface area of 13.68 m^2/g. The functionalized cellulose nanofibers were tested as adsorbent for lead and cadmium from water. In particular, the adsorption performances of functionalized cellulose nanofibers were compared with those of functionalized cellulose raw fibers having a surface area of 3.22 m^2/g. The adsorption capacity for functionalized cellulose nanofibers were 1.21 and 0.59 mmol/g for Pb and Cd, respectively, compared to 0.24 and 0.13 mmol/g for functionalized cellulose raw fibers. Moreover, oxolane-2,5-dione-modified electrospun cellulose nanofibers can be easily regenerated by using a 3M HNO$_3$ solution, without affecting their adsorption performances.[51]

Among the cellulose derivatives, CA is an important cellulose ester, widely used in fabrication of membranes for UF and RO.[52] Upon functionalization with –COOH, –SO$_3$H, and –NH$_2$ groups, CA is able to bind heavy-metal ions.[53]

The CA nanofibrous membranes were successfully prepared by electrospinning the biopolymer dissolved in acetone/dimethylacetamide (DMAc) solvent system. The electrospun membranes, functionalized with poly(methacrylic acid), have been demonstrated to be potentially inexpensive and highly efficient adsorbents for heavy metals in water. In fact, those membranes showed a good adsorption capacity toward metal ions such as Cu(II), Hg(II), and Cd(II), and higher adsorption occurred in higher initial pH. This is due to the pH-responsive behavior of PMMA chain: at higher pH, the carboxyl groups of PMMA are ionized to COO$^-$ acting as binding sites for cationic species, while at lower pH, the carboxyl groups are deionized to COOH, and the uncharged chains are aggregated on the surface losing their ability to bind metal ions. Moreover, they showed high adsorption selectivity for Hg(II) ions, and they could be easily regenerated with ethylene-di-nitrilo tetra-acetic acid solution.[54] At the same time, A. A. Taha et al. developed novel NH$_2$-functionalized

CA/silica composite nanofibrous membranes with high porosity (73%) and high surface area (126.5 m^2/g), by combining sol–gel and ES methods. These composite membranes have a significant potential application in wastewater treatment showing good adsorption performances of Cr(VI) in both static and dynamic conditions (in a continuous flow filtration model).

In situ polymerization of fluorinated polybenzoxazine layer incorporated with silica nanoparticles, carried out on acetate nanofibers surface, produced superhydrophobic (water contact angle 161°) and superoleophilic (oil contact angle of 3°) nanofibrous membranes that showed an excellent separation of oil–water mixtures. These membranes could find interesting applications for practical oil-polluted water.[55]

Another widely considered polysaccharide for water treatment is Chitosan (CS). This biopolymer is the *N*-acetylated derivative of chitin, which, after cellulose, is the most abundant organic material produced by biosynthesis. A stable CS-based nanofibrous membrane would have several advantages over currently used polymers, not only because CS is an environmentally friendly natural material but also because it has numerous polar and ionizable groups and antibacterial properties. Therefore, the CS-based nanofibrous membrane can be used, at the same time, for fine filtration and removal of toxic metal ions and microbes.[56] However, as with cellulose, also CS displays solubility problems in solvents suitable for ES.

Haider and Park have successfully electrospun CS dissolved in trifluoroacetic acid. The CS electrospun membranes (235 nm mean diameter) neutralized with potassium carbonate showed good stability in water and high adsorption affinity for Cu(II) and Pb(II). In fact, the maximum adsorption capacities were 484.4 mg/g for Cu(II) and 263.1 mg/g for Pb(II), about 10 times higher than those of CS microspheres. This is due to the high porosity and high specific surface area of CS nanofibers compared to CS microspheres.[57] In another work of Desai et al., electrospun CS/poly(ethylene oxide) nanofibers were tested for physical particle separation, heavy-metal adsorption, and microbes removal. The authors demonstrated that filtration efficiency of the nanofibers was strongly related to the size of the fibers and their surface CS content. CS-based nanofibrous filter media exhibited a binding ability up to 35 mg of hexavalent chromium/g CS along with a 2–3 log reduction in *Escherichia coli* bacteria colony forming unit (cfu).[58]

Other electrospun membranes at high surface area per unit mass (22.4 m^2/g), made of CS nanofibers with a mean diameter of 42 nm, were produced by electrospinning of 0.4 wt% CS in 1,1,1,3,3,3-hexafluoroisopropanol. This environmentally friendly CS-based nanofibrous filter media showed high sorption percentage (90%–100%) for Fe(III), Cu(II), Ag(I), and Cd(II) and high selectivity for copper ions.[59]

9.4.2 PROTEIN

Despite being a major category of biopolymers, the process of electrospinning proteins into fibers is still challenging, due to protein's complex macromolecular and 3D structures joined by strong inter- and intramolecular bonds. However, several proteins have been electrospun into nanofibers, such as collagen,[60] gelatin,[61] silk,[62] keratin,[63] soy,[64] gluten,[65] and zein.[66] Among those proteins, fibroin- and keratin-based nanofibrous membranes are proposed as biodegradable filter media for water depuration from heavy metals and cationic dyes. Despite the encouraging results about the fabrication of fibroin- and keratin-based nanofibrous membranes, one of the major concerns is the poor stability and structural integrity of these membranes in the aqueous environment. Several strategies have been proposed to overcome this drawback such as treatments with formaldehyde vapor,[67] aqueous alcohol solution,[68] water vapor treatment,[69] or treatment with cross-linking agents as genipin.[70] Even if the eco-friendly stabilization methods based on treatments with alcohols, such as ethanol, work well with fibroin, they are not so efficient with keratin, for which stabilization treatments with formaldehyde or glutaraldehyde are often required. Recently, a novel water stabilization treatment of pure keratin nanofibers, which entails its thermal treatment at 180°C for 2 h, was proposed.[63] Unlike the water-soluble untreated membranes, the thermally treated keratin nanofibrous membranes not only do not dissolve in water, but also they maintain their nanostructure after immersion in water for 24 h. The improved stability of thermally treated keratin nanofibers seems to be ascribed to the formation of amide bonds between acid and base groups of some amino acids within the side chains.

Silk fibroin (SF) is a protein mainly composed of amino acetic acid, alanine, serine, and tyrosine, and it is an excellent candidate for heavy-metal adsorption due to a large amount of functional groups able to bind cationic species. SF/CA electrospun blended membranes made of randomly oriented nanofibers having diameters between 100 and 600 nm and water stabilized with ethanol were tested for Cu(II) ion adsorption from water. Practically, 200 mL of a Cu(II) solution (100 mg/L) was pumped through the electrospun blended membranes at a flow rate of 2.0 mL/min for 3 h, and the adsorption capacity was evaluated using Equation 9.2. The higher adsorption capacity of pure SF nanofibrous membranes (10.8 mg/g), compared to that of pure CA nanofibrous membranes (4.3 mg/g), is due to the higher number of ionizable groups (such as carboxyl, amino, carbonyl, and hydroxyl) of fibroin compared to CA. However, the blend nanofibers containing the 20% of CA showed the highest adsorption capacity (22.8 mg/g) indicating that SF and CA have a synergistic effect. This means that the carbonyl and the hydroxyl groups in CA molecular chains adsorb Cu(II) together with the carboxyl groups in SF molecular chains. The Cu(II) adsorption capacity of SF/CA blend nanofibrous membranes substantially decreased when the content of CA was more than 20%; this is because of the significant decrease of the number of ionizable groups.[71]

The adsorption capacity of pure SF nanofibrous membranes toward Cu(II) ions was also compared with that of blend nanofibers made of SF/keratose blend nanofibers, where keratoses were obtained by oxidative extraction of keratin from wool using performic acid.[72] Through the oxidative extraction, the cysteine residues of keratin are converted into cysteic acid (SO_3H) as shown in Figure 9.5a.

Keratin, being the major component of wools, hairs, horns, nails, and feathers, is an abundant nonfood protein, widely considered as adsorbent of ionic species because of its great number of functional groups.

An interesting advantage of keratin compared to fibroin is that it can be extracted from low-cost biomasses such as feathers, horns, and nails from butchery, poor quality raw wools from sheep breeding, and by-products from textile industry. Even if

(a) Keratin-S-S-Keratin + HCO$_3$H $\longrightarrow$ Keratin -SO$_3$H

(b) Keratin-S-S-Keratin + Na$_2$S$_2$O$_5$ $\longrightarrow$ Keratin-SH + Keratin-S-SO$_3$H

FIGURE 9.5   Keratin extraction by (a) oxidation and by (b) sulfitolysis.

keratin proteins have been demonstrated to show good adsorption performances toward cations such as heavy metals or cationic dyes,[73–75] some works underline how the adsorption capacity improves when keratin proteins are transformed into nanostructured membranes.[55,61]

In the work of Ki et al., pure SF and wool keratose/SF (WK/SF) electrospun membranes made of nanofibers of 300 nm mean diameter were water stabilized by immersion in methanol followed by a treatment with formaldehyde vapor. For the evaluation of the adsorption capacity toward Cu(II) ions, 30 mL of the stock solution at initial copper ion concentration of 3.5 mg/L was forced to circulate through the membranes using a flow rate of 13mL/min for 30 min. The WK/SF blend nanofibrous membranes exhibited a higher adsorption capacity (2.88 mg/g) than SF membrane (1.65 mg/g), and this is mainly attributed to different amino acid compositions of the two proteins. The nonpolar amino acids of fibroin such as alanine, glycine, and serine, which occupy more than 80% of the total SF amino acid content, hardly play a role as Cu(II) binding sites. On the other hand, WK contains more hydrophilic amino acids, which have polar side residues such as aspartic acid, glutamic acid, and cysteic acid that can act as binding sites for copper ions.

The Cu(II) adsorption capacity, evaluated on pure cysteine-*S*-sulfonate keratin (SSO$_3$H) nanofibrous membranes, in dynamic conditions using 70 mL of the stock solution at initial copper ion concentration of 3.5 mg/L, a flow rate of 18 mL/min, and a circulation time of 3 h, increased up to 5.82 mg/g.[76] The cysteine-*S*-sulfonate keratin is obtained when the sulfitolysis reaction is used as keratin extraction method. In this case, the disulfide covalent bonds are broken into cysteine residues (SH) and cysteine-*S*-sulfonate residues as described in Figure 9.5b.[77,78]

The maximum Cu(II) adsorption capacity of the aforementioned membranes, obtained using a copper ion starting solution at concentration of 20 mg/L, was about 11 mg/g (Figure 9.6a). Moreover, the selectivity studies carried out on pure cysteine-*S*-sulfonate keratin electrospun membranes showed that the Cu(II) > Ni(II) > Co(II) was followed (Figure 9.6b).

The membranes of cysteine-*S*-sulfonate keratin are tested also as adsorbents for methylene blue (MB), chosen as an example of cationic dye.[77] The adsorption test, carried out in batch conditions (simply by shaking the nanofibrous membranes immersed in 10 mL of MB starting solution), revealed that the maximum MB adsorption capacity obtained at 20°C, pH 6, and an adsorbent dosage of 1 g/L was about 170 mg/g. Moreover, in order to evaluate the effect of the specific surface area, the MB adsorption capacity of keratin nanofiber membranes was compared with that of a wool fabric, which is chemically similar, but it has a specific surface area of two orders of magnitude lower than that of the nanostructured membrane. As shown in Figure 9.7, due to its higher specific surface area, the adsorption capacity of keratin nanofiber membrane is two orders of magnitude higher than that of the wool fabric.

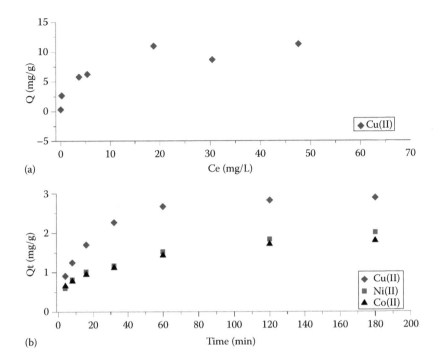

FIGURE 9.6 Adsorption capacities of heavy-metal ions onto keratin nanofibrous membranes: (a) copper ions and (b) copper, nickel, and cobalt ions. (Adapted from Aluigi, A. et al., *Text. Res. J.*, 83, 1574, 2012.)

9.4.3 POLYVINYL ALCOHOL

Polyvinyl alcohol (PVA) is a nontoxic and biocompatible polymer of synthetic origin that can be easily processed in solution due to its solubility in water. In the field of electrospun membranes for filtration purpose, PVA is used to improve the mechanical properties of biopolymer-based nanofibers.

The PVA was blended with CS and cyanobacterial extracellular polymeric substances (EPS) and successfully electrospun using a solution of acetic acid 1% as common solvent.

Electrospun PVA/CS and PVA/EPS blend nanofibers with uniform and smooth morphology and narrow diameter distribution from about 50 to 130 nm were obtained.

The electrospun blended PVA/EPS membrane showed better tensile mechanical properties when compared with PVA and PVA/CS and displayed higher resistance against disintegration in the temperature range between 10°C and 50°C. Moreover, the blended membranes showed a great potential for water depuration purposes displaying a 5% increase capacity in binding chromium.[79] In another work of Keshtkar et al., polyvinyl alcohol/tetraethyl orthosilicate/aminopropyltriethoxysilane (PVA/TEOS/APTES) electrospun membranes made of nanofibers having a mean diameter

Properties	Keratin nanofibers	Wool fabric
Fiber diameter (nm)	223 ± 74	$19\ \mu m$
Thickness (μm)	50	500
Porosity (%)	90	69.7
Specific surface area (m^2/g)	13.6	0.2

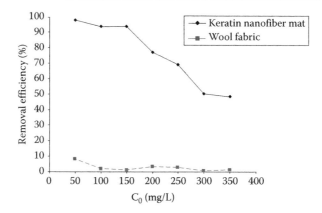

FIGURE 9.7 Comparison of methylene blue adsorption capacities between keratin nanofibers and wool fibers. (Adapted from Aluigi, A. et al., *J. Hazard. Mater.*, 268, 156, 2014.)

of approximately 200 nm and a specific surface area of 282 m^2/g were developed by the sol–gel/ES method. Their adsorption capacity of uranium from aqueous solutions was compared with PVA/TEOS/APTES membrane prepared by sol–gel/casting method having a specific surface area of 153 m^2/g. The maximum adsorption capacity of uranium ions onto the PVA/TEOS/APTES hybrid nanofibrous membrane was found to be 168.1 mg/g, and it was more than fivefold greater than that of uranium sorption onto the PVA/TEOS/APTES hybrid membrane (33.6 mg/g). Moreover, those membranes could be regenerated by a simple wash with a 1M HNO_3 solution and reused many times without any significant loss in adsorption performance.[80]

Wu et al. demonstrated that surface modification of PVA nanofibers could be a useful tool for improving the performance of an adsorbent. The authors proposed PVA/SiO_2 composite nanofibers functionalized with mercapto groups suitable for the heavy-metal removal from water. The surface area of these membranes depended on the PVA content: when PVA content was in the range 1.43%–4.55%, the surface area of PVA/SiO_2 electrospun membranes was higher than 290 m^2/g and decreased by increasing the PVA content. Moreover, the adsorption capacity toward Cu(II) ions of thiol-functionalized mesoporous PVA/SiO_2 electrospun membranes was greater than that of pure PVA nanofiber membrane. The higher Cu(II) uptake was 489 mg/g with the capacity maintained through six recycling processes of adsorption and desorption. Consequently, this membrane can be a promising material for heavy-metal ions removal and water recovery.[81]

9.4.4 BIODEGRADABLE POLYESTERS

Among biodegradable polyesters, aliphatic polyesters from synthetic and natural origins have been intensively investigated.

The poly-(lactic acid) (PLA) is an aliphatic polyester that can be derived from renewable agricultural resources such as sugar beets, rice, or corn.[44] This polymer is an innovative eco-efficient material that degrades by hydrolysis after several months of exposure to an active compost environment.[82] Moreover, unlike other natural polymers, PLA is thermoplastic; therefore, it can be transformed into various shapes through melt processes.[83] Even if, from the point of view of mechanical and physical properties, PLA has been demonstrated to be comparable to PET and polystyrene,[84] the electrospun PLA membranes showed poor mechanical strength that limited their applications. The poor mechanical properties of PLA nanofibers are due to the low crystallinity degree of the polymer associated to the fast solvent evaporation or polymer solidification that occurs during ES process.[85] Recently, annealed electrospun PLA membranes were proposed as eco-friendly membranes for MF by Li et al.[86] The annealing is an environment-friendly treatment that involves heating the material to a temperature above its glass transition temperature and below its melting temperature, followed by slow cooling of the material to induce changes in molecular structure and leading to a higher crystallinity degree.[87,88] The nanofibrous membranes were prepared by ES solutions (13 wt%) of PLA dissolved in acetone and submitted to annealing at 90°C for 30 min. Compared to as-spun PLA membrane, the annealed membrane shows smaller average pore size, lower porosity, and a higher crystallinity degree, which in turn provides an increased stiffness (increased Young modulus) and a decreased ductility (decreased elongation at break). Moreover, after annealing treatment, the formation of interfibrous fusions increases the structural integrity of the membrane performing a higher stress at break. As regards the filtration properties, the annealed membrane shows a higher particle rejection (85%) than the as-spun one (10%).

Polycaprolactone (PCL) is another biodegradable polyester of synthetic origin widely considered for ES. In a study of Cooper et al., CS–PCL fibrous membranes were prepared with different amounts of CS to impart an antibacterial property to the membrane. The authors found that at 25% and 50% of CS content, the highly porous membranes supported high water permeability, while with increasing the CS content, the membranes were more susceptible to swelling, preventing their application as filters. The CS–PCL fibrous membrane was able to remove about 100% of particles having a diameter of about 300 nm demonstrating their ability to act as a pre-filter. Moreover, the blend nanofibers containing 25% of CS reduced *Staphylococcus aureus* bacterial colonization by 50% compared to membranes made of pure PCL fibers demonstrating promising antibiofouling properties.[89]

Polyhydroxyalkanoates (PHAs) are a family of biodegradable polyesters produced in nature by bacterial fermentation of sugar or lipids. They are stored intracellularly as energy storage inclusions by various bacteria during environmental stress.[90] Over the years, PHAs have gained much commercial interest due to their thermoplastic, nontoxic, biodegradable, and renewable nature.[91] Among the known PHAs, the poly-3-hydroxybutyrate (P(3HB)) is the most widely studied biopolyester because of its high crystallinity and availability from a large number of Gram-negative and Gram-positive bacteria.[92,93]

Sridewi et al. developed an electrospun membrane made of P(3HB)–TiO$_2$ nanocomposite nanofibers, which show simultaneous adsorption and photocatalytic degradation of malachite green (MG) dye.

Even if the electrospun P(3HB) membrane exhibited excellent MG adsorption capability by removing approximately 80% of the dye in less than 2 h, the same membrane functionalized with titanium dioxide (P(3HB)-50 wt%) gave the highest decolorization activity under solar irradiation. Moreover, the aforementioned membrane was found to be usable, by removing nearly 98% of the dye at every cycle for 10 consequent rounds of decolorization experiments.[94]

9.5 CONCLUSION

The increasing air and water pollution represents undeniable threats to the world's environment. In particular, the global research in the field of water remediation is directed toward new, sustainable, affordable, safe, and robust methodologies enabling improved water disinfection, decontamination, and desalination.

Membrane-based technologies are favored over many other technologies for water treatment since they do not require chemical treatment or thermal input; they do not cause secondary pollution, while requiring low-energy working conditions. Therefore, the next generation of membrane-based technologies for environmental applications will be based on energy-saving and cost-effective membranes.

In this regard, due to their extraordinary structural characteristics such as their high surface area and interconnected porosity, which implies higher permeability to fluids streams, together with their tunable selectivity and energy/cost efficiency, the ENMs will play an important role in the replacement of conventional membranes for water remediation.

ENMs have been shown to hold a great potential in different water purification processes such as prefiltration, UF/NF, desalination, and removal of microbes and other toxic substances such as heavy metals and dyes. Nevertheless, although several ENMs based on synthetic polymers have been developed, the circumvention of persistent plastics is one of the greatest challenges in order to reduce the environmental impact. Therefore, the development of materials based on biodegradable polymer is necessary for the coexistence of the human society with the nature. Several classes of biopolymers have been successfully electrospun into membranes having good performances in MF, heavy-metal and dye adsorption, and microbes removal. The advancement in fabricating nonwoven biopolymer nanofibers is appreciable, but much work still remains to be done. One of the greatest challenges for polymer scientists is to manufacture, at a sustainable cost, biodegradable polymers having well-balanced biodegradability and mechanical properties. Ultimately, the ES technology needs to make a step further, allowing the large-scale production of reproducible and uniform nanofibers with suitable mechanical properties.

REFERENCES

1. Urban Urgency. 2007. Water caucus summary. Marseille, France: World Water Council (WWC).
2. Seckler, D. et al. 1999. Water scarcity in the twenty-first century. *Int. J. Water Res. Develop.* 15:29–42.
3. Montgomery, M. A. and Elimelech, M. 2007. Water and sanitation in developing countries: Including health in the equation. *Environ. Sci. Technol.* 41:17–24.
4. WHO. 2007. Combating waterborne disease at the household level/International Network to Promote Household Water Treatment and Safe Storage, World Health Organization, NLM classification: WA 675. Geneva, Switzerland: WHO, pp. 7–34.
5. Lima, A. A. M. et al. 2000. Persistent diarrhea signals a critical period of increased diarrhea burdens and nutritional shortfalls: A prospective cohort study among children in northeastern brazil. *J. Infect. Dis.* 181:1643–1651.
6. van der Kooij, D. 2003. Managing regrowth in drinking-water distribution systems. In: Bartram, J., Cotruvo, J., Exner, M., Fricker, C., and Glasmacher, A. (eds.). *Heterotrophic Plate Counts and Drinking-Water Safety: The Significance of HPCs for Water Quality and Human Health.* Geneva, Switzerland: World Health Organization/ IWA Publishing, pp. 199–232.
7. Li, L. et al. 2007. Transport of water and alkali metal ions through MFI zeolite membranes during reverse osmosis. *Sep. Purif. Technol.* 53:42–48.
8. Molinari, R. et al. 2000. Study on a photocatalytic membrane reactor for water purification. *Catal. Today* 55:71–78.
9. Ao, Y. et al. 2008. Deposition of anatase Titania onto carbon encapsulated magnetite nanoparticles. *Nanotechnology* 19:405604.
10. Jeong, B. H. et al. 2007. Interfacial polymerization of thin film nanocomposites: A new concept for reverse osmosis membranes. *J. Membr. Sci.* 294:1–7.
11. Corry, B. 2008. Designing carbon nanotube membranes for efficient water desalination. *J. Phys. Chem. B* 112:1427–1434.
12. Phillip, W. et al. 2010. Self-assembled block copolymer thin films as water filtration membranes. *ACS Appl. Mater. Interfaces* 2:847–853.
13. Mohanty, A. K. et al. 2002. Sustainable bio-composites from renewable resources: Opportunities and challenges in the green materials world. *J. Polym. Environ.* 10:19–26.
14. Mulder, M. 1996. *Basic Principles of Membrane Technology.* London, U.K.: Kluwer Academic Publishers.
15. Ulbricht, M. 2006. Advanced functional polymer membranes. *Polymer* 47:2217–2262.
16. Ma, Z. W. et al. 2005. Electrospun cellulose nanofiber as affinity membrane. *J. Membr. Sci.* 265:115–123.
17. Zhu, M. X. et al. 2007. Removal of an anionic dye by adsorption/precipitation processes using alkaline white mud. *J. Hazard. Mater.* 149:735–741.
18. Robinson, T. et al. 2001. Remediation of dyes in textile effluent: A critical review on current treatment technologies with a proposed alternative. *Bioresour. Technol.* 77:247–255.
19. Lin, J. and Murad, S. 2001. A computer simulation study of the separation of aqueous solutions using thin zeolite membranes. *Mol. Phys.* 99:1175–1181.

20. Lu, N. et al. 2010. Organic fouling and regeneration of zeolite membrane in wastewater treatment. *Sep. Purif. Technol.* 72:203–207.
21. Kondo, M. et al. 1997. Tubular-type pervaporation module with Zeolite NaA membrane. *J. Membr. Sci.* 133:133–141.
22. Law, M. et al. 2004. Semiconductor nanowires and nanotubes. *Annu. Rev. Mater. Res.* 34:83–122.
23. Gudiksen, M. S. and Lieber, C. M. 2000. Diameter-selective synthesis of semiconductor nanowires. *J. Am. Chem. Soc.* 122:8801–8802.
24. Wang, X. and Li, Y. 2002. Selected-control hydrothermal synthesis of alpha- and beta-MnO_2 single crystal. *J. Am. Chem. Soc.* 124:2880–2881.
25. Kovtyukhova, N. I. et al. 2002. Layer-by-layer self-assembly for template synthesis of nanoscale devices. *Mater. Sci. Eng. C* 19:255–262.
26. Doshi, J. and Reneker, D. H. 1995. Electrospinning process and applications of electrospun fibers. *J. Electrostat.* 35:151–160.
27. Fridrikh, S. V. et al. 2003. Controlling the fiber diameter during electrospinning. *Phys. Rev. Lett.* 90:144502.
28. Theron, S. A. et al. 2004. Multiple jets in electrospinning: Experiment and modeling. *Polymer* 45:2017.
29. Donaldson Company Inc. 2011. Home page. Pennsylvania, PA. http://www.donaldson.com. Accessed August 10, 2011.
30. Elmarco. 2011. Elmarco home page. Liberec, Czech Republic. http://www.elmarco.cz. Accessed August 10, 2011.
31. Ahlstrom Corporation. 2011. Ahlstrom corporation home page. Helsinki, Finland. http://www.ahlstrom.com. Accessed August 10, 2011.
32. Dotti, F. et al. 2007. Electrospun porous mats for high efficiency filtration. *J. Ind. Text.* 37:151–162.
33. Huang, C. et al. 2006. High-strength mats from electrospun poly(p-phenylene biphenyltetracarboximide) nanofibers. *Adv. Mater.* 18:668–671.
34. Shao, M. et al. 2013. Preparation and surface modification of electrospun aligned poly(butylene carbonate) nanofibers. *J. Appl. Polym. Sci.* 130:411–418.
35. Lijo, F. et al. 2010. Electrospun polyimide/titanium dioxide composite nanofibrous membrane by electrospinning and electrospraying. *J. Nanosci. Nanotechnol.* 10:1–6.
36. Gopal, R. et al. 2007. Electrospun nanofibrous polysulfone membranes as pre-filters: Particulate removal. *J. Membr. Sci.* 289:210–219.
37. Aussawasathien, D. et al. 2008. Separation of micron to sub-micron particles from water: Electrospun nylon-6 nanofibrous membranes as pre-filters. *J. Membr. Sci.* 315:11–19.
38. Yoon, K. et al. 2006. High flux ultrafiltration membranes based on electrospun nanofibrous PAN scaffolds and chitosan coating. *Polymer (Guildf)* 47:2434–2441.
39. Curcio, E. and Drioli, E. 2005. Membrane distillation and related operations—A review. *Sep. Purif. Rev.* 35:151–160.
40. Feng, C. et al. 2008. Production of drinking water from saline water by air-gap membrane distillation using polyvinylidene fluoride nanofiber membrane. *J. Membr. Sci.* 311:1–6.
41. Ma, Z. et al. 2006. Surface modified nonwoven polysulphone (PSU) fiber mesh by electrospinning: A novel affinity membrane. *J. Membr. Sci.* 272:179–187.
42. Kaur, S. et al. 2006. Oligosaccharide functionalized nanofibrous membrane. *Int. J. Nanosci.* 5:1–11.
43. La Mantia, F. and Morreale, M. 2011. Green composites: A brief review. *Compos. Part A Appl. Sci. Manuf.* 42:579–588.

44. Vink, E. T. H. et al. 2003. Applications of life cycle assessment to natureworks (TM) polylactide (PLA) production. *Polym. Degrad. Stabil.* 80:403–419.
45. Doi, Y. 1994. *Biodegradable Plastics and Polymers.* Studies in Polymer Science, Vol. 12. Amsterdam, the Netherlands: Elsevier.
46. Hartmann, M. H. 1998. High molecular weight polylactic acid polymers. In: Kaplan, D. L. (ed.). *Biopolymers from Renewable Resources.* Chapter 15. Berlin, Germany: Springer-Verlag, pp. 367–411.
47. Johnson, D. C., Nicholson, M. D., and Haigh, F. C. 1969. U.S. Patent 3,447,939.
48. Hata, K. 1969. U.S. Patent 3,424,702.
49. Kim, C. W. et al. 2005. Preparation of submicron-scale, electro-spun cellulose fibers via direct dissolution. *J. Polym. Sci. Part B Polym. Phys.* 43:1673–1683.
50. Frey, M. W. 2008. Electrospinning cellulose and cellulose derivatives. *Polym. Rev.* 48:378–391.
51. Stephen, M. et al. 2011. Oxolane-2,5-dione modified electrospun cellulose nanofibers for heavy metals adsorption. *J. Hazard. Mater.* 192:922–927.
52. Edgar, K. J. et al. 2001. Advances in cellulose ester performance and application. *Prog. Polym. Sci.* 26:1605–1688.
53. Liu, C. X. and Bai, R. B. 2006. Adsorptive removal of copper ions with highly porous chitosan/cellulose acetate blend hollow fiber membranes. *J. Membr. Sci.* 284(1–2): 313–322.
54. Tian, Y. et al. 2011. Electrospun membrane of cellulose acetate for heavy metal ion adsorption in water treatment. *Carbohydr. Polym.* 83:743–748.
55. Shang, Y. et al. 2012. An in situ polymerization approach for the synthesis of super-hydrophobic and superoleophilic nanofibrous membranes for oil–water separation. *Nanoscale* 4:7847–7854.
56. Desai, K. et al. 2008. Morphological and surface properties of electrospun chitosan nanofibers. *Biomacromolecules* 9:1000–1006.
57. Haider, S. and Park, S. Y. 2009. Preparation of the electrospun chitosan nanofibers and their applications to the adsorption of Cu(II) and Pb(II) ions from an aqueous solution. *J. Membr. Sci.* 328:90–96.
58. Desai, K. et al. 2009. Nanofibrous chitosan non-wovens for filtration applications. *Polymer* 50:3661–3669.
59. Horzum, N. et al. 2010. Sorption efficiency of chitosan nanofibers toward metal ions at low concentrations. *Biomacromolecules* 11:3301–3308.
60. Matthews, J. A. et al. 2002. Electrospinning of collagen nanofibers. *Biomacromolecules* 3:232–238.
61. Huang, Z. M. et al. 2004. Electrospinning and mechanical characterization of gelatin nanofibers. *Polymer* 45:5361–5368.
62. Fan, S. et al. 2013. Electrospun regenerated silk fibroin mats with enhanced mechanical properties. *Int. J. Biol. Macromol.* 56:83–88.
63. Aluigi, A. et al. 2013. Morphological and structural investigation of wool-derived keratin nanofibres crosslinked by thermal treatment. *Int. J. Biol. Macromol.* 57: 30–37.
64. Thirugnanaselvam, M. et al. 2013. SPI/PEO blended electrospun matrix for wound healing. *Fibers Polym.* 14:965–969.
65. Dong, J. et al. 2010. Aqueous electrospinning of wheat gluten fibers with thiolated additives. *Polymer (Guildf)* 51:3164–3172.
66. Li, J. et al. 2013. Coaxial electrospun zein nanofibrous membrane for sustained release. *J. Biomater. Sci. Polym. Ed.* 24:1923–1934.
67. Aluigi, A. et al. 2011. Adsorption of copper(II) ions by keratin/PA6 blend nanofibres. *Eur. Polym. J.* 47:1756–1764.

68. Servoli, E. et al. 2005. Surface properties of silk fibroin films and their interaction with fibroblast. *Macromol. Biosci.* 5:1175–1183.
69. Min, B. M. et al. 2006. Regenerated silk fibroin nanofibers: Water vapor-induced structural changes and their effects on the behavior of normal human cells. *Macromol. Biosci.* 6:285–292.
70. Silva, S. S. et al. 2008. Genipin-modified silk-fibroin nanometric nets. *Macromol. Biosci.* 8:766–774.
71. Zhou, W. 2011. Preparation of electrospun silk fibroin/cellulose acetate blend nanofibers and their applications to heavy metal ions adsorption. *Fibers Polym.* 12:431–437.
72. Ki, C. S. et al. 2007. Nanofibrous membrane of wool keratose/silk fibroin blend for heavy metal ion adsorption. *J. Membr. Sci.* 302:20–26.
73. Hartley, F. R. 1968. Studies in chrome mordanting II. The binding of chromium (III) cations to wool. *Aust. J. Chem.* 21:2723–2735.
74. Kar, P. and Misra, M. 2004. Use of keratin fiber for separation of heavy metals from water. *J. Chem. Technol. Biotechnol.* 79:1313–1319.
75. Monier, M. et al. 2010. Adsorption of Cu(II), Hg(II) and Ni(II) ions by modified natural wool chelating fibers. *J. Hazard. Mater.* 176:348–355.
76. Aluigi, A. et al. 2012. Wool-derived keratin nanofiber membranes for dynamic adsorption of heavy-metal ions from aqueous solutions. *Text. Res. J.* 83:1574–1586.
77. Aluigi, A. et al. 2014. Study of methylene blue adsorption on keratin nanofibrous membranes. *J. Hazard. Mater.* 268:156–165.
78. Tonin, C. et al. 2007. Thermal and structural characterization of poly(ethylene-oxide)/keratin blend films. *J. Therm. Anal. Calorim.* 89:601–608.
79. Santos, C. et al. 2014. Preparation and characterization of polysaccharides/PVA blend nanofibrous membranes by electrospinning method. *Carbohydr. Polym.* 99:584–92.
80. Keshtkar, A. R. et al. 2012. Removal of uranium (VI) from aqueous solutions by adsorption using a novel electrospun PVA/TEOS/APTES hybrid nanofiber membrane: Comparison with casting PVA/TEOS/APTES hybrid membrane. *J. Radioanal. Nucl. Chem.* 295:563–571.
81. Wu, S. et al. 2010. Effects of poly (vinyl alcohol) (PVA) content on preparation of novel thiol-functionalized mesoporous PVA/SiO$_2$ composite nanofiber membranes and their application for adsorption of heavy metal ions from aqueous solution. *Polymer (Guildf)* 51:6203–6211.
82. Auras, R. A. et al. 2004. An overview of polylactides as packaging materials. *Macromol. Biosci.* 4:835–864.
83. Hufenus, R. et al. 2012. Biodegradable bicomponent fibers from renewable sources: Meltspinning of poly (lactic acid) and poly [(3hydroxybutyrate)co(3hydroxyvalerate)]. *Macromol. Mater. Eng.* 297:75–84.
84. Auras, R. A. et al. 2005. Evaluation of oriented poly (lactide) polymers vs. existing PET and oriented PS for fresh food service containers. *Packag. Technol. Sci.* 18:207–216.
85. Zong, X. et al. 2002. Structure and process relationship of electrospun bioabsorbable nanofiber membranes. *Polymer* 43:4403–4412.
86. Li, L. et al. 2013. Development of eco-efficient micro-porous membranes via electrospinning and annealing of poly (lactic acid). *J. Membr. Sci.* 436:57–67.
87. Inai, R. et al. 2005. Structure and properties of electrospun PLLA single nanofibres. *Nanotechnology* 16:208–213.
88. Tsuji, H. and Ikada, Y. 1995. Properties and morphologies of poly (L-lactide): 1. Annealing condition effects on properties and morphologies of poly (L-lactide). *Polymer* 36:2709–2716.
89. Cooper, A. et al. 2013. Chitosan-based nanofibrous membranes for antibacterial filter applications. *Carbohydr. Polym.* 92:254–259.

90. Lakshman, K. and Shamala, T. R. 2006. Extraction of polyhydroxyalkanoate from Sinorhizobium meliloti cells using *Microbispora* sp. culture and its enzymes. *Enzyme Microb. Technol.* 39:1471–1475.

91. Valappil, S. P. et al. 2008. Polyhydroxyalkanoate biosynthesis in *Bacillus cereus* SPV under varied limiting conditions and an insight into the biosynthetic genes involved. *J. Appl. Microbiol.* 104:1624–1635.

92. Manna, A. and Paul, K. 2000. Degradation of microbial polyester poly(3-hydroxybutyrate) in environmental samples and in culture. *Biodegradation* 11:323–329.

93. Tokiwa, Y. and Ugwu, C. U. 2007. Biotechnological production of (R)-3-hydroxybutyric acid monomer. *J. Biotechnol.* 132:264–272.

94. Sridewi, N. et al. 2011. Simultaneous adsorption and photocatalytic degradation of malachite green using electrospun P(3HB)-TiO$_2$ nanocomposite fibers and films. *Int. J. Photoenergy* 2011:1–11.

10 Electrospinning for Air Pollution Control

Antonella Macagnano, Emiliano Zampetti,
Andrea Bearzotti, and Fabrizio De Cesare

CONTENTS

10.1 INTRODUCTION

The need of monitoring and cleaning the environment from chemical and biological contaminants is an ever-growing problem for underdeveloped countries as well as for the rest of the industrialized countries, where industry, anthropogenic, and agriculture activities (e.g., pesticides) have been causing significant dangers to environmental safety and human health.[1-3] There is a strict relationship between the climate change and the pollution. Processes such as fossil fuel burning in industry, motor vehicles, and buildings emit pollutants that cause local and regional pollution. These pollutants include particulate matter (PM) and ground-level ozone (O_3), nitrogen oxides (NO_x), sulfur oxides (SO_x), volatile organic compounds (VOCs), and carbon monoxide (CO). The same processes are responsible, too, of the release of greenhouse gases (GHGs), mainly carbon dioxide (CO_2), methane (CH_4), and nitrous oxide that are linked to global climate change. Since up to date high levels of pollutants were produced, their effects were found on a global environmental scale. Air pollution and climate change influence each other through complex interactions in the atmosphere. Increasing levels of GHGs modify the energy balance between the atmosphere and the earth's surface that, in turn, can cause temperature changes altering the chemical composition of the atmosphere.[4] Climate change is a progressively crucial problem with serious consequences for life on earth. Humans and wildlife are also exposed to chemical, physical, and biological stressors. One of the consequences of climate change that has recently attracted attention is its ability to change the environmental distribution and biological effects of chemical toxicants. For instance, the continuous emission of fine particles in air has provoked both acute and chronic effects on human health, ranging from minor upper respiratory to chronic respiratory inflammation, even lung cancer, bronchitis in adults and respiratory infections in children, aggravating pre-existing heart and lung diseases, or asthma spasms.[5–7] To protect the environment from the adverse effects of pollution, many countries worldwide enacted legislation to regulate various types of pollution as well as to mitigate the adverse effects of contamination. In 1998, a total of 141 countries ratified the Kyoto Protocol (*Kyoto Protocol to the United Nations Framework Convention on Climate Change: http://unfccc.inunt/kyoto_protocol/items/2830.php*), one of the most significant international treaty on global warming, committing themselves to observe a series of specific rules, such as decreasing GHG emissions up to defined concentration levels, using more advanced cleaning systems, and improving the monitoring systems. The environmental monitoring sounds as the prerequisite for the handling of the environmental pollution, strictly linked to the strategies of cleaning and remediation. In order to achieve this objective with affordable investment and running costs, there is a need to develop advanced sensors for air-quality monitoring and more in general for environmental monitoring capable to produce data that are comparable to those provided by adopting standard methods and technologies. Similarly, the conventional strategies used to clean the environmental compartments need to be improved in efficiency and convenience, using materials and technologies capable of inflicting the minimal or no harm at all upon ecosystems. Due to the huge advance of nanoscience and nanotechnology over the last two decades, great progress has been made not only in nanomaterials engineering and characterization but also in their functional applications. Because of their reduced size and large radii of curvature, the nanomaterials have a surface that is particularly reactive (mainly due to the high density of low-coordinated atoms at the surface, edges, and defects).[8] These unique properties can be applied to degrade and scavenge pollutants in air and water. Khin et al.[9] reported in an overview the mechanism and the efficiency of variously shaped nanomaterials (nanoparticles, tubes, wires, rods, fibers, etc.) in removing/detecting gas (SO_2, CO, NO_x, etc.), chemical (heavy metals, nitrate, arsenic, etc.), organic (aliphatic and aromatic hydrocarbons, VOCs, etc.), and biological (virus, bacteria, antibiotics, etc.) contaminants in the environment. Among nanostructured materials, nanofibrous structures offer unique properties, such as a high specific surface area (SSA) (depending on the diameter of fibers and intrafiber porosity), good interconnectivity of the pores, and the potential to combine active chemistry or functionality at a nanoscale.[10] The working principle of these nanomaterials involved in remediation may be used to design and fabricate nanosensors with improved detection of the pollutants. Functionalized and engineered nanofibers (NFs) have been investigated and used as smart materials for a plethora of advanced environmental applications. This chapter would be a short overview on any of the most performing nanofibrous materials obtained by electrospinning technology concerning both air filtration and gas/VOC detection. Can one technology solve the global environmental problems?

10.2   NANOFIBER-BASED AIR FILTRATION MEMBRANES

Air filtration technology is of great interest because of its low equipment cost, low energy consumption, high performance, and wide application range. Nowadays, fibrous membranes have been used in different filtration applications, comprising disposable respirators for citizens, industrial gas cleaning equipment, and air purifiers for cars, clean rooms, and indoor environments. Depending on the application, they can be made of a wide variety of materials, such as cellulose, glass, plastics, ceramics, metals, or composites. It is known that conventional fibrous structures suffer from many structural disadvantages, like large fiber diameter, nonuniform fiber diameter and pore size, low filtration efficiency, and poor high-temperature resistance.[11] On the other hand, it is well documented that the smaller the fiber diameter, the better the filtration performances.[12] Generally, a filter is made of packed fibers placed perpendicularly to the direction of the air flow. Efficiency and pressure drop are the two key parameters used to assess the performances of a filter. The efficiency is defined as the fraction of the entering particles (or mass concentration) that are collected by the filter:

$$E = \frac{N_0 - N}{N_0}$$

where N_0 and N are the particle number (or mass) concentrations at the filter inlet and outlet, respectively. Pressure drop is a parameter related to the resistance of individual fibers to gas flow, and it is proportional to the filter thickness and inversely proportional to the square of the fiber diameter. NFs are able to increase the filtration efficiency by providing an enhanced *slip effect* that should cause less resistance against airflow, consequently leading to a reduced pressure drop across the fibrous media. Template synthesis, phase separation, drawing, self-assembly, solvothermal synthesis, and electrospinning are just some techniques used to design nanofibrous filters.[13–16] Regarding electrospinning, the first patent for production of polymer fibers was issued in 1902 in the United States,[17] and practical results were obtained by A. Formhals in Germany (patented in the United States in 1934).[18] Electrospun nanofibrous media quickly found a successful use in filtration of air and liquids in the USSR (Petranov filters)[19] from the 1940s. Further advances in electrospinning process allowed organization of the industrial production of numerous types of fiber filter materials.[20] In the United States, the production of nanofibrous materials gained momentum in 1980 with the efforts of "Donaldson." In Europe, the commercial production of fibers by electroforming method started in the 1990s by "Freudenberg" (*http://www.krunk.ee/eng/2/21.shtml*). In 2007, Barhate and Ramakrishna[21] reported that more than 20 enterprises were keeping interest in either production or use of NF filter media. In the last decade, applications of nanofibrous filter media have been increasing rapidly. Further improvements in melt blowing and electrospun technology are allowing to facilitate mass-scale production of NFs. A more recent research on global NF market (2012–2016) reported an increased list of enterprises devoted to produce electrospun filtering systems (*http://www.efytimes.com/e1/fullnews.asp?edid=133596*). Key vendors dominating this market space, reported as follows, include *Ahlstrom Corp., Donaldson Co. Inc., E. I. du Pont Nemours and Co., Hollingsworth and Vose Co.,* and *Johns Manville.* Other vendors mentioned in the report are *Kuraray Co. Ltd., Mitsubishi Rayon Co. Ltd., Teijin Fibres Ltd., Toray Industries Inc., Argonide Corp., Biomers, C-Polymers GmbH, Carbon NT and F 21, Catalytic Materials LLC, Catalyx Nanotech Inc., Clearbridge NanoMedics, Electrovac AG, Elmarco Sro., Espin Technologies Inc., Esfil Tehno AS, Espin Technologies Inc., Fiberio Technology Corp., Finetex Technology, FutureCarbon GmbH, Grupo Antolin, Helsa-automotive, HemCon Medical Technologies Inc., Hills Inc., Irema-Filter GmbH, Japan Vilene Co. Ltd., Kertak Nanotechnologies S. R. O., Metpro Corp., Nanofiber Solutions, NanoMas Technologies Inc., Nanotechnics Co. Ltd., NanoTechLabs Inc., Nanoval GmbH and Co. KG, Pyrograf Products Inc., Revolution Fibres Ltd., Shenzhen Nanotech Port Co. Ltd., SNS Nanofiber Technology LLC,* and *US Global Nanospace Inc.*

The versatility of the process, the low equipment cost, the tunable porosity, and the high surface area per unit volume as well as the open-pore structure interconnections and the good permeability for gases are some of the features of electrospun nanofibrous membranes that have made them very appreciated in separation technology.[22–24] Additionally, they can be chemically designed and properly functionalized for a more selective filtration, not limiting their actions to the mere physical entrapment related to size and shape of the samples. Some examples of recent progresses in the development of electrospun nanofibrous membranes (e.g., organic, hybrid, inorganic) for both fine particle filtration and selected adsorption of gases/VOCs defined as potentially hazardous to environment and human health have been here reported.

10.2.1 Particle Matter Filtration

Inhalation exposure to airborne nanoscale particles is one big concern related to human health: epidemiological studies reported a strict relationship between the incidence of chronic obstructive pulmonary and cardiovascular diseases and workers exposed to microscale–nanoscale particles.[25,26] Anyhow, exposure to PMs can occur both indoor and outdoor. The size of the particle is a crucial parameter: larger particles are generally filtered in the nose and throat via cilia and mucus, but PM smaller than about 10 µm, referred to as PM10, can get stuck in the bronchi and lungs and cause health problems. Similarly, particles smaller than 2.5 µm, called PM2.5, tend to penetrate into the gas exchange regions of the lung (alveolus), and very small particles (<100 nm) may pass through the lungs to affect other organs. Penetration of particles is not wholly dependent on their size; shape and chemical composition also play a part. Additionally, atmospheric PM can both interfere with photosynthesis functions clogging stomatal openings and causing mortality in some plant species and affect the climate of the earth by altering the amount of incoming solar radiation. Some particulates occur naturally, originating from dust storms, volcanoes, forest and grassland fires, living vegetation, and sea spray. Human activities, such as the burning of fossil fuels in vehicles, power plants, and various industrial processes, also produce substantial amounts of particulates. Smoking, cooking, and other combustion sources are instead the predominant activities related to indoor contamination. Thus, successful attempts of 3D and functionalized electrospun membranes for air filtration have been proposed and fabricated in organic, inorganic, and composite materials according to several strategies of particles capturing and the final working temperature. Wang et al. recently reported an overview on electrospun NFs for air filtration.[27] The basic approach to capture particulate by a fibrous layer is reported in Figure 10.1.

Depending on the used polymer, a range of fabric properties, such as strength, weight, and porosity, can be achieved. Experimental results of Gibson et al.[28] showed that electrospun fiber mats were extremely efficient in trapping aerosol particles. Their high capturing efficiencies were due to the submicron-sized fibers, and the reduction of high pressure drop was due to the comparable values between the fiber diameter and the air molecules' mean free path flowing through the filter. Ahn et al.[29] prepared nylon 6 nanofilters by a solution of nylon 6 pellets ultrasonicated in formic acid. In order to get homogeneous and comparable fibrous layers, the concentration was increased up to 24 wt%. The calculated filtration efficiency was 99.993% with an average pore size of 0.24 μm (200 nm diameter size), better than a commercialized high-efficiency particulate air filter (99.97%, pore size 1.7 μm) at a face velocity of 5 cm/s using 300 nm test particles. Zhang et al.[30] investigated the contribution of some spinning parameters, such as polymer concentration, feed rate, and distance, to the filtration efficiency. According to Darcy's law,[31] since a volume Q flows in

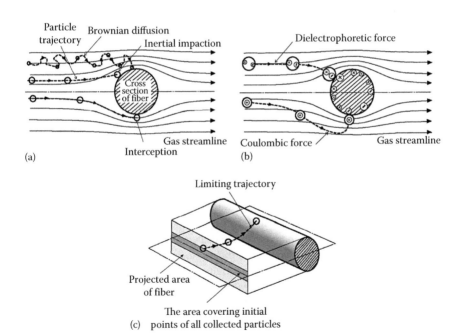

FIGURE 10.1 Sketch of several mechanisms occurring when particles approach a fiber: (a) particle collection by inertial impaction, interception, and convective Brownian diffusion; (b) particle collection by Coulombic and dielectrophoretic forces; and (c) efficiency of a single fiber. (Reprinted with permission from Wang, C. and Otani, Y., *Ind. Eng. Chem. Res.*, 52, 5–17, 2013. Copyright 2013 American Chemical Society.)

t seconds throughout a cross-sectional area A, the apparent linear rate of flow is $q = Q/At$; thus, the latter is linearly dependent on the pressure difference (Δp), the flow permeability coefficient (K_{fp}), and inversely on the thickness of the filter (L):

$$q = K_{fp}\frac{\Delta p}{L}$$

Flow permeability coefficient, the key parameter in filtering materials, can be measured by experimental or theoretical methods.[32] The authors experimentally observed lower normalized pressure drop at higher polymer (nylon 6) concentration and larger tip-to-distance and higher feed rate. Larger fibers (>100 nm) showed a tendency to decrease the normalized pressure drop. Additionally, these fibrous filters showed relatively better filtering efficiency (E) at very small particles (50 nm) than at larger ones, suggesting a possible application in producing membranes for capturing particles smaller than 50 nm. Barhate et al.[33] reported the effects of the applied electric field (drawing rate), the rotational speed of collector (collection rate), and the tip-collector distance, respectively, on particle transport properties, demonstrating that a control over the pore size distribution can be achieved by coordinating the drawing and the collection rates. Generally, morphological parameters such as fiber diameter, orientation, and thickness are estimated by electron microscopy and porosity by measuring weight and volume of a sample according to the following equation:

$$\varepsilon = \frac{V_0}{V_0 + V_s} = \left(1 - \frac{\rho_{mat}}{\rho_{fibers}}\right) = \left(1 - \frac{(w/(v_0 + v_s))}{\rho_{fibers}}\right) = \left(1 - \frac{w/A_z}{\rho_{fibers}}\right)$$

where

V_0 and V_s are the volumes of void and fibers, respectively

ρ_{mat} and ρ_{fibers} are the densities of the nanofibrous mat and fibers, respectively

w, A, and z are the weight, the area, and the thickness of the nanofibrous mat, respectively

Therefore, the authors reported that increasing the electric field strength caused a shifting trend in the fiber distribution toward a lower fiber size, reducing, too, the number of beaded fibers (Figure 10.2), probably due to the suppression of the axisymmetric instabilities. An increase in the drawing rate (by increasing the electric field strength) generated an increase in the number of fiber crossings, reduced the pore size, and improved the interconnectivity of pores. Consequently, the permeability, as well as the pore size, decreased as the electric field strength increased. Otherwise, charge accumulation on fibers also due to high electric field values can enhance repulsion between the fibers and change the trend in the arrangement as evinced in Table 10.1. The effects of the rotational speed of the collector were also reported: packing density and fibers' alignment significantly increased when the collection rate increased (ranged between 0.9 and 4.5 m/s) determining a decrease in the porosity, due to the better alignment of fibers. Thus, the authors stated that it is possible to define a collection speed (in their work, it was equal to 2.7 m/s) to get the best arrangement of the fibers. In fact, below that speed rate, NFs were randomly

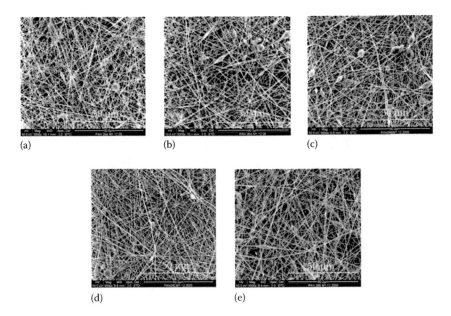

FIGURE 10.2 Morphological changes in an electrospun polyacrylonitrile mat due to increasing electric field strength. (a) $E = 93,000$ V/m (M), (b) $E = 111,000$ V/m (M), (c) $E = 148,000$ V/m (M), (d) $E = 167,000$ V/m (M), and (e) $E = 130,000$ V/m (M). (Reprinted from *J. Membr. Sci.*, 283, Barhate, R.S., Loong, C.K., and Ramakrishna, S., 209–218. Copyright 2006, with permission from Elsevier.)

collected providing a reduced control of the orientation and then the porosity. Sambaer et al.[34] suggested a novel approach to create a 3D membrane model based on actual performances of polyurethane (PU) filtering membranes.

Two PU membranes, having a difference only in the average pore size, were combined to form multilayers with different thicknesses. Using scanning electron microscopy (SEM) image-based 3D structure model (Figure 10.3), transition/free molecular flow regime, Brownian diffusion, particle–fiber interactions, and aerodynamic slip, the model was able to predict the measured filtration efficiency. The authors investigated the four different forces mainly involved in capturing particles, the drag force (F_D), the lift force (F_L), the adhesion force (F_A), and the friction force (F_F) (Figure 10.4a), and concluded that the particle–fiber friction coefficient had the highest effect on the filtration efficiency.

Considering the proposed slip-path modeling by using the local coordinate system-based force balance and geometrically complex NF shapes, fully 3D character of the particle crossing of the filter was modeled (Figure 10.4b and c).

The possibility of combining a variety of polymers, particulate inorganic nanofillers, and biological agents through electrospinning leads to the development of nanocomposite/hybrid membranes with a better filtration efficiency and a much broader domain of environmental applications than their neat counterparts. Several strategies can be adopted to combine nanostructures, such as nanoclusters, nanoparticles, nanotubes, and nanosheets with the electrospun NFs. Commonly,

TABLE 10.1

Effects of Electrospinning Process Parameters on Structure and Property of Filtering Membranes

Electric Held Strength (V/m)	Darcy Permeability (m²)			Maximum Pore Diameter (μm)			Minimum Detectable Pore Diameter (μm)			Average-Flow Pore Diameter (μm)		
	L	M	R	L	M	R	L	M	R	L	M	R
Effect of Applied Electric Field on Permeability and Pore Size												
93,000	5.00×10^{-13}	5.00×10^{-13}	5.00×10^{-13}	291.20	58.24	5.94	2.17	2.04	2.24	4.12	4.13	4.12
111,000	5.00×10^{-13}	2.50×10^{-13}	5.00×10^{-13}	58.24	24.27	11.65	1.90	1.76	1.97	4.15	3.00	4.13
130,000	1.43×10^{-13}	2.00×10^{-13}	2.50×10^{-13}	18.20	11.65	3.51	1.53	1.57	1.89	2.33	2.70	2.96
148,000	2.50×10^{-13}	2.00×10^{-13}	2.00×10^{-13}	2.09	3.94	58.24	1.40	1.37	1.57	2.96	2.67	2.68
167,000	3.33×10^{-13}	5.00×10^{-13}	5.00×10^{-13}	4.48	6.62	13.24	2.21	2.14	2.31	3.47	4.16	4.13

Speed of Collector (m/s)	Fiber Diameter (μm)			Deposition (Thickness) (μm)			Porosity (ε) (%)		
	L	M	R	L	M	R	L	M	R
Effect of Rotational Speed of Collector on the Fiber Diameter, Deposition, and Porosity									
0.9	0.2 (0.045)	0.2 (0.043)	0.2 (0.098)	147 (11.82)	138 (6.89)	131 (13.46)	94.37	95.03	97.66
1.8	0.2 (0.049)	0.2 (0.036)	0.2 (0.042)	118 (9.41)	236 (9.51)	69 (2.25)	96.41	95.19	91.04
2.7	0.2 (0.062)	0.2 (0.037)	0.2 (0.036)	166 (6.91)	120 (8.30)	110 (5.91)	96.42	94.48	96.06
3.6	0.3 (0.022)	0.3 (0.046)	0.3 (0.037)	211 (8.12)	149 (8.15)	148 (6.57)	96.89	95.79	97.86
4.5	0.2 (0.035)	0.2 (0.084)	0.2 (0.064)	38 (2.44)	155 (11.43)	128 (20.34)	88.73	94.82	92.62

Source: Reprinted from *J. Membr. Sci.*, 283, Barhate, R.S., Loong, C.K., and Ramakrishna, S., 209–218. Copyright 2006, with permission from Elsevier.

Values in parenthesis are S.D.; L, M, and R: the left, middle, and right portion of membrane, respectively.

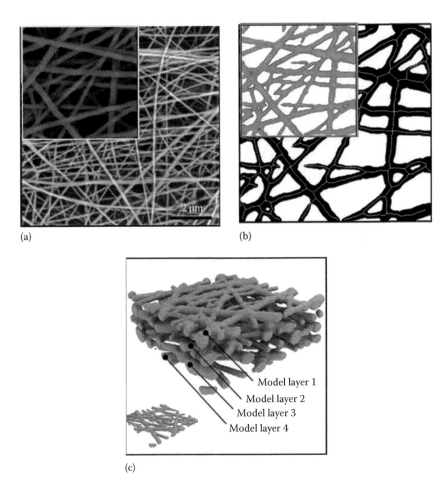

FIGURE 10.3 Three-dimensional structure model creation: (a) SEM top view of a polyure-thane fibrous layer with the zoomed section used to provide the model (inset); (b) application of the average threshold level to extract the top fibers resulting in a black and white image, with nanofiber center line determination (inset); and (c) a full 3D model of a nanofibrous layer with inset a perspective view of one layer. (Reprinted from *Chem. Eng. Sci.*, 82, Sambaer, W., Zatloukal, M., and Kimmer, D., 299–311. Copyright 2012, with permission from Elsevier.)

nanocomposite synthesis can be done through the blending of insoluble nanopar-ticles with the polymer solution to be processed, thereby encapsulating nanopar-ticles, like soluble drugs, bacterial agents, and metal-oxide sol–gel solutions, in the solidified NFs. The combination of an inorganic–organic composite can benefit the light weight, flexibility, and moldability of the polymers together with high strength, chemical resistance, and thermal stability of inorganic additives. Thus, since electrostatic attraction is a mechanism of capturing of submicron par-ticles, Yeom et al.[35] fabricated a nanocomposite filter made of nylon 6 (73 nm diam-eter), boehmite nanoparticles (3%), and electrostatic charging agents. As a result,

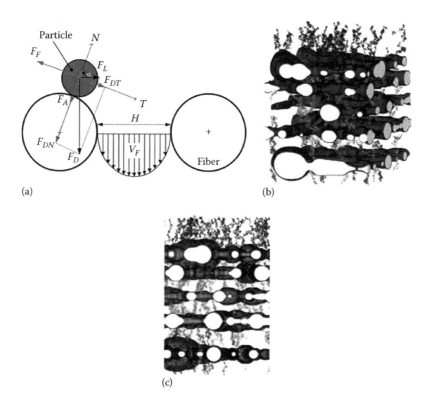

(a)

(b)

(c)

FIGURE 10.4   (a) Sketch of the system-based force balance used for particle–fiber interaction modeling and flow path lines through a part of the filter for 100 nm diameter particles, (b) perspective view, and (c) side view. (Reprinted from *Chem. Eng. Sci.*, 82, Sambaer, W., Zatloukal, M., and Kimmer, D., 299–311. Copyright 2012, with permission from Elsevier.)

a high capturing efficiency (96.8%) of submicron aerosol was obtained, attributed to the increase of surface potential of the NFs. Similarly, Cho et al.[36] demonstrated that, despite the less fiber coverage density on the cellulose substrate, the filtration efficiency of the polyacrylonitrile (PAN) filter media with TiO_2 nanoparticles added was much greater than that of those made with just pure PAN fibers, even with reduced pressure drop. Such enhancement was supposedly due to the added electric charge for the interaction between TiO_2 nanoparticles (7.5 vol%) in nanofibrous filter media and the simulated dust particles. According to the pore sizes and the pore size distribution, measured by capillary flow porometry, the hybrid nanofibrous filter media showed larger pore sizes and broader pore size distribution than the pure PAN nanofibrous filter media. The pressure drop of the cellulose filter media covered with the PAN/TiO_2 hybrid NFs (0.5 g/ft^2) was four times less than the counterpart filter media. Additionally, the pure PAN nanofibrous filter media with the same fiber mass showed higher fiber coverage density on the substrate than the PAN/TiO_2 hybrid counterpart filter media. However, the filtration

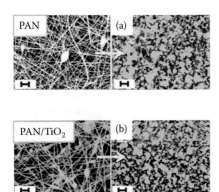
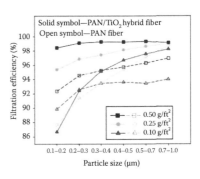

FIGURE 10.5 SEM micrographs of polyacrylonitrile (0.5 g/ft²) and composite polyacrylonitrile/TiO₂ filters before (on the left) and after aerosol filtration (a and b); (on the right) filtration efficiency (%) of the various fibrous filters with different mass as a function of particle size. Scale bars are 3 μm. (Reprinted from *Polymer*, 54(9), Cho, D., Naydich, A., Frey, M.W., and Joo, Y.L., 2364–2372. Copyright 2013, with permission from Elsevier.)

efficiency of the PAN filter media with TiO₂ nanoparticles added was found to be greater (4%–6%) than those made with just pure PAN fibers, especially for smaller particles in the range 0.1–0.5 μm (Figure 10.5).

When environmental conditions for filtration are extreme, like close to waste incinerators, cement plants, and coal-fired boilers, ceramic fibrous membranes with excellent resistance against corrosion, chemical erosion, and thermal stability are required. These membranes, joining the thermal stability to nanometer fiber diameter, can be used (without expensive pretreatments) for hot waste gases. However, despite efforts devoted to fabricate ceramic NFs, their brittleness significantly limits their practical applications. Recently, several strategies[37–39] to overcome that issue have been proposed. For instance, Mao et al.[40] proposed a new class of flexible and thermally stable silica nanofibrous membranes for high-performance filtration media by the combination of electrospinning and sol–gel processes. Starting from precursor solutions based on various concentrations of polyvinyl alcohol (PVA) aqueous solutions (ranging between 8 and 14 wt%) and silica gel (tetraethyl orthosilicate [TEOS], phosphoric acid, and water), the authors successfully developed robust and flexible membranes with excellent thermal stability (up to 1000°C). The authors reported that morphology and crystalline phase, related to PVA concentration and calcination temperature value, played an important role in the mechanical properties, including flexibility and tensile strength, as well as in filtration performances toward NaCl aerosols of nanometer range. Specifically, membranes with 12 wt% PVA, annealed at 800°C, with 303 nm average fiber diameter (Figure 10.6), showed the best compromise between mechanical properties and the filtering quality factor $QF = 0.089$ (calculated as the ratio between the filtration efficiency measure and the pressure drop as $QF = -\ln(1-E)/\Delta P$, where E and ΔP are the efficiency of the filter and the pressure drop across the filter, respectively).

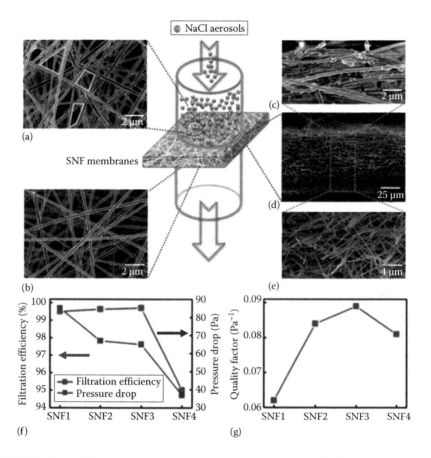

FIGURE 10.6 Field-emission scanning electron microscope (FE-SEM) micrographs of SNF3 membranes after the filtration process: (a) top surface; (b) bottom surface; and (c through e) cross section with different magnifications, respectively; filtration efficiency and pressure drop and (f) QF values (g) of silica nanofibrous filters (SNF) with increasing amount of polyvinyl alcohol and named as SNF1, SNF2, SNF3, and SNF4 membranes. (Mao, X., Si, Y., Chen, Y. et al., *RSC Adv.*, 2(32), 12216, 2012. Reproduced by permission of The Royal Society of Chemistry.)

10.2.2 VOC and Gas Removal

In addition to PMs, numerous VOCs and gases can be dangerous to human health or cause harm to the environment. Significant quantities of gases and VOCs[41] are produced and emitted into the environment each year, both indoor and outdoor, by different sources, natural (forests, oceans, wetlands, volcanoes) and anthropogenic, including the emissions from industrial plants, vehicles, aircraft, petroleum handling, chemicals, painting operations, etc. With regard to VOCs, a great concern arises for their ability of long-range transport, distribution, and accumulation in several environmental compartments. Each chemical has its own toxicity, depending on the reactivity (chemical structure) and lifetime. Furthermore, VOCs are mainly responsible for indoor air pollutants, able to cause significant effects on health, including mucous membrane irritation, headache, and fatigue (symptoms of a pathology called sick building syndrome). Long-term exposure to VOCs can damage the liver, kidney, and central nervous system and cause cancer.[42] Thus, VOC removal from a huge diversity of sources is required to decrease the negative health risks in both outdoor and indoor environments. Adsorption is one effective way to decrease VOC level. Pinto et al.[43] reported that membranes of PUs adsorbed VOCs with performances comparable to activated carbon. Scholten et al.[44] demonstrated that suitably selected PUs can be designed to prepare robust fiber and elastic mats with high sorption capacities, especially for toluene, since it is chemically similar to the polymer building block (4,4-methylenebis(phenyl isocyanate)–based PU [named MDI based]), and chlorinate compounds, since it is polar in nature. The absorption plots for chloroform and hexane were linear (described by Henry's law, $C = K_H p$, where C is the capacity of the fibers at partial gas pressure p and K_H is the Henry's Law constant), showing instead for toluene a pronounced deviation at higher concentrations, probably due to both van der Waals dispersion forces and π–π interactions between toluene and MDI-based PU (Figure 10.7). Changing some segments of the building blocks of the polymers, the selectivity of VOC can be changed, too.

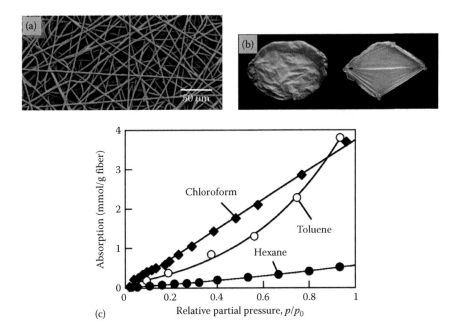

FIGURE 10.7 (a) SEM micrograph of a 4,4-methylenebis(phenyl isocyanate)–based non-woven fibrous mat; (b) normal and stretched fiber mats, respectively, held in place with push pins; and (c) absorption curves of the fibrous membrane for toluene, chloroform, and hexane at 25°C. (Reprinted with permission from Scholten, E., Bromberg, L., Rutledge, G.C., and Hatton, T.A., *ACS Appl. Mater. Interfaces*, 3, 3902–3909. Copyright 2011 American Chemical Society.)

Furthermore, such fibrous membranes could be easily regenerated by desorption under ambient conditions (under nitrogen flow), in contrast with the harsh thermal treatments (extensive heating and washing procedures) needed to remove the VOCs chemisorbed on the exposed surfaces of activated carbon adsorbents.[45] The incorporation of commercially available fly ash (FA) (SiO_2–Al_2O_3–$CaCO_3$), commonly known as waste material as well as adsorbent of different toxics, within PU fibrous mats, confirmed the improvement of both their adsorption and mechanical properties.[46] For instance, the physicochemical features of the FAs, such as bulk density, porosity, and water-holding capacity, make them interesting for absorbing mechanisms. The absorbed VOC amount on electrospun composites hugely increased when FAs were blended with the polymer up to 30 wt%. This result was attributed to the related decrease in the composite fiber diameter (with a consequent increase in surface area) and the presence of inorganic particles on fibers' surfaces. The VOC sorption kinetics of different aromatic hydrocarbons at room temperature (styrene > xylene > toluene > benzene) was strictly related to their electronic and steric structures forming π-complexes on the surface of the composite fibers. Styrene is the most highly absorbed VOC (Figure 10.8).

Moreover, VOC desorption was easily achieved by placing the membrane under the suction of a rotary pump at room temperature after each VOC sorption cycle, demonstrating potential of regeneration of the same membrane for more applications. VOC sorption process has been also investigated to develop activated carbon nanofibers (CNFs). The most common polymer for the preparation of CNFs is PAN, mainly due to its high melting point and carbon yield after thermal treatments.[47] Furthermore, the surface of PAN-based CNFs may be modified using a relatively simple activation process. Other precursors also used to fabricate CNFs are PVA, polyimide, cellulose, and pitch. Bai et al. described the adsorption of some VOCs on CNFs generated by PAN.[48] NFs were obtained by dissolution of PAN in dimethylformamide (DMF) (range 10–15 wt%), electrospinning deposition, gradual stabilization at 280°C (generally between 200°C and 300°C, 5°C/min, useful to prevent the decay of fibrous shape during pyrolysis), and carbonization at 800°C in an inert atmosphere (nitrogen). The activation was carried out by adding 30 vol% steam at the same temperature (50 min). When NFs were activated by different conditions, the microporosity increased with increasing burn off, thus resulting in a general higher adsorption capacity for both benzene and ethanol. NFs with higher oxygen content were reported to have a higher adsorption capacity for ethanol. Composite activated carbon nanofibrous mats with TiO_2 particles enclosed (powder together PAN solution in DMF) were investigated by Gholamvand et al.[49] as a practical indoor air purification system, where the combination of titania photocatalytic features, together with the adsorbing capacity of the CNFs, provided a very promising material for VOC degradation and sorption. In addition to the adsorption of a series of VOCs, comprising formaldehyde, a typical indoor pollutant, activated CNFs,[50] looked promising candidate to adsorb and catalytically oxidize NO into NO_2 at room temperature, when subjected to a graphitization process at higher temperature (1900°C–2400°C). Wang et al.[51] supposed that the enhancement of the catalytic activity of NO oxidation by graphitization depended on the increased

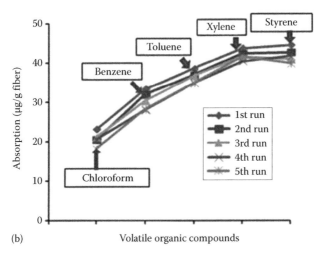

FIGURE 10.8 (a) FE-SEM image of a polyurethane fibrous mat containing 30 wt% fly ashes (FAs), with transmission electron microscopy (TEM) micrograph of an FA aggregated on a fiber (inset); (b) cyclic sorption of different volatile organic compounds on the composite membrane. (Reprinted from *Chem. Eng. J.*, 230, Kim, H.J., Pant, H.R., Choi, N.J., and Kim, C.S., 244–250, Copyright 2013, with permission from Elsevier.)

number of active sites, generated by the liberation of groups O–H and N–H and the increase of topological defects. In another study,[52] the synergic cooperation between ceria (CeO_2) and CuNPs, enclosed in PAN fibers and dispersed on activated CNFs, gave rise to NO oxidation at room temperature and atmospheric pressure increasing the filtering performances of the electrospun mat. Porous carbon fibers, named also PCF, can be applied to adsorb SO_2.[53,54] Song et al.[55] described that the ultraporous carbon fibers (UPCF) obtained by electrospinning technology were highly sulfur dioxide adsorbing. The conventional PAN fibers were electrospun from a solution with 10% concentration, degassed, and carbonized at 800°C in a flow of 30 vol% steam/CO_2 (N_2 carrier). Adsorption capacities of UPCF for SO_2 could be improved by dipping the UPCF membrane in a nitrogenated compound solution and drying at 100°C under vacuum. Adsorption capacity, in fact, is supposed to be determined not only by the filter textural structure but also by its surface chemistry: the nitrogen-containing functional groups over UPCF membrane showed alkaline behavior with a strong affinity to acidic SO_2 gas, which could be oxidized into SO_3 or into H_2SO_4, when in dry or humid environment, respectively. Also, carbon dioxide (CO_2) emission into the atmosphere has attracted great attention due to its relationships with the global climate change phenomenon. In fact, CO_2 concentration has increased from 310 to 395 ppm (*http://www.esrl.noaa.gov/gmd/ccgg/trends/*) over the last half century, mainly due to anthropogenic sources; thus, several strategies devoted to capture and store that gas have been developed.[56] Currently, zeolitic imidazolate frameworks (ZIFs[57]) are known as the best porous materials for the selective capture of CO_2; thus, Ostermann et al.[58] synthesized an electrospun composite membrane based on a mixture of polyvinylpyrrolidone (PVP, final concentration of 3.5 wt%) in methanol with several amounts of a dispersion of ZIF-8 (50 nm in size).[59] The resulting textile (Figure 10.9) revealed a homogeneous distribution of the nanoparticles inside the fibers, a smooth polymeric surface (150–300 nm diameter), and good adsorbing properties, even if the polymer layer constituted an additional diffusion barrier. In fact, changing polymers (polystyrene [PS] and polyethylene oxide [PEO]), as well as diameter of fibers, changed the adsorption kinetics, too. To overcome this issue, Wu et al.[60] proposed to use suitable electrospun polymer scaffolds in order to produce a new class of adsorbent materials based on hierarchically nanostructured metal-organic frameworks (MOFs) (comprising ZIF-8). They enclosed MOF–crystal seeds as nucleation sites within polymer fibers and, by a sequential deposition procedure, carried out a controlled crystal growth over the fibers, obtaining a 3D network of MOF–crystal assemblies extremely appealing for gas adsorption ad desorption. The existence of macroporous vacancies in the composite polymer fibers facilitated the gas transfer and therefore accelerated the preferred CO_2 adsorption and desorption rate in the whole process.

Further results in designing CO_2-adsorbing membranes have been described by Yoo et al.,[62] by electrospinning Nafion/PEO fibrous layers capable to entrap a low-melting salt (1-hexyl-3-methylimidazolium tetrafluoroborate), a chemical compound well known in literature[63] for its capacity of capturing carbon dioxide. More specifically, the gas permeance electrospun fibrous mat maintained the characteristics of the membranes containing the liquid salt, while the selectivity to CO_2 increased, implying that the high adsorption of the salt into Nafion determined a definite contribution in CO_2 separation. Metal-oxide fibrous layers have been also described,[64] where the proposed mechanism was supposed to be due to the numerous structural defects and oxygen vacancies of the fibers, capable of trapping and then binding carbon dioxide. Supplementary strategies based on electrospun membranes for capturing carbon dioxide have been collected and described in a recent review of Wang and Li.[65]

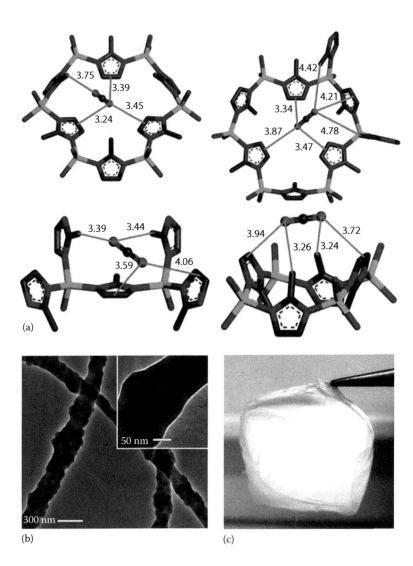

FIGURE 10.9 (a) Schematic visualization of CO_2 interactions with four-, five-, and six-membered rings in ZIF-8, according to density-functional theory calculations (Fisher, M. and Bell, R.G., *Cryst. Eng. Commun.*, 16, 1934–1949, 2014. Reproduced by permission of The Royal Society of Chemistry), (b) TEM micrographs (with inset of zoomed-out particles) of ZIF-8 nanoparticles in polyvinylpyrrolidone, and (c) a digital photograph of the 1.5 cm diameter fibrous mat. (Ostermann, R., Cravillon, J., Weidmann, C., Wiebke, M., and Smarsly, B., *Chem. Commun.*, 47, 442–444, 2011. Reproduced by permission of The Royal Society of Chemistry.)

10.3 NANOFIBROUS MATERIALS FOR GAS SENSORS

The most recent efforts to develop advanced sensors and smart sensing systems for air-quality control (AQC) cover several aspects joined to the general need to develop global-scale monitoring systems in order to track the evolution of environmental quality with changing environmental pressures. The progress made in developing nanotechnologies, based on nanostructured materials and advances in microelectronics, allows to design and develop sensors that have high performance in terms of sensitivity, detection limits, response time, and selectivity for different molecules. The most recent efforts to develop advanced sensors and smart sensing systems for AQC cover several aspects including the designing and developing of new sensor materials; the manufacturing of new gas sensors, new transducers, portable gas sensor systems, wireless technology, and sensors networking supported by intelligent algorithms and distributed computing; and the definition of new protocols, standards, and methods comprising conventional analyzers together with sensor technologies. Therefore, the innovative approaches to design sensor materials take in consideration many factors such as the surface area, interfacial characteristics, nanocrystallite size, and fast, simple, and inexpensive deposition processes. Today, a plethora of functional nanostructured materials includes a very large set of structures, textures, and chemical compositions. In addition to their high flexibility with regard to shaping, chemical, and physical properties, nanostructured materials provide an evident advantage of enabling both integration and miniaturization. At present, conductometric chemical sensors based on nanocrystalline metal-oxide semiconductors are the most promising among solid-state gas detectors due to their reliability and ease of manufacture. Their sensitivity is related to change in conductivity resulting from chemical interaction of adsorbed target gas with active centers on the surface of the sensor. The main disadvantage of such sensors due to the low selectivity to gases having similar chemical characteristics and the relatively high operating temperature can be overcome by using composite materials in which nanoparticles of metal-oxide semiconductor (TiO_2, SnO_2, VO_2) or metal (Pt, Pd, Au, Ag) are embedded.[66] They play both passive and active roles in sensing, increasing the active surface area, improving gas diffusion inside the layer, changing the conductivity, optical absorption of the hosting metal oxide, and showing catalytic properties. The doping of metal-oxide structures with other metal oxides offers better performance than single oxides for sensitive and selective detection of various gases in the environment.[67,68] Conducting polymers are able to work at low temperature in comparison to most of the metal-oxide sensors and can behave like semiconductors due to their heterocyclic compounds. Polymer nanocomposites combine the advantages of both organic polymer material and inorganic metal or semiconductor oxide and also show significant improvements in optical and mechanical properties for gas sensor applications. For instance, by mixing metal oxides with polyaniline to form nanocomposites, performances like high sensitivity, fast responses, and recovery rates of NH_3 have been reported.[69] Recently, electrospun technology allowed the fabrication of very attractive nanofibrous layers for gas and VOC detection, as well as for biosensing. Thus, a sensor based on titania fibers coated with a very thin film of poly(3,4-ethylendioxythiophene)/polystyrene sulfonate (PEDOT/PSS) was reported to have a detection limit of about 1 ppb to NO_2, compared to the performances of some semiconductor nanostructured sensors, working at room temperature and at very low power consumption.[70] Further efforts have been addressed to design, fabricate, test, and characterize new cost-effective sensor systems at a level of proof of concept for enhanced air-pollutant detection up to trace levels by means of laboratory experiments and field campaigns. Today, several European networks (*http://www.cost.eu*), involving interdisciplinary platforms (COST Actions TD1105, MP0901, MP1206, ES1004, ES1002, ES0602), are devoted to explore and improve several aspects of the advanced monitoring systems, keeping in mind protocols and standardization methods (*European Directive 2008/50/ EC: Ambient Air Quality*). Therefore, advanced transducers (chemoresistors, field-effect transistors, electrochemicals, surface acoustic wave, optical fibers, quartz crystal microbalance [QCM], cantilevers, hybrid transducers, etc.) have been investigated for high-performance environmental sensors with new functionalities of advanced electronic interfaces and wireless communications at low power consumption.[71] Furthermore, many portable sensor systems are currently designed and investigated not only as potential personal sensors and/or wearable sensors in the everyday life of an individual but also as air-quality monitoring for indoor applications (green buildings, low CO_2 emissions, air-ventilation systems, indoor energy efficiency, etc.).[72] Smart devices with pattern recognition algorithms and artificial neural networks are today studied, and microsensors at low cost and low power consumption are expected to be joined in a distributed wireless sensor network with distributed computing in multiple nodes for measuring concentration of air pollutants in the real scenario.[73] Fully autonomous systems for gas sensing, from both technological point of view and budgetary reasons, are strongly desired. The present paragraph is not intended to be exhaustive, rather a brief overview on the key strategies applied to electrospinning to develop extremely smart materials for sensing applications. Therefore, some of most recent applications for environmental monitoring of electrospun nanofibrous chemical sensors, classified according to the detecting mechanisms, will be introduced. In 2010, a review of Ding et al.[74] collected for the first time the most significant results about electrospun nanomaterials for ultrasensitive sensors, updated in 2014 as a book chapter.[75] Furthermore, a book entitled *Electrospinning for High Performance Sensors* (Macagnano et al., 2015) focused on the most recent and encouraging results (presented to the International Workshop within COST Action MP1206 in Rome 2014) concerning the advanced nanostructured (bio)sensors, ranging from high sensitivity to extreme operating conditions and covering a wide range of requirements, investigating and debating the different aspects of electrospinning to design and fabricate high-performing materials for sensors applied in monitoring gaseous and liquid environments, and especially emphasizing the electrospinning sensor market perspective.[76]

The convenience of sensors fabricated employing electrospinning technology lies in the combination of the extreme versatility of the technology with the benefits from the high surface-area-to-volume ratio, compatibility with semiconductor processes and adaptability to numerous surfaces, practically unlimited choice of molecules (electronic structure, configuration, size, shape, etc.), and finally cost-effectiveness of manufacturing. When proper setup parameters are selected, a broad range of complex architecture ranging from porous, core shell, and multicore, to tubular or cable-like structures can be designed and developed. Thus, simple tunings of the electrospinning process, as well as posttreatment operations, are able to modulate the properties of the resulting sensors. Porosity of both fibers and layers deposited onto transducers can be designed on purpose, and layer thicknesses can be modulated. Novel sensing materials can be achieved by mixing organic polymers in blends or organic polymers with nanoparticles (e.g., metals, graphene, metal oxide, and carbon nanotubes) during electrospinning to create nanocomposites. In fact, these processes can enlarge interacting capacity of NFs with analytes or can provide nano(bio)sensors with greater resistance to sustain extreme environmental conditions. Finally, in situ processing, such as chemical and biochemical functionalizations and plasma treatments, can supply greater specificity and sensitivity to nanosensors.

10.3.1 MASS SENSORS: QCM-BASED DEVICES

Among the most investigated sensing systems,[77] the QCM-based ones have attracted great attention for their inexpensiveness, compact volume, easy portability, and capability to respond to mass changes on a nanogram scale.[78] It comprises a thin slice of piezoelectric quartz placed between two metal electrodes that, applying an alternating electric field across the crystal, are able to induce oscillations related to its resonant frequency (electromechanical coupling). The resonance frequency is a function of the gravitating mass: for small amount of mass, a linear relationship between variations of both mass and frequency exists (Sauerbrey law: $\Delta f = -f_0^2 \Delta m / SN\rho$, where f_0 is the fundamental frequency, Δm is the gravitating mass on the surface, S is the area of the electrode, and ρ and N are density and constant of the quartz plate, respectively). As mass sensor, QCM can detect the guest molecules binding the host molecules (the sensing layer) immobilized on the electrode by decreasing its frequency linearly to the adsorbed mass. The change in QCM frequency determines the mass of analyte adsorbed in ng/cm². The sensing ability of the QCM toward an environmental toxic chemical is indubitably influenced by the characteristics of the coating material that works as an adsorbing surface. Up to now, a variety of materials such as polymers, macromolecules, ceramics, carbon nanotubes, and nanocomposites have been investigated both as thin and compact films as well as arranged within well-defined nanostructures having greater adsorbing surface (Figure 10.10). Ding et al.[79] described the sensing features of QCMs covered with electrospun nanofibrous blend of polymers with promising sensitivity to ammonia (NH₃), which is the main component of fertilizers, pharmaceuticals, surfactants, and coolants widely used in industry and agriculture and strongly irritating the human skin and mucosa as well as causing protein degeneration. The authors reported that diameter size and rigidity of NFs increased when polyacrylic acid (PAA) content is increased. Viscosity and conductivity of the homogeneous blend solution of PAA and PVA increased, too. They observed that the average resonance frequency changes were gradually increased by increasing the content of PAA mixed to PVA, due to the increase in the amount of ammonia adsorption sites (the carboxyl groups), thus tuning the sensitivity, as

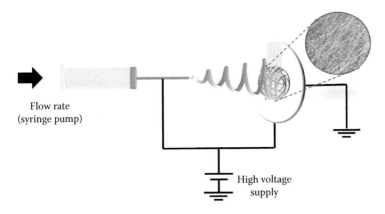

Flow rate
(syringe pump)

High voltage
supply

FIGURE 10.10 Schematic illustration of an electrospun deposition on a quartz crystal microbalance that can be either motionless or rotated at various speeds to achieve different fiber orientations.

well as selectivity, to the sampled gas. Although the resulting measures are strongly affected by relative humidity, at appropriated humidity values, the sensors detected up to hundreds of ppb of ammonia. More recently, Jia et al.[80] described a one-step electrospun process based on a suitable combination of the solvents (acetone/DMF) for diluting poly(styrene-block-maleic acid). The authors reported that the sensitivity of the sensor to ammonia (25°C and 40%–50% RH) was dependent linearly on the SSA of the electrospun membrane, with higher results when acetone/DMF ratio was 1:3 (wt/wt) and NFs had a diameter of about 364 nm. The sensing process, supposed to be ascribed to the reversible interaction between the ammonia molecules (injected into the measuring chamber at different concentrations) and the functional groups (–COOH) of the NFs, exhibited excellent reversibility and a limit of detection of 1.5 ppm by a 5 MHz QCM. Thereafter, Ding's group prepared successfully QCM ultrasensitive sensors based on the combination of electrospun polymers functionalized with a simple solution dropwise or nanofibrous netting (NFN). These QCM sensors were applied to detect noxious gases, such as formaldehyde and HCl.[81,82]

Specifically, dispersions of titania fibers into ethylene glycol (EG) were dropped onto QCM surfaces by a micropipettor and heated under vacuum at 120°C until complete evaporation of EG (1300–400 Hz of coating loads). Therefore, polyethylenimine (PEI) solution drops, diluted into 1 wt% water/ethanol (1:2 wt%) and coated the fibers achieving an increase in frequency shifts. The authors confirmed that the highly porous layer of PEI–TiO_2 supplied high-diffusivity paths for formaldehyde molecules, enhancing the diffusion through the sensing material and providing faster response time. Additionally, the rigidity of the glassy fibers of titania facilitated the oscillation transmission, resulting in much faster responses. The decrease in QCM frequency when formaldehyde was injected into the sensing chamber was presumably due to the reversible nucleophilic addition reaction between primary amine groups of PEI and formaldehyde molecules. The large Branauer-Emmett-Teller (BET) surface area (68.72 m^2/g) and porosity of the fibers enhanced their sensitivity to formaldehyde due to the internal surface area accessible in them. The low limit of detection (1 ppm), the selectivity over various vapors, and the simplicity of preparation suggested attractive applications for real-time monitoring gas sensors (Figure 10.11).

When the polymer scaffold comprises the unique structure made of the combination of NFN (nanofibrous nets) with nonwoven fibers, the QCM sensors can achieve excellent performances. Thus, Ding's group employed polyamide-6 NFN membranes to be coated by PEI solution drops. Benefiting from the extremely large SSA, high porosity, and strongly tight adhesive force to the device of NFN, the as-prepared QCM sensors were reported to have a remarkably low detection limit of 50 ppb to formaldehyde vapors.[83]

10.3.2 CONDUCTOMETRIC SENSORS

Certain kinds of conducting materials cause a change in their electrical resistance in response to an interaction with gases and vapors. These measurements are easily

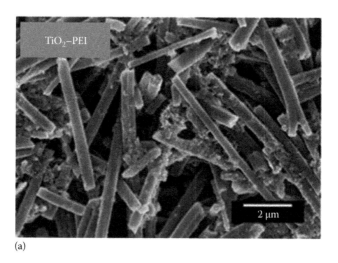

(a)

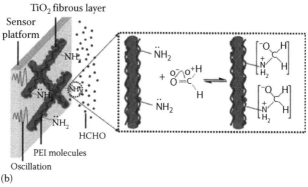

(b)

FIGURE 10.11 (a) FE-SEM micrograph of PEI–TiO$_2$ composite fibers, (b) a scheme of formaldehyde vapor–detecting mechanism based on the designed fibers. *(Continued)*

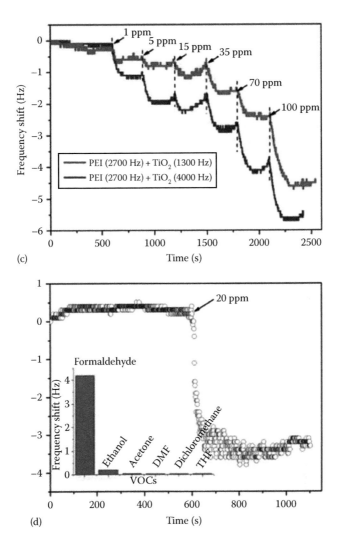

FIGURE 10.11 (Continued) (c) frequency shifts of two differently titania-loaded quartz crystal microbalances (QCMs) to increase concentration of formaldehyde, and (d) a bar plot with the comparison of the QCM responses to 20 ppm of several chemicals. (Wang, X.F., Wang, J.L., Si, Y., Ding, B., Yu, J.Y., Sun, G., Luo, W.J., and Zheng, G., *Nanoscale*, 4, 7585–7592, 2012. Reproduced by permission of The Royal Society of Chemistry.)

performed by connecting two adjacent electrodes (often microelectrodes) with the interacting layer and then measuring its conductivity (or resistivity) as a function of the gas/VOC concentration.[84] Among these materials, the most used for gas sensing are semiconductors organics and inorganics: such materials, in fact, are inexpensive, and their electrical measurements require simple DC electronics.

Organic semiconductors (OS), such as phthalocyanines and conducting polymers, can respond to a wide range of polar and nonpolar gases and vapors depending on several interaction mechanisms. For instance, the analyte can affect the charge transfer between the OS and the electrode contact or change the density of charge carriers by oxidizing or reducing reactions; the analyte can interact with the mobile charge carriers at the SO backbone changing the carrier mobility or alter the probability of carrier hopping between chains thus affecting the resistivity of the film. One more mechanism is based on polymer swelling under the effect of an organic vapor analyte that changes the charge density. Sensor response is generally rapid and reversible at room temperature and can be represented as follows:

$$\frac{1}{\Delta R_g} = \frac{K + c}{K(R_{sat} + R_0)}$$

where

ΔR_g is the change in resistance in the presence of the analyte $(R_g - R_0)$
c is the concentration of the analyte
K is the binding constant
R_0 is the resistance without the analyte
R_{sat} is the resistance at saturation

Conducting polymers (CPs) can also be used in the form of NFs where either a single fiber or the whole nonwoven membrane can be employed as sensing element.[85] The most common electrodes used for manufacturing conductometric sensors are interdigitated electrodes (IDEs) that are implemented over insulating and flat substrates (like alumina, silicon dioxide wafers) in various geometrical layout (one or several pair of electrodes) using photolithography and metal deposition techniques. A constant potential is applied, and the output signal strength of IDEs is controlled by the design of the active area, width, and spacing of the electrode fingers. For instance, if the number of electrode fingers and their finger widths is increased, the resulting electric signal will increase, too. The change in any parts of the sensor will cause a consequential change of overall resistance of the device according to the following rule: $R = (1/(2N-1))\rho(w/(h \cdot L))$, where N and L are the number and size of the fingers, h is the electrode thickness, and r is the resistivity of the overlying material. The whole resistance is depending on both the quantity and quality of the texture covering the electrodes, such as fiber density, shape, and dimension of fibers (individual resistance value), assembling over the electrodes (aligned, nonwoven, etc.). Since charge transport is dominated via hopping mechanism, the change in resistance, when a gas reacts with the surface sites of electrospun polymer fibers, occurs mainly for doping/dedoping or acid/base reactions as conformational changes due to changes of crystallinity, H-bonds forming, dipolar–dipolar interactions, etc. Some modifications of the transducer design allow to measure further electrical parameters (i.e., capacitance, work function, field effect) obtaining additional information. During the last 10 years, organic NF-based sensors have been also investigated for the detection of environmental contaminants showing very attracting sensing features (i.e., fast responses, easy tuning of chemical selectivity, high sensitivity) despite their short lifetime (few months), possible poisoning at high concentration of the analyte or/and interferents (causing changes in morphology, hysteresis, etc.), and their poor solubility in common solvents. Macagnano's group[86–88] designed and fabricated various blends of polyaniline electrospun with insulating host polymers (PS, PVP, PEO) to investigate about their sensing features to toxic gases such as NH_3, NO, and NO_2 at extreme temperature and relative humidity values and also according to peculiar biomimetic arrangements. Host polymer carriers are responsible for great modifications of the topology of the interacting surface (diameter and length of the fibers, roughness, porosity, presence of beads and grains, nonwoven framework and branched junctions, adhesion, etc.), in addition to the different affinity to the analytes tested. The linear shape (ohmic behavior) within the applied voltage indicated constant and low resistance values for all the polymer blends, thus suggesting a good adherence of the NFs over the electrodes and providing the possibility to work at very low voltage. The transient responses to several gases reported the achievement of 90% equilibrium responses after 30 s. Specifically, polyaniline-polyethylene oxide(PANi–PEO) was the most porous layer with good electrical performance and high sensitivity to ammonia but was the most fragile sensor when subjected to environmental stress (such as high humidity and temperature). PANi–PVP seemed to be a very good sensor for NO_2, but it was strongly influenced by humidity even at low values. Electrical and morphological characteristics of PANi–PS seemed to be less affected by humidity (up to 50% RH) and temperature, thus suggesting that it could be a stable and promising sensor. Surely, the implementation of humidity protection strategies (such as filters, cartridges) should provide the means of overcoming some of the problems found in other materials, taking advantage of the attractive sensing properties of PANi–PVP. However, all of the tested sensors are very attractive due to their low cost, ease of measurement, low power consumption, and good performance at room temperature (Figure 10.12).

The combination of nanosized metal, metal oxides, and polymers has the potential to increase the sensitivity and the selectivity, as well as the stability, of the conducting polymers. Such composites can in fact operate at room temperature (or decreased temperature), and the selectivity toward different gas species can be controlled by the volume ratio of nanosized compounds. In addition, the composite fibers may have a better long-term stability. The properties of these materials depend not only on the features of their constituents but also on their combined morphology and interfacial characteristics. Wang et al.[89] fabricated TiO_2–ZnO nanofibrous scaffold (the supporting material), which, after dipping in a $FeCl_3$/EtOH solution (both oxidant and doping agent), is overlayered with a thin shell of polypyrrole (PPy), generated by the exposure of fibers to saturated pyrrole vapor with consequent vapor phase polymerization. Thus, the substrate changed its color from white to black and showed a very thin shell (7 nm) on the rough surface with numerous nanoaggregates. The resulting increased electrical resistance of the sensor after gas exposure was due to the charge transferring between electron-donating NH_3 and p-type PPy, with a decrease in both charge carrier density and number of holes. The high sensitivity (with limit of detection [LOD] = 60 ppb) and fast recovery time was supposed to be due to the highly porous and stable structure and the ultrathin PPy-coating layer that reduced the diffusion resistance of the gas. Similarly, Zampetti et al.[70] reported that an electrospun TiO_2 nanofibrous framework, grown up directly onto the IDE transducer and then covered by dipping with a homogeneous ultrathin film of PEDOT–PSS, was an excellent NO_2 sensing material. The final nanocomposite structure induced the increase in conductivity of the chemosensor upon exposure to a few ppb of NO_2, at room temperature and 40% RH. The n–p contacts between TiO_2 NFs and PEDOT–PSS layer, indeed, gave rise to CP's electronic structure changes and allowed the gas sensitivity increment. Exposure to NO_2, maybe causing reduction of the depletion region, increased the nanocomposite conductivity. TiO_2–PEDOT/PSS NFs exhibited good, reversible, and reproducible concentration-dependent response to NO_2. The resulting sensor showed several suitable features, such as a small size, low power consumption, stability over time, and high sensitivity (LOD was estimated ≤ 1 ppb) to the analyte. The very good limit of detection could be related to both the background noise reduction and free access of gaseous molecules to the interacting sites of the highly nanoporous structure (Figure 10.13).

Alternatively, nonconducting polymers (insulators) possess a relatively better stability and easy processability than CPs and hence can be considered as potential resistive sensors when provided with electrical conductivity. Research activities have been intensified on tailoring the electrical property of nonconducting polymer composites by loading with electrically active fillers. Fillers have been used successfully in imparting electrical conductivity to nonconducting polymers, but the conductivity is dependent on the type of filler used. Lala et al.[90] demonstrated the feasibility of using nylon 6 NFs with homogeneous dispersions of multi-wall carbon nanotubes (MWCNTs) (by dipping) on their surfaces for gas sensing, functioning at room temperature. The sensors are particularly selective to polar molecules capable of withdrawing electrons from MWCNTs. Since the key parameters influencing the

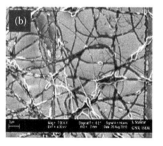
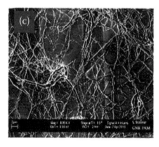

FIGURE 10.12   FE-SEM micrograph of interdigitated electrodes coated with (a) PANi–polystyrene, (b) PANi–polyvinylpyrrolidone, (c) PANi–PEO nanofibrous layers.   (*Continued*)

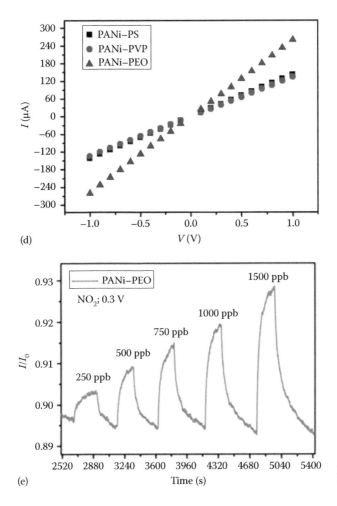

FIGURE 10.12 (*Continued*) FE-SEM micrograph of interdigitated electrodes coated with (d) current–voltage curves of the chemoresistors between −1 and +1 V, and (e) normalized transient responses of PANi–PEO sensor to NO_2 increasing concentrations. (Reprinted from *Thin Solid Films*, 520, Macagnano, A., Zampetti, E., Pantalei, S., De Cesare, F., Bearzotti, A., and Persaud, K.C., 978–985, Copyright 2011, with permission from Elsevier.)

responsiveness of sensor device were supposed to be both dipole moment, nature of functional groups and vapor pressure of analytes, when MWCNTs were embedded within the fibers, gas detection decreased, suggesting that the analyte interaction with the MWCNT was a fundamental mechanism. However, up to date, the most popular gas sensors commercially available are semiconducting sensors based on metal oxides (inorganic semiconductors), operating at high temperatures, where conductivity changes occur when specific gases are adsorbed. The gas sensitivity arises from a diffusion-reaction mechanism: the analyte gases diffuse into the layer

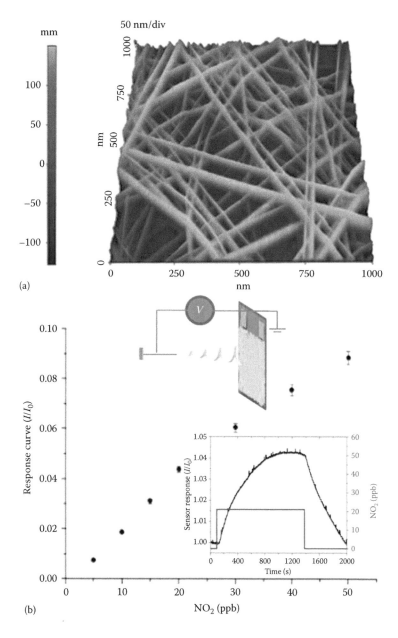

FIGURE 10.13   (a) A 3D image of a TiO$_2$ nanofibrous layer topography, NC mode, 1 µm×1 µm area; (b) current changes vs. increasing concentration of oxidized NO$_2$ normalized to I_0 (starting current) with inset of the current variation reported when 20 ppb of oxidized NO$_2$ interacted with the sensor, at 25°C and 40% RH, normalized to the starting current value (I_0) in the gas carrier; and (upper) the sketch of the nanofibrous scaffold deposition on interdigitated electrode. (Reprinted from *Sens. Actuators. B*, 176, Zampetti, E., Pantalei, S., Muzyczuk, A. et al., 390–398, Copyright 2013, with permission from Elsevier.)

and chemisorb or interact with preadsorbed species (oxygen atoms in different states) on the surface of the nanosized metal-oxide particles, leading to significant variations in the electrical resistance of the sensing layer. Sol–gel solutions containing different polymers and inorganic precursors can be electrospun on various substrates. During the electrospinning process, reactions such as hydrolysis, condensation, and gelation of the precursors are involved in the morphological and microstructural evolution of the fibers. Upon subsequent calcination at elevated temperatures (450°C), the organic components decompose, while the inorganic precursors oxidize and crystallize to form metal-oxide nanoparticles aligned through the fiber. This unique morphology facilitates easy penetration of the surrounding gas phase into the layer and effective distribution of the reactants to the surface of the metal-oxide nanoparticles inside the layer. This is probably the underlying key for the exceptionally high gas sensitivity observed in metal-oxide gas sensors produced by electrospinning. Benefiting from the enhanced surface activity, fast diffusion, and great space-charge modulation depth, the SnO_2 hollow NF-based NO_2 gas sensor achieved a LOD $\leq$ 0.5 ppm, which resulted to be four times lower than that of an analogue thin film.[91] Another strategy to increase the sensing features of the inorganic nanofibrous sensors for gas detection has been suggested by Kim et al.[92]: TiO_2–poly(vinyl acetate) composite NF mats could be directly electrospun onto interdigitated Pt electrode arrays, hot pressed at 120°C, and calcined at 450°C. The network is composed of sheaths of 200–500 nm diameter cores filled with readily accessible gas ~10 nm thick single-crystal anatase fibrils. TiO_2 NF sensors exhibited exceptional sensitivity to NO_2 (833% increase in sensor resistance when exposed to 500 ppb NO_2 at 300°C) and a LOD below 1 ppb. Additionally, both attaching nanoparticles with catalytic function (e.g., Pd, Ag, Pt) onto the surface of NFs and choosing metal salts such as KCl, LiCl, NaCl, and $MgCl_2$ as dopant agents to be added into the NFs can improve the sensitivity and recovery speed of the prepared sensors. Thus, Ag nanoparticle–coated ZnO–SnO_2 NFs were able to detect formaldehyde up to 9 ppb[93]: response signals were increased by about six times with respect to the undoped ones, presumably due to *spillover effects* between the AgNPs and ZnO–SnO_2 nanotubes; meanwhile, the working temperature was decreased by about 140°C. Such features propose the sensor as a promising effective indoor HCHO detector. Among sensors for the environment, obviously, humidity sensor is a very significant tool for practical applications in environment monitoring, industrial process control, and human daily life. LiCl-doped TiO_2 NFs[94] exhibited ultrafast response and recovery (3 and 7 s, respectively) when humidity changed from 11% to 95%, with the impedance value changing from 10^7 to 10^4 Ω. Presumably, during water adsorption, LiCl could be dissolved and then Li(I) ions be transferred along with the fibers, thus leading to a decrease in the impedance. The operating temperature is another important issue for several applications, especially for sensors. A room temperature sensor simplifies the device design by eliminating the heater component. Moreover, this feature allows to save electrical power and assemble the sensor potentially on flexible polymer substrates. Another feature of semiconductor materials is photoconductivity, which is an optical and electrical phenomenon in which materials can become more electrically conductive, at room temperature, due to the absorption of electromagnetic radiation such as visible, ultraviolet (UV), or infrared light. When the UV light hits the sample, a couple of electron–hole pairs are generated. The holes migrate to the surface along the potential gradient produced by the band bending and are neutralized by the negatively charged adsorbed oxygen ions. Consequently, oxygen is desorbed from the surface, and the free carrier concentration is increased (with a consequent decreasing of the depletion layer). The remaining unpaired electrons become the major contributors to the current. Because of the wide bandgap of 3.39 eV and the capability of working at extreme conditions (temperature and voltage bias), photoconductive GaN materials have been the most investigated devices,[95] but nanofibrous layers of ZnO (wide bandgap of 3.37 eV) and TiO_2 (3 and 3.2 eV, for anatase and rutile phases, respectively) have been investigated, too, as promising photoconductive sensors for gas and vapor detection. Zhu et al. reported that NFs coated with a shell of PVP got enhanced photoconductivity, specifically at a UV light intensity of 12 mW/cm^2, and the current of the PVP-coated ZnO NF device was approximately two times higher than that of the pristine ZnO NF device.[96] Such an improvement in current could be explained since under UV irradiation, the electrons, trapped by PVP at the grain boundary and the surface of the NFs, could recombine with the photon-generated holes, leaving the photon-generated electrons for conduction free. Exploiting such a sensor design, it will be possible to develop several polymers–ZnO-based sensors having different functional groups for selective VOC adsorbing, such as amines, ketones, acids, and aromatic hydrocarbons, from the environment. Doping fibers with metal NPs is an additional strategy used to improve photoconductivity as well as selectivity and sensitivity in gas detection. The cause of the enhancement may be related to the increased separation of photon-generated couple of electron–hole and the resulted efficient discharge of surface states. Thus, the ZnO NF decoration with NPs makes them conductive even under visible light and at room temperature. The effect of the photoresponse enhancement by Au nanoparticle decoration can be explained according to the band diagram of both the materials. The work function of Au is about 5.3 eV, while the work function of ZnO is 5.1 eV; thus, the Fermi level of zinc oxide is higher than that of gold. So the electron in ZnO will transfer to Au until equilibrium, and finally, a Schottky junction should be formed at the interface: it is known that the Schottky junction enhances the photogenerated electron–hole separation, extending the life of free electrons. Additionally, the increase of light adsorption efficiency due to the Au nanoparticle–induced light scattering further enhances the photoinduced current. The selection of several metals will confer several selectivity and sensitivity to the photoconductive sensor. Indeed, when PtNPs have been used to decorate nonwoven NFs of titania,[97] the resulting devices were able to combine the properties of photoconduction of titania (1 V voltage polarization) with the sensitivity of platinum to H_2 (LOD = 75 ppm), under UV (UV LED 1 = 365 nm) and room temperature (25°C) (Figure 10.14). The I/V curves pointed out that the conductance of the IDE devices increased with the increasing particle concentration. Without the UV irradiation, the measured current of the devices was more than three orders of magnitude lower than the measured current under UV irradiation. Platinum nanoparticles significantly affected the dynamic behavior of the absorption and desorption processes. The dynamic responses, during the hydrogen adsorption phase, were dominated by both the hydrogen adsorption on PtNPs and the ionic titanium–hydride bonds formation after dissociation at surface defect sites. In fact, the time constant was comparable with the dynamic characteristic of the undecorated device response.

10.3.3 OPTICAL SENSORS

Among the output signals, the colorimetric sensor array system, implemented by Rakow and Suslick,[98] seems to be the most convenient sensing platform because of its effectiveness and simplicity in using a simple naked eye to detect the color change after chemical interaction. A minor limitation of this method comes from the fact that a software program is needed to deconvolute the color pattern changes taking place on exposure to VOCs in order to get a colorimetric digital fingerprint of the analyte concentration as well as more complex environment monitored. Yoon et al.[99] investigated about some commercially available diacetylene (DA) monomers, known in literature for changing blue to red color due to stress factors (such as thermal,

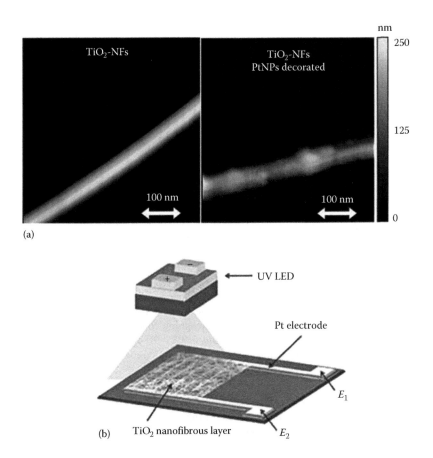

FIGURE 10.14 (a) Atomic force microscope (AFM) images of a titania nanofiber (on the left) before and (on the right) after doping with PtNPs; (b) layout of the photoconductive system.   (*Continued*)

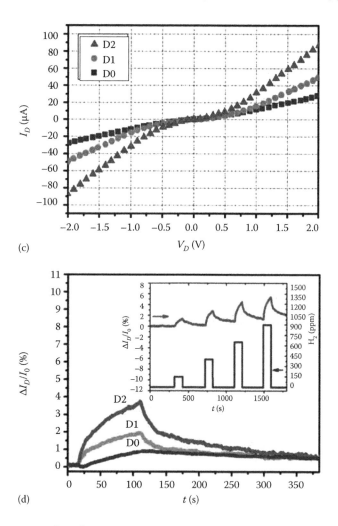

(c)

(d)

FIGURE 10.14 (Continued) (c) *I/V* curves of nonwoven nanofibrous layer of (D0) pristine TiO$_2$, with (D1) slight and (D2) improved PtNPs doping, performed under ultraviolet irradiation and N$_2$ stream; and (d) comparison of the dynamic responses to 700 pm dihydrogen (120 s) and (inset) the D2 electrical signal to increasing concentrations of H$_2$. (Fratoddi, I., Macagnano, A., Battocchio, C. et al., *Nanoscale*, 6, 9177–9184, 2014. Reproduced by permission of The Royal Society of Chemistry.)

mechanical, chemical). PEO (hosting polymer) and TEOS (stability enhancer) were mixed with different DA monomers generating several viscous solutions suitable for electrospinning deposition. In order to generate conjugated polydiacetylene (PDA), the fibers were irradiated with 254 nm UV light for 1 min (blue color). When used in combination, these systems allowed a visual differentiation of several organic solvent by simply monitoring the color patterns that they generated under vapor exposure. The advantages, when the electrospun layers were compared to flat films, can be resumed as (a) a higher intensity of blue to red color change recognizable by the naked eye, (b) more fibrous structures embedding PDA, and (c) porous membrane more desired for sensing and filtering applications. An effective application has been reported by Lee et al.[100] that revealed, using PS and PAA matrices of NFs embedding PDA, fake or adulterated gasoline, often responsible of noxious and polluting tailpipe emissions. Figure 10.15 shows the representation of the mechanism inducing PDA

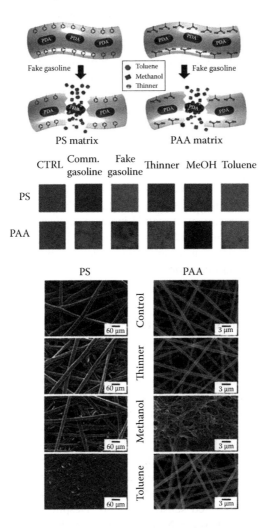

FIGURE 10.15 (a) Representation of the mechanism for the fake gasoline-induced polydiacetylene chromic transition in polystyrene and polyacrylic acid matrices and the related photographs after exposure to vapors of commercial gasoline, fake gasoline, thinner, methanol, and toluene. (b) SEM micrographs of the electrospun fibers before and after exposure to the selected analytes. (Lee, J., Balakhrishnan, S., Cho, J., Jeon, S.-H., and Kim, J.-M., *J. Mater. Chem.*, 21(8), 2648–2655, 2011. Reproduced by permission of The Royal Society of Chemistry.)

chromic transition and photographs showing the shades of color of fibrous matrices after exposure to commercial and fake gasoline as well as several solvents. Recently, electrospun polymer NFs have been investigated, too, to develop efficient chemosensors for the detection of vapors of explosive agents,[101,102] exploiting the fluorescence quenching caused by nitroaromatic electron transfer due to the interaction between the vapor molecules and a synthesized fluorescent polymer (polymer probe) blended with electrospun PS. To improve the porosity and then the surface-area-to-volume ratio, a surfactant was added to the electrospun solution.

The nanofibrous sensing film adequately maintained the fluorescence property of fluorescent polymer and showed higher sensitivity toward traces of several explosive vapors (like picric acid, 2,4,6-trinitrotoluene) compared to conventional spin-casting film. The proposed explosive sensor also exhibited satisfactory reversibility with less than 5% loss of signal intensity after four quenching–regeneration cycles and good reproducibility (SD 2.8%). More recently, nanoparticles called quantum dots (QDs) have been investigated for their unique optical properties. Electrospinning technology was used to get uniformly distributed QDs of cadmium selenide/zinc sulfide (CdSe/ZnS) core shell within NFs in polymeric mixture of PS and polystyrene-co-maleic anhydride (PSMA) (4:1 ratio). Specifically, the introduction of QDs only into the PS–PSMA NFs, in addition to fluorescence, induced electrical conductivity, allowing the user to investigate, simultaneously, more parameters. Tatavarty et al.[103] also showed a two- to threefold increase in electrical conductivity in the presence of VOCs. The obtained quantum-dot nanofibers (Qd-NFs) were photostable for more than 6 months and promising for designing both optical and conductive chemosensors for vapor detection (chlorides, amides, furans) with a response time of less than 1 s.

10.4   TECHNICAL CHALLENGES AND PERSPECTIVES

From recent and more and more growing number of peer-review publications, electrospinning technology emerges as one of the best strategies, among the various nanotechnologies, for designing and developing smart and ultrasensitive systems for meeting the challenges caused by anthropogenic pollution of the environment. Thus, some recent advances in manufacturing nanofibrous filter media via electrospinning have been reported, in an effort to enhance the several approaches used to improve both filtration performances and mechanical properties. The uniqueness of the resulting nanostructures and their direct growth onto various microtransducers as well as the low cost of the equipment and materials confirmed the electrospinning techniques as a promising candidate for manufacturing advanced sensing systems. Despite the great interest on the clear potentials of electrospun NFs by both scientific and industrial communities, many practical problems have to be solved, like a major control of the whole process and industrial scale-up of the laboratory-based electrospinning. However, it is expected that the continuous efforts on investigating electrospun nanofibrous structures, inspired by their unique features, will address the current challenges and upgrade a rapid development of the electrospun nanofibrous-based devices aiding the environment. Therefore, presumably, electrospinning will be able to solve the global environmental problems.

10.5   CHAPTER SUMMARY

To protect life on earth from pollution, environmental monitoring is undoubtedly the necessary prerequisite to proceed to cleaning and remediation. In order to achieve these objectives with affordable investment and running costs, there is a need to develop smart sensing devices for air-quality monitoring as well as to improve the filtrating systems in efficiency and convenience, using materials and technologies capable of inflicting minimal or no harm at all upon ecosystems. Recently, electrospinning technology has been hugely investigated for environmental pollutant detection and removal. This chapter is a brief overview, surely not exhaustive, on the main strategies used over the last decade to improve the electrospun nanofibrous materials in several aspects (i.e., chemical, physical, structural) depending on the final target. Thus, in turn, some adopted approaches in tuning mechanical properties and filtering efficiency of the electrospun fibers have been described for gas/VOC abatement systems, as well as the fibers' lifetime, stability, accuracy, sensitivity, selectivity, and response time for sensing.

REFERENCES

1. Noyes PD, McElwee MK, Miller HD et al. *Environ Int* 35(6) (2009):971–986.
2. Nriagu JO, Pacyna JM. *Nature* 333(6169) (1988):134–139.
3. Jarup L. *Environ Health Perspect* 112(9) (2004):995–997.
4. Law K. *Sci Environ Policy* 24 (2010):1–3. http://ec.europa.eu/environment/integration/research/newsalert/pdf/24si_en.pdf
5. Rodríguez S, Querol X, Alastuey A et al. *Sci Total Environ* 328(1–3) (2004):95–113.
6. Querol X, Alastuey A, Rodriguez S, Plana F, Mantilla E, Ruiz CR. *Atmos Environ* 35(5) (2001):845–858.
7. Kampa M, Castanas E. *Environ Pollut* 151(2) (2008):362–367.
8. Sanchez A, Recillas S, Font X, Casals E, Gonzalez E, Puntes V. *TrAC—Trends Anal Chem* 30 (2011):507–516.
9. Khin MM, Nair AS, Babu VJ, Murugan R, Ramakrishna S. *Energy Environ Sci* 5 (2012):8075–8109.
10. Fang J, Wang X, Lin T. In Tong L (ed.) *Nanofibers - Production, Properties and Functional Applications.* InTech Shangai China (2011), pp. 287–326.
11. Adiletta JG. Fibrous nonwoven web. U.S. Patent No. 5,954,962 (1999).
12. Wang C, Otani Y. *Ind Eng Chem Res* 52 (2013):5–17.
13. Tanaka S, Doi A, Nakatani N, Katayama Y, Miyake Y. *Carbon* 47(11) (2009):2688–2698.
14. Wang D, Sun G, Chiou BS. *Macromol Mater Eng* 292(4) (2007):407–414.
15. Qiu P, Mao C. *ACS Nano* 4(3) (2010):1573–1579.
16. Wang X, Li Y. *J Am Chem Soc* 124(12) (2002):2880–2881.
17. Cooley JF. Apparatus for electrically dispensing fluids. U.S. Patent No. 692,631 (1902).
18. Formhals A. Process and apparatus for preparing artificial threads. U.S. Patent No. 1,975,504 (1934).
19. Lushnikov A. *J Aerosol Sci* 28(4) (1997):545–546.
20. Shepelev AD, Rykunov VA. *J Aerosol Sci* 26(1) (1995):S919–S920.

21. Barhate RS, Ramakrishna S. *J Membr Sci* 296 (2007):1–8.
22. Ding B, Kim HY, Lee SC, Lee DR, Choi KJ. *Fibre Polym* 3(2) (2002):73–79.
23. Ding B, Kimura E, Sato T, Fujita S, Shiratori S. *Polymer* 45(6) (2004):1895–1902.
24. Wu J, Wang N, Zhao Y, Jiang L. *J Mater Chem A* 1(25) (2013):7290–7305.
25. Borm PJA, Kreyling W. *J Nanosci Nanotechnol* 4 (2004):521.
26. Mills NL, Donaldson K, Hadoke PW et al. *Nat Clin Pract Cardiovasc Med* 6 (2009):36.
27. Wang N, Mao X, Zhang S, Yu J, Ding B. In Ding B, Yu J (eds.). *Electrospun Nanofibres for Energy and Environmental Applications.* Springer-Verlag, Berlin, Germany (2014), pp. 299–324.
28. Gibson P, Schreuder GH, Rivin D. *Colloid Surf A: Physicochem Eng Asp* 187–188 (2001):469–481.
29. Ahn YC, Park SK, Kim GT, Hwang YJ, Lee CG, Shin HS, Lee JK. *Curr Appl Phys* 6 (2006):1030–1045.
30. Zhang S, Shim WS, Kim J. *Mater Des* 30 (2009):3659–3666.
31. Carmen PG. *Flow of Gases through Porous Media.* Butterworth Scientific Publications, London, U.K. (1956).
32. Bhattacharyya RK. *ASTM* 718 (1980):272–286.
33. Barhate RS, Loong CK, Ramakrishna S. *J Membr Sci* 283 (2006):209–218.
34. Sambaer W, Zatloukal M, Kimmer D. *Chem Eng Sci* 82 (2012):299–311.
35. Yeom B, Shim E, Pourdeyhimi B. *Macromol Res* 18(9) (2010):884–890.
36. Cho D, Naydich A, Frey MW, Joo YL. *Polymer* 54(9) (2013):2364–2372.
37. Guo M, Ding B, Li X, Wang X, Yu J, Wang M. *J Phys Chem C* 114(2) (2009):916–921.
38. Zhao F, Wang X, Ding B et al. *RSC Adv* 1(8) (2011):1482–1488.
39. Yang L, Raza A, Si Y. et al. *Nanoscale* 4(20) (2012):6581–6587.
40. Mao X, Si Y, Chen Y. et al. *RSC Adv* 2(32) (2012):12216.
41. Srivastava A, Dipanjali Mazumdar D. In Mazzeo N (ed.). *Air Quality Monitoring, Assessment and Management.* InTech (2011), pp. 137–148, ISBN: 978-953-307-31.
42. Wang S, Ang HM, Tade MO. *Environ Int* 33 (2007):694–670.
43. Pinto ML, Pires J, Carvallo AP, de Carvalho MB, Bordado JC. *J Phys Chem B* 108 (2004):13813–13820.
44. Scholten E, Bromberg L, Rutledge GC, Hatton TA. *ACS Appl Mater Interfaces* 3 (2011):3902–3909.
45. Kim HJ, Pant HR, Choi NJ, Kim CS. *Chem Eng J* 230 (2013):244–250.
46. Ahmaruzzaman M. *Energy Fuel* 23 (2009):1494–1511.
47. Raza A, Wang J, Yang S, Si Y, Ding B. *Carbon Lett* 15 (2014):1–14.
48. Bai Y, Huang ZH, Wang MX, Kang F. *Adsorption* 19 (2013):1035–1043.
49. Gholamvand Z, Aboutalebi SH, Keyanpour-Rad M. *Int J Mod Phys: Conf Ser* 5 (2012):622–629.
50. Lee KJ, Shiratori N, Lee GH et al. *Carbon* 48 (2010):4248–4255.
51. Wang M-X, Huang Z-H, Shen K, Kang F, Liang K. *Catal Today* 201 (2013):109–114.
52. Talukdar P, Bhaduri B, Verma N. *ACS Ind Eng Chem Res* 53(31) (2014):12537–12547.
53. Raymundo-Piñero E, Cazorla-Amorós D, Salinas-Martinez DLC, Linares-Solano A. *Carbon* 38 (2000):335.
54. Davini P. *Carbon* 41 (2003):277.
55. Song X, Wang Z, Li Z, Wang C. *J Colloid Interfaces Sci* 327 (2008):388–392.
56. Li B, Duan Y, Luebke D, Morreale B. *Appl Energy* 102 (2013):1439–1447.
57. Phan A, Doonan CJ, Uribe-Romo FJ, Knobler CB, O'keeffe M, Yaghi OM. *Acc Chem Res* 43 (2010):58–67.
58. Ostermann R, Cravillon J, Weidmann C, Wiebke M, Smarsly B. *Chem Commun* 47 (2011):442–444.
59. Cravillon J, Muenzer S, Lohmeier S, Feldhoff A, Huber K, Wiebcke M. *Chem Mater* 21 (2009):1410–1412.

60. Wu Y, Li F, Liu H et al. *J Mater Chem* 22 (2012):16971–16978.
61. Fisher M, Bell RG. *Cryst Eng Commun* 16 (2014):1934–1949.
62. Yoo S, Won J, Kang SW, Kang YS, Nagase S. *J Membr Sci* 363(1–2) (2010):72–79.
63. D'Alessandro DM, Smit B, Long JR. *Angew Chem Int Ed* 49(35) (2010):6058–6082.
64. Bender ET, Katta P, Lotus A, Park SJ, Chase GG, Ramsier RD. *Chem Phys Lett* 423(4–6) (2006):302–305.
65. Wang X, Li B. In Ding B, Yu J (eds.). *Electrospun Nanofibres for Energy and Environmental Applications.* Springer-Verlag (2014), pp. 249–263.
66. Sun T, Donthu S, Sprung M et al. *Acta Mater* 57 (2009):1095–1104.
67. Shouli B, Dianqinga L, Dongmei H, Ruixian L, Aifan C, Liu C. *Sens Actuators B* 150 (2010):749–755.
68. Liu J, Guo Z, Meng F, Jia Y, Liu J. *J Phys Chem C* 112 (2008):6119–6125.
69. Tai H, Jiang Y, Xie G, Yu J, Chen X. *Sens Actuators B* 125 (2007):644–650.
70. Zampetti E, Pantalei S, Muzyczuk A et al. *Sens Actuators B* 176 (2013):390–398.
71. Courbat J, Yue L, Raible S, Briand D, de Rooij NF. *Sens Actuators B* 161 (2012):862–868.
72. Hue J, Dupoy M, Bordy T et al. *Sens Actuators B* 189 (2013):194–198.
73. Mead M, Popoola O, Stewart G et al. *Atmos Environ* 70 (2013):186–203.
74. Ding B, Wang M, Wang X, Yu J, Sun G. *Mater Today* 13 (2010):16–27.
75. Wang X, Li Y, Ding B. Electrospun nanofiber-based sensors. In Ding B, and Yu J (eds.). *Electrospun Nanofibers for Energy and Environmental Applications.* Springer-Verlag, Berlin, Germany (2014), pp. 267–297.
76. Macagnano A, Zampetti E, Kny E. Electrospinning for high performance sensors. Springer International Publishing Switzerland (2015), pp. 1–329 .
77. Janata J. *Principles of Chemical Sensors.* Springer Science + Business Media (2009), pp. 1–373.
78. Vashist SK, Vashist P. *J Sens* (2011), pp. 1–13, ID 571405.
79. Ding B, Kikuchi M, Yamazaki M, Shiratori S. *Sens Proc IEEE2* 2 (2004):685–688.
80. Jia Y, Yan C, Yu H, Chen L, Dong F. *Sens Actuators B* 203 (2014):459–464.
81. Wang X, Cui F, Lin J, Ding B, Yu J, Al-Deyab SS. *Sens Actuat B* 171–172 (2012):658–665.
82. Wang XF, Wang JL, Si Y, Ding B, Yu JY, Sun G, Luo WJ, Zheng G. *Nanoscale* 4 (2012):7585–7592.
83. Ding B, Wang X, Yu J, Wang M. *J Mater Chem* 21 (2011):12784–12792.
84. McQuade DT, Pullen AE, Swager TM. *Chem Rev* 100 (2000):2537–2574.
85. Banica F-G. In *Chemical Sensors and Biosensors.* John Wiley & Sons Ltd., Chichester, U.K. (2012), pp. 246–257.
86. Macagnano A, Zampetti E, Pantalei S, De Cesare F, Bearzotti A, Persaud KC. *Thin Solid Films* 520 (2011):978–985.
87. Zampetti E, Pantalei S, Scalese S et al. *Biosens Bioelectron* 26 (2011):2460–2465.
88. Zampetti E, Pantalei S, Pecora A et al. *Sens Actuators B: Chem* 143 (2009):302–307.
89. Wang Y, Jia W, Strout T, Schempf A, Zhang H, Li B, Cui J, Lei Y. *Electroanalysis* 21 (2009):1432–1438.
90. Lala NL, Thavasi V, Ramakrishna S. *Sensors* 9 (2009):86–101.
91. Cho NG, Yang DJ, Jin MJ, Kim HG, Tuller HL, Kim ID. *Sens Actuators B: Chem* 160(1) (2011):1468–1472.
92. Kim ID, Rothschild A, Lee BH, Kim DY, Jo SM, Tuller HL. *Nano Lett* 6(9) (2006):2009–2013.
93. Xu L, Xing R, Song J, Xu W, Song HJ. *Mater Chem C* 1(11) (2013):2174–2182.
94. Li Z, Zhang H, Zheng W et al. *J Am Chem Soc* 130 (15) (2008):5036–5037.
95. Wu H, Sun Y, Lin D, Zhang R, Zhang C, Pan W. *Adv Mater* 21 (2009):227–231.
96. Zhu Z, Zhang L, Howe JY et al. *Chem Commun* 18 (2009):2568–2570.
97. Fratoddi I, Macagnano A, Battocchio C et al. *Nanoscale* 6 (2014):9177–9184.
98. Rakow NA, Suslick KS. *Nature* 406 (2000):710.

99. Yoon J, Chae SK, Kim J-M. *J Am Chem Soc* 129 (2007):3038–3039.
100. Lee J, Balakhrishnan S, Cho J, Jeon S-H, Kim J-M. *J Mater Chem* 21(8) (2011): 2648–2655.
101. Long Y, Chen H, Yang Y et al. *Macromolecules* 42 (2009):6501–6509.
102. Long Y, Chen H, Wang H et al. *Anal Chim Acta* 744 (2012):82–91.
103. Tatavarty R, Hwang ET, Park JW, Kwak JH, Lee JO, Gu MB. *React Funct Polymers* 71 (2011):104–108.

Index